The Highwaymen

The Highwaymen

Warriors of the Information Superhighway

Ken Auletta

RANDOM HOUSE NEW YORK

All of the essays in this work were originally published in *The New Yorker* in different form.
A small portion of the introduction was originally published in the *Media Studies Journal* of
the Freedom Forum.
Grateful acknowledgment is made to *U.S. News and World Report* for permission to reprint
an excerpt from an article by John Leo (March 27, 1995). Copyright © 1995 by *U.S. News
and World Report*. Reprinted by permission.

Library of Congress Cataloging-in-Publication Data
Auletta, Ken.
The highwaymen : warriors of the information superhighway / Ken Auletta. — 1st ed.
p.   cm.
ISBN 0-679-45738-0
1. Businessmen—United States—Biography.   2. Telecommunication—United States—
Biography.   3. High technology industries—United States—Biography.   I. Title.
HC102.5.A2A95   1997
3388.04′092273—dc21   96-39714

Random House website address: http://www.randomhouse.com/
Printed in the United States of America on acid-free paper
24689753
First Edition

For Kate

CONTENTS

INTRODUCTION

The information revolution has changed the way we work, play, learn, shop, bank, retrieve information, and govern ourselves. In 1980, few people could purchase a VCR. They were too expensive, too complicated. Today VCRs are in 90 percent of American homes. As recently as 1979—less than two decades ago—there were no PCs, no fax machines, no cellular phones or CDs, no MTV or CNN. And no one had yet invoked the term "information superhighway."

Over the Internet or through interactive television, we can or will soon be able to summon news, data, books, catalogues, magazines, movies. We will use computers to pay our bills, order groceries, buy and sell stocks, simulate sex, play the lottery, and register to vote.

That science will make this possible is certain. Whether consumers will want—or can afford—these services remains a mystery. No one can foresee the social consequences of these changes, and we don't yet know what impact instantaneous access to information (or disinformation) will have on deliberative democracy.

Nor do we know what delivery system will be ascendant. Will information and entertainment be delivered by a cable or telephone wire? Or through a direct-broadcast satellite dish? Will more-powerful computers be the TV screens of the future? Will telephone calls be made over the Internet? How will traditional broadcast networks compete with five hundred or more channel choices? Will networks survive because they are familiar life rafts—brands—in a churning sea of choices?

No one can predict the final contours of this electronic revolution, but we do know this: Changes in the way we communicate will be profound. Elementary-school children will not only read and be told about ancient Egypt, they will experience it by summoning virtual images of pyramids and figuratively floating along the Nile. With electronic tools, more citizens will work at home. Corporations will become less hierarchical as middle managers and others begin to talk directly to CEOs through e-mail. Many layers of management will become superfluous. File cabinets will be discarded, as information will be stored in computer files. In unpredictable ways, computers will alter the importance of geographic proximity—and reshape cities. Work, and play, will be transformed.

This collection of sixteen *New Yorker* articles spans a four-year period, from 1992 to 1996—a time of tumultuous change in the communications industry, in which, in the words of *The Economist*, "the introduction of the PC caused the largest creation of wealth in the history of the planet." The net worth of Microsoft, Intel, and Compaq—$130 billion in 1995—exceeded that of all of Hollywood's film studios. And the burgeoning Internet may soon dwarf this. Today, the entertainment/information business, with $350 billion in domestic and overseas sales, has supplanted jet engines as America's foremost export. And this market is only half the size of the worldwide telephone business. Those who strive to dominate this volatile new world—the Highwaymen—are the subjects of this book.

The Highwaymen begins with Barry Diller's 1992 quest to discover the future, using his PowerBook as a guide. Diller found that his laptop freed him from secretaries, meetings, and memos, and would help him envision the possibilities of a new interactive electronic democracy. By following Diller's journey we discover how a tiny chip can transform dumb TV sets into smart ones, how viewers may soon be able to program for themselves rather than rely on network programmers to tell them when to watch *60 Minutes, Seinfeld,* or an HBO movie. With a click of a switch, viewers may soon have instant access to a video jukebox—a hit movie, a shopping channel, a local council meeting, tomorrow's *New York Times*. The television—or computer—could become the equivalent of a personal secretary or agent, setting aside favorite programs or news clippings, gathering information on stocks or bonds.

With advances in compression technology, digitalization, and microprocessing, the intoxicating dream of five hundred or more channels was soon off the drawing boards and ready for testing. By 1993, Diller had decided to link his fate with QVC, a low-budget home-shopping network. His principal partner was John Malone, the CEO of Tele-Communications, Inc., the nation's and the world's largest cable company, and a part owner of QVC. Malone, an enigmatic man who avoided media interviews, was among the most powerful businessmen in America. In early 1994, when my profile of him ("The Cowboy") appeared, he was championing a merger with another powerful distributor, the cash-rich Bell Atlantic telephone company.

To combat Malone's capacity to collect tolls from program suppliers, men like Sumner Redstone of Viacom and Michael Eisner of Disney strove to become so indispensable as brand-name suppliers that no distributor could deny them access to customers. In September 1993, Redstone announced that he would swallow Paramount. Although challenged by Diller, whose QVC had soared in value, Redstone won the bidding war, and Viacom entered the front ranks of worldwide entertainment software companies—at least on paper.

With the communications business promising growth and riches and glamour, new players joined the fray. Among the new Highwaymen was Edgar Bronfman, Jr., whose Seagram Company purchased 15 percent of Time Warner's stock in 1993. This set up a conflict, explored in "No Longer the Son Of," between two men who were seeking to escape from the shadows of dominant personalities—Bronfman from his father's shadow, and Time Warner chairman Gerald Levin from the shadow of a predecessor, Steve Ross.

Words like "synergy," "leverage," "branding," "convergence," and "partnership" soon became the rage of the communications business. The proliferating communications-company mergers, sales, and alliances were spurred by investment bankers. Among these, the foremost matchmaker was Herbert Allen ("The Consigliere"), who shepherded the sale of Columbia Pictures to Coca-Cola and then to Sony, and who sold MCA first to Matsushita and then to Seagram.

Within communications, government is often viewed as an impediment, but when Newt Gingrich's Republican majority claimed Congress, companies felt they had a potent ally. The relationship between the government and the communications/entertainment conglomerates is explored in "The Referee," a piece on the FCC.

Government regulations are not the only potholes corporations encounter. Ted Turner's CNN collided with the public's appetite for local news presented by local anchors, which technology can now satisfy. Instead of the uniform global village envisioned by Marshall McLuhan, satellite technology has spawned what is now called localism. Meanwhile, companies are vying to become more global. Starting in 1994, worldwide competitors to CNN began to emerge. As American studios and communications companies transform overseas markets—where the studios now generate half their profits—the race for global hegemony has become vicious.

And technology leaves its own potholes. By 1994, we were learning that the magic box envisioned by Barry Diller wasn't so magical. The chasm between vision and reality is explored in a portrait of Time Warner's video-on-demand system in Orlando ("The Magic Box"), which experienced unanticipated technical and financial problems and whose success remains dubious.

And then there are the human potholes, especially the toxic mix of vanity, pride, and ambition. Primal forces drove Disney's Michael Eisner and Jeffrey Katzenberg toward a business divorce each knew could be harmful yet neither could prevent ("The Human Factor").

By 1995, the buzz of the previous year had subsided. Cable looked weaker, as did the telephone companies. To bet all one's chips on content, as Disney and Viacom were doing, no longer seemed such a good idea. A blend of content and distribution seemed the shrewder choice, and Rupert Murdoch was now the Highwayman whom everyone was watching ("The Pirate"). Murdoch's successes with the Fox network and with satellite systems that blanketed the globe helped provoke a frenzy of activity, as first Barry Diller, then Ted Turner, and finally Michael Eisner chased after a network distribution system, while Michael Ovitz and others searched for new jobs. Insecurity and uncertainty abounded, and such talented executives as Frank Biondi and Michael Fuchs soon found themselves out of a job ("The Beheadings").

If these moves diminished the Highwaymen, so did their handling of the "family values" issue. The entertainment industry—and especially Time Warner—had no idea how to respond to President Clinton's V-chip or William Bennett's attack on gangsta rap. The clash between Bennett and Time Warner is chronicled in "The Power of Shame."

This book begins in 1992 with Barry Diller, interactive TV, and the birth of an "information superhighway"; it ends in 1996 with Michael Kinsley and a new interactive paradigm, cyberspace.

These have been years of dizzying change. When the decade began, Rupert Murdoch's News Corporation neared bankruptcy. Today Murdoch sets the pace among the global communications superpowers. A few years ago, video on demand was the interactive rage. Today interactivity usually means the Internet. One year, broadcast stations and networks are said to be dinosaurs; the next they are valued as brand names. One day, Bill Gates disparages the Internet; the next he transforms Microsoft into an Internet provider. One moment companies proclaim that software is king, and the next they fret that distributors control worldwide access.

Our Information Age is filled with agitation as journalists frantically declare winners and losers, Wall Street scrambles to find the next Netscape, and stock prices fluctuate like cardiograms. It's a thrilling, and terrifying, time to be involved in communications. Broadcasters worry about cable, who worry about the Baby Bells, who worry about long-distance telephone carriers, who worry about Microsoft and Intel, who worry about the Internet. Companies fret over their advertising dollars; publishers and journalists fret over what will happen to the print medium. On the surface, nothing seems certain in the communications revolution.

Yet there are some certitudes—maxims—that can serve as guideposts for this Information Age.

One: All companies strive to crush competitors and to take the risks out of capitalism. The media battlefield today resembles Europe in the nineteenth century, when there were potent nation-states but no single superpower. At least seven distinct groups vie to become information/entertainment superpowers—cable, telephone, film studios, and broadcast networks, as well as the computer, publishing, and consumer-electronics industries. They seek to become vertically integrated, to leverage their power in one business in order to assist another, as Disney plans to leverage ABC to throw business to its TV production arm, or to use the popularity of ESPN to muscle more cable channels for its other cable networks. Companies struggle to minimize risks by controlling every aspect of their business, from the creation of an idea to its manufacture and distribution to copyright ownership and thus its afterlife.

Two: There will be more concentration of corporate power and more competition. On the one hand, each industry says it is ready to fight: phone companies prepare to enter the cable business as cable readies to challenge Baby Bells and broadcast stations and computer companies prepare to offer phone service on the Internet. Yet companies are hedging their bets by seeking partners who share financial risks, provide absent skills or services, and sometimes minimize political exposure. Film studios share costs—and profits—with producers. MCI, which doesn't own software to distribute over its long-distance wires, invests two billion dollars in News Corporation, a software company. AT&T enters the video business by buying a stake in DirecTV, the nation's leading direct-broadcast satellite service. Microsoft invests five hundred million dollars to help fund a twenty-four-hour cable news network and on-line service (MSNBC) with NBC.

For citizens, these maneuvers evoke memories of the 1890 Sherman and 1914 Clayton antitrust acts, which were designed to curb corporate dominance over markets and suppliers. The dangers of monopoly today are often more subtle than they once were. Powerful corporate allies could cede spheres of influence to each other, choosing not to compete but to adopt instead the Japanese model known as *keiretsu*, whereby mutual back-scratching closes markets and restricts competition.

Three: Technology threatens bigness and ideology. About this, Orwell was wrong: governments cannot control information. Try as it may, the government of China cannot block access to the Internet, any more than the Communist governments of Eastern Europe could seal their borders against faxes and telephones and satellites. And television viewers have more sources of news today than they did fifteen or more years ago, including CNN, CNBC, C-Span, and MSNBC.

Four: As Palmerston once said of nations, corporations have no eternal allies and no perpetual enemies, only permanent interests. Thus an ally in one venture or country can be an adversary elsewhere. One week, MCI enters a joint Internet effort with News Corporation. The next, it dumps this deal to join with Microsoft in a similar venture. Yet MCI's long-distance competitor, AT&T, also signed with Microsoft to reciprocally market products on the Internet. Universal and Paramount compete in the studio business yet jointly own the USA cable network, while Hearst and Capital Cities/ABC compete in the magazine business yet are partners in ESPN.

Five: The media battlefield is global. As the American market matures and other nations duplicate our deregulated environment, U.S. companies often see greater growth opportunities overseas. Since roughly half of our movie revenues come from overseas, Hollywood continues to spend vast sums on action-adventure or sex-driven movies that transcend language barriers and assure global sales. Disney expects that within five years half the revenues from its diverse activities, including theme parks and stores, will come from outside the United States. Viacom's chairman, Sumner Redstone, has announced that though only 17 percent of his revenues came from outside the United States in 1995, he expects that by the year 2000 this will rise to 40 percent. As the revenues of music companies flattened in the United States, they exploded overseas. The same is true of video stores and movie theaters.

Six: While the contest is global, localism is key. We are witnessing the rise not of a single global village but of hundreds of local villages. American tourists may like to turn on their hotel TV sets in Europe or Asia and learn the results of American football games, but a resident of New Delhi wants local cricket scores, local weather, and local anchors. And the concern many governments have about American "cultural imperialism" will only speed this desire to localize, while sending media giants scrambling to locate local partners.

Seven: Familiar brand names can sometimes beat localism. That's why NBC puts its brand on its global news service (a move News Corporation failed to make when it launched Sky-TV in Europe without a Fox label). The goal is to have a name that instantly triggers an association in the consumer's mind—as MTV means youth and cool, Disney means child-friendly, *The Wall Street Journal* means business news, and IBM once meant impeccable service.

Eight: Distribution is not dead yet. We commonly hear from folks like Eisner and Redstone that software is king, that as competition increases, distribution systems will lose leverage to strong brand names like MTV. Yet today four studios own their own broadcast networks and thus exercise some control over the distribution of their television product. Rupert Murdoch started a twenty-four-hour news service, but he knew that without a cable distribution system he could not succeed. So he used his overseas distribution systems for leverage over cable operators here. And software factories like Viacom learned that in order to get the best overseas distribution of Paramount movies on television, or of MTV, they needed to pay gatekeepers like Murdoch's satellite systems in Europe, Asia, and South America.

Nine: The companies that will survive are those that define themselves broadly. CBS, NBC, and ABC lost a third of their audience over the past fifteen years, in part because they defended their single channels and blindly fought cable, failing to understand that they owned a brand and not just a single channel. Thus the danger for, say, *The New York Times*, would be to think that it is in the newspaper business rather than the information business. As long as customers pay for information from the *Times*, and advertisers pay the freight, it doesn't matter whether readers receive news on paper or from computer screens.

Ten: The human factor counts, though it can't be quantified. Gerald Levin, Time Warner's chairman and CEO, fired Robert Morgado and then Michael Fuchs as chief of Warner Music not because they were failures—they were spectacularly successful—but because he felt uncomfortable with them. Similarly, Viacom's chairman, Sumner Redstone, fired CEO Frank Biondi because, depending on whose story one believes, he either questioned his gumption or envied his press clippings. Disney's Michael Eisner and Jeffrey Katzenberg divorced for personal rather than business reasons. The proposed merger of Bell Atlantic and TCI collapsed as much because of a culture clash—a cowboy culture banged heads with a staid corporate culture—as for business reasons. Viacom has yet to achieve the much-touted synergies from its acquisition of Paramount, partly because synergies require teamwork and the sublimation of ego.

Eleven: The great mystery is the consumer. We know some things about consumers: that they want more choice and convenience; but there is much more we don't know. Do people want to program for themselves, or would they rather collapse in front of their TV sets and let network programmers make the decisions? Will they entrust personal information—such as credit card numbers—to computers? Will there be enough customer demand to justify the enormous investment companies must make?

A reminder of how baffling it is to predict winners and losers came for me in January 1997, during a visit to the Consumer Electronics convention in Las Vegas. Like others, I was awed by thousands of inventive appliances and products. But I was also struck by the notion that few of these products were as simple to use as, say, a television remote-control device. Keypads were too small, the lighting was too dim, the instructions were too dense. The appliances were not user-friendly, and until they are, there is no accurate way to gauge whether they will be accepted by the consumer.

Twelve: Beware the social and political consequences. Will we create two classes of citizens, information haves and have-nots? What will be the consequences of a society in which 40 percent of homes have computers but only 5 percent of low-income homes have them? What should we do, if anything, for the 98 percent of classrooms without access to telephones, the essential link to cyberspace? (Half the world's population lives more than one hundred kilometers from a telephone line.) Who should pay for providing this link—taxpayers or corporations? With unlimited communication and entertainment choices, will more citizens isolate themselves? Will the rise of virtual Internet communities increase or decrease our tribal tendencies, generate greater harmony or an electronic Bosnia? Because e-mail and instant polls allow instant opinions, will they permit the cool deliberation James Madison and the other Founding Fathers envisioned when they chose to divide power among the branches of government to slow down decision-making?

Thirteen: Synergy is no friend of journalism. As companies strive to boost their stock price, they inevitably notice that news divisions rarely make as much money as other divisions. So to cut costs they replace foreign correspondents and bureaus and contract out for cheaper video news pictures. They ask junior employees to re-process material from their own video libraries as documentaries. They insist that news programs conform to what marketing surveys say the public wants. Parent companies striving for synergies will often ask their news divisions to promote their products: Disney asked *Good Morning America* and ABC to promote *The Hunchback of Notre Dame*; each network expects its local stations to promote its entertainment shows by scheduling interviews on local newscasts with their stars. This team culture, though common in business, is a menace to independent journalism.

Fourteen: The marketplace alone will not determine the winners and losers; government decisions matter. The government decides whether to bring antitrust action against Microsoft (so far, it has declined), whether to approve the merger of Time Warner and Turner Broadcasting (it did), whether to lift restrictions on foreign ownership, and whether to let broadcasters inherit the extra spectrum space made possible by digital compression, or to extract a "public trust" commitment from broadcasters, or to put it up for auction (it has not yet decided). Overseas, the government of China will decide whether to regulate the Internet, Singapore licenses Web-site providers, and France and Spain, among others, will decide whether to sustain their barriers to foreign competition. Will China and other Asian nations ignore copyright piracy? And will this trigger a trade war?

Government has a legitimate role to play as referee between communications rivals. If local and long-distance telephone companies, for instance, are to compete, the government must assure that both companies rent their wires to competitors. Without government regulation, companies may have little incentive to cooperate, and if they don't, prices will not come down.

Fifteen: Beware of maxims. If synergy were as magical as promised, Sony and Matsushita would successfully have mated their hardware and software, as they expected when they overpaid a decade ago for two Hollywood studios. Nor has Viacom achieved the promised synergies of its merger with Paramount Communications. Perhaps the foremost example of successful synergy is Disney's development of products from its movies for sale in Disney stores. After little more than a decade, sales now total over $2 billion annually, or twice the revenues from Disney movies. But this dollar success is an exception.

Similarly, a brand name didn't prevent Apple—despite its brilliant software—from slipping because it refused to allow (until 1995) any other computer maker to use its superb software. Two computer pioneers—Ken Olsen of the Digital Equipment Corporation (DEC) and An Wang of Wang Laboratories—faltered because each failed to foresee the multiple uses of PCs.

Sixteen: It's OK to be confused. Many of us are understandably nervous about how technology will alter our lives. As investors guess what stocks to buy or sell, so consumers can only guess whether it makes sense to buy a combined printer/fax, or a Newton personal digital assistant, or a TV set dedicated to the Internet, or a "dumb" but cheap computer that retrieves software from the Internet.

Cyberspace has attracted the swagger of know-it-all-ism. Yet many of the certitudes of seers and CEOs about "interactivity" and "multimedia" and "synergy" bring to mind that old *New Yorker* cartoon in which the psychiatrist says to the patient, "I have good news for you."

"Yes?" says the expectant patient.

"You have no conflicts," says the psychiatrist. "You are inferior!"

When one looks closely at the decision-making process in giant communications companies, one learns that even the "experts" are guessing. Think of the many dumb bets smart people have made. Remember how Gerald Levin and John Malone promised we would all be surfing five hundred channels by now? Remember how the

Baby Bells said they would dominate the video-dial-tone business? Several hundred million dollars later, they are not in sight of their first customer. Remember when Bill Gates and Microsoft ignored the Internet? Or when IBM passed on buying Microsoft? Or when AT&T vowed to dominate the computer business? With an annual R&D budget of $200 million, Microsoft's technology seer, Nathan Myhrvold, nevertheless concedes, "Most decisions are seat-of-the-pants judgments. You can create a rationale for anything. In the end, most decisions are based on intuition and faith." On jumping-off-cliff guesses.

Journalists also jump off cliffs. We travel to strange places to meet strangers. We struggle to sort the truth from a muddle of sometimes contradictory and always incomplete information. We trust that the person on the other end of the phone is who he says he is. We never have enough time, or space. The journalistic form—gimme a lead, a headline, a scoop, and write it in a few minutes or hours and in no more than five hundred words—unavoidably shapes the content. In the fever of daily reporting, it is easy to lose perspective. Swept up by the coverage of the O. J. Simpson trial, we forget that there were other "trials of the century," including the Scopes trial, the trial of Bruno Hauptmann, and the Nuremberg trials. Today we think Rush Limbaugh has power, but fifty years ago, when the population was smaller, Walter Winchell's Sunday-night radio show reached many more listeners and generated more headlines.

One of the joys of writing for *The New Yorker* is that you have time, and space, to pursue Christopher Morley's sage aphorism: "Truth is a liquid, not a solid." My task as a journalist is to understand the people and issues I write about, not to prosecute nor to celebrate them. Anytime a journalist gains access to decision-makers, whether the White House staff or Rupert Murdoch, there is a danger of being co-opted, of forgetting that the audience is not the person interviewed but anonymous readers or viewers. But it's also important to remember that moguls are not stick figures. They may be ruthless or diabolic or shallow, but they are rarely just that.

It is tempting, when writing about business, to treat decisions as if they were coldly rational. Often they are—companies do, after all, strive to maximize profits. But decisions are often composed of many strands, and people are not always rational. Egos can get in the way. Although journalists don't have subpoena power, vanity can become our ally. People talk to journalists for a variety of reasons—to promote their interest, because they think the reporter will be fair, will tell their story, will set the record straight. They talk if the reporter is persistent, and doesn't make the interview experience as unpleasant as a trip to the dentist. And they talk because vanity is a great equalizer. One doesn't become a CEO or a senator without confidence in one's ability to sell anything, to convince anyone.

The Highwaymen

1

DILLER PEEKS INTO THE FUTURE
Intimations of Interactivity
(The New Yorker, February 22, 1993)

When Barry Diller, the former chairman of Fox, Inc., speaks of his Apple PowerBook, a laptop computer, he grows rhapsodic. "My odyssey began with the PowerBook," Diller says of the months he spent after leaving Fox. The odyssey was Diller's ten-month search among the seven industries hoping to dominate global communications—studios, TV networks, cable companies, telephone companies, computer companies, consumer-electronics companies, and publishers—to decide where he should stake out his own future. In fact, the search began in the fall of 1991, several months after Diller and Rupert Murdoch, whose News Corporation owns Fox, met to discuss Diller's future. After being head of prime-time television for ABC Entertainment in the early seventies, and chief of Paramount Pictures for ten years—into the first half of the eighties—and then guiding Fox into the nineties, Diller told Murdoch that he wanted to be a principal, not just a well-paid employee. Murdoch, who had moved to Hollywood and plunged into Fox's business—Diller's business—asked for a few days to consider Diller's request. When he had done so, he responded bluntly, Diller recalls, saying, "There is only one principal in this company."

Diller was grateful for the candor, he concedes, but, with his fiftieth birthday approaching, he began to think about leaving. Lifting up his PowerBook, he explains, "I learned it to leave Fox."

A tutor taught him how to use it. The machine's allure was that it promised a certain kind of freedom—from secretaries, meetings, memos, press leaks. Diller used it to compose his resignation statement; to fax draft copies of the statement to Murdoch and to his closest friend, the clothing designer Diane Von Furstenberg; to list things he must do before issuing the statement; to sort from his copious address book the three hundred people he wanted to have receive the resignation statement before they heard or read about it; to jot down notions of what he might like to do next and whom he might consult. The PowerBook went with him everywhere. Diller punched keys in the middle of meetings, while others were left to stare at the top of his bald head or to listen as he related the many extraordinary feats his machine could perform. "He's had an unbelievable love affair with his computer," Von Furstenberg says. "It has expanded his horizon. No question that his relationship with his little screen—which is irritating to everybody in the room—has altered his life."

Among other things, the machine helped Diller better understand the new video democracy. Through it he could see how technology, with incredible speed, was transforming dumb television sets into smart ones, making it possible for viewers to select, organize, and interact with programming and information rather than passively consuming what was offered on fifty, or even five hundred, channels. The PowerBook became for him a means of peering into the future, for he uses the laptop the way Apple Computer, which makes it, hopes that people will use a book-size machine, referred to as a "personal digital assistant," that Apple is developing. Just as Diller could convert his laptop into a word processor, a fax, a file cabinet, a spreadsheet, a conveyor of commands, or a link to various networks of news or data, so in the next few years, he came to believe, viewers will receive video on demand—be able to watch what they want when they want. With a click of a remote control or a telephone button, they will summon up movies from the equivalent of a video jukebox. In an instant, they will send for and receive a paperless newspaper, a program they missed last night, a weather report.

Still, as the day Diller would announce his departure from Fox neared, he felt vulnerable. "I liked my life," he says. "I liked power." He worried that no one would call, that he might lose his conspicuous seat at restaurants, that it might appear Murdoch had dismissed him, that the story would leak before he was ready.

Diller issued a statement on February 24, 1992, which came as a complete surprise to Fox employees and to Hollywood. In it Diller said he "yearned" to be an entrepreneur like Murdoch, whom he called an "inspiration." Murdoch responded with wide praise for Diller's many accomplishments, from creating a fourth TV network, which in 1991 made more money than either NBC or CBS, to engineering the comeback of the Fox film studio. "It's nice to read your obituary while you're living," friends told him. Diller left Fox in good financial condition. He received a severance, bonus, and stock-payment package that a Diller intimate says was worth $140 million; upon his departure, he bought a thirteen-seat Fox Gulfstream jet for some $5 million.

Within days, he recalls, thirty-three executives offered either to back Diller or to become his partner in any future venture, and their names were immediately filed in his PowerBook. Among the most persistent of these suitors in the months that followed were John Malone, the chief executive officer of Tele-Communications, Inc. (TCI), which is America's and the world's largest cable company, and Brian L. Roberts, the president of the Comcast Corporation, which is the nation's fourth-largest cable company. Both men believed that cable's weakness—programming—was Diller's strength. They knew that technology would one day allow a virtually unlimited number of channels, but they also knew that viewers watched programs, not technology.

For months, rumors chased Diller. It was said that he might be acquiring NBC from General Electric, and that he might be asked to run Time Warner. Whenever he was seen lunching with a wealthy financier, the news would appear in the columns, increasing speculation that Diller was about to launch something big.

Ten months after his Fox announcement, Diller's plans were still a mystery. People in the entertainment and communications worlds began to say that he couldn't be a principal, because he had no capital. "A guy who's an employer is a guy who writes checks and he can't," one Hollywood power broker said. Then on December 10, 1992, it was announced that Diller had become a partner of Malone's and Roberts's in something called the QVC Network, a twenty-four-hour home-shopping cable network. People passing Diller at his regular table in the Grill Room of the Four Seasons had almost embarrassed expressions as he kept looking around, as if for applause. Those who knew Diller waved a greeting, but they seemed to be thinking: Barry Diller's going to run *what*? A home-shopping network? You've got to be kidding!

"All they care about is status," Diller said some weeks later, in his suite at the Waldorf Towers. "That's why they can't understand why I'm doing this. They say, 'It's not very glamorous.' " Every now and then, he glanced over at his PowerBook, which was sitting on a coffee table relaying minute-to-minute QVC sales. In 1991, QVC made a net profit of nearly $20 million; in the first nine months of 1992, it already showed a profit of $36 million. It was recently rated America's "fastest-growing small public company" by the magazine *Inc.*

The people who found QVC insufficiently glamorous probably weren't aware of those numbers. Or, if they were, they wondered why Diller, who had done so much to attract mass audiences, was now in the narrowcasting business. Worse, they wondered why he was hawking "tacky" merchandise. Diller pretends to be unfazed, but friends admit that he is sensitive. "Remember, the royalty of America is Hollywood," one friend says. "We are the place of fantasy and intrigue and overnight rags to riches. For the time being, Barry is concerned that he's out of that limelight. He feels that what he's in is a bit undignified."

Diller seeks to counter the skepticism in several ways. After ten months of searching for the future, he thinks that he has become a pioneer in a new form of interactive television—one that began with home shopping and will soon include news and programming. He thinks of QVC not as just two televised shopping channels, which is what it is now, but as a springboard to a universe beyond the limited world of channels. He also thinks of QVC as a gold mine. Within two weeks of the announcement that he would run QVC, its stock rose from about thirty dollars a share to forty. On paper, Diller's seven million shares climbed in worth by $70 million, yielding half as much in two weeks as he got when he left Fox. "If he fulfills his vision at QVC, he'll be the richest person any of us know," says Jeffrey Katzenberg, chairman of Walt Disney Studios, who is a close friend. "Barry Diller will be worth many billions of dollars."

A year ago, as Diller contemplated his future, he said to himself, "There are a couple of concrete possibilities—NBC is one. The others are all large enterprises, the kind I have always run. OK, but what else is interesting? I don't know very much. I know topic headings, that's all. Some part of me is urging me to take advantage of this gift of time."

The PowerBook taught Diller he could work at home, so he had his pool house, in Beverly Hills, converted into an office for himself, constructed an additional office for his secretaries, and moved to his beach house, in Malibu. Meanwhile, his phone was not quiet. Of all those in Diller's wide circle of well-known intimates—the music impresario David Geffen, the designer Calvin Klein, the producer Sandy Gallin, the actor Warren Beatty—the friend he confided in was Diane Von Furstenberg. She says they spoke, and still speak, at least three times a day. "He's like a husband, really," she says.

On February 29, 1992, Von Furstenberg visited QVC's headquarters, in West Chester, Pennsylvania, and the trip amazed her. As soon as she returned, she told Diller on the phone that she and several other people, including Marvin Traub, a former chairman of Bloomingdale's, had just gone on a "field trip" to a place outside Philadelphia to see something called QVC, which stands for "Quality, Value, and Convenience." When they arrived, she said, they sat in a studio behind several banks of telephone operators and watched the soap-opera star Susan Lucci pitch a hair-care product carrying her name.

Suddenly, the phones lit up. "It was amazing. She sold four hundred and fifty thousand dollars' worth of hair products in an hour," Von Furstenberg told Diller. She described how the operators had punched in the orders on computer screens in front of them, charged each order to a credit card, and then pressed a key to send the order to a warehouse, which promised delivery usually within seven days. "There you have the potential of talking to millions of people all at once, and you don't have to rely on an in-between," she said. "It's more honest." Sure, much of the merchandise was cheesy, she told Diller. But that was correctable. What mattered was the directness of the system. "Barry, you've got to go out there," she said.

Diller wanted to go, if for no other reason than that over the years he had personally negotiated most of Von Furstenberg's business contracts. But home shopping did sound—well, *small*, for his ambition. He wouldn't physically leave Fox until spring, and he thought, he said later, that then "I'd have this big, long holiday."

Through the winter and into the spring, Diller fiddled with his PowerBook for at least a couple of hours every day, typing options, lists of things he might do; he raised money for his fellow Democrats Bill Clinton and Bob Kerrey; he read a few books; and by May he was exhausted with leisure. He was ready to begin his odyssey. "I don't need a wall to stare at," he remembers thinking. "I need new data." He decided to combine a lifelong wish to travel across the country with stops at such potential windows onto the future as the Massachusetts Institute of Technology's Media Lab, William Gates's Microsoft, near Seattle, and Bill Clinton's governor's mansion, in Little Rock.

Diller knew from the entertainment business the lightning impact of technology. He remembered how the studios had fought the videocassette recorder and, before that, television, and how both had eventually swelled studio profits. The PowerBook gave Diller some further technical insights. He came to call it "an enabler," because it allowed him to do many different things at once—to be alone and still be able to communicate instantly. In the spring, Brian Roberts, of Comcast, called Diller to chat. He wanted to invite Diller once again to join the cable team. The two men wound up talking about the PowerBook, since Roberts had just bought one and was still trying to figure it out. "I wanted to understand it for the cable business," Roberts says. Diller shared his impressions of what computers mean for the future. One day, he speculated, the computer screen might become a TV set, and the keyboard would be a mechanism for summoning anything. The speed would be astonishing. A billion bits of information per second would travel over a wire, contrasted with only a few thousand bits a second sent by a PowerBook fax. Diller had read that these bits of information would perhaps be retrieved by a powerful microprocessor in a cable-converter box inside or beside the screen, and perhaps computer software would make remote-control devices user-friendly, like a computer mouse, permitting viewers to choose what they watch not only by surfing among five hundred or more channels but also by specifying categories—movies, comedy, sports, books, news.

As Diller thought about the competing interests of the cable companies, Hollywood studios, TV networks, computer hardware and software companies, publishers, telephone companies, and assorted consumer-electronics powerhouses, like Sony—many of which already did business with one another—he realized that each one hoped someday to control either the wire highway to each home or the switching mechanism that would someday direct video traffic or the computer databases that would serve as a library or the technology that converted pictures and programming to digital signals and back again. He knew that the current system of sending analog signals to TV sets would eventually be replaced, because these electrical impulses took up too much space on the highway, or bandwidth, that they traveled over, limiting the number of channels. What he didn't understand was the digital-compression technology that would replace it. Nor did Diller know—in fact, no one knew, or knows now—whether the future means of delivering television signals would be through backyard or rooftop dishes (direct-broadcast satellite) or over expensive but almost limitless fiber-optic cables or through some hybrid of fiber-optic cables and existing coaxial cables, or whether those cables would be owned by cable companies or the telephone company or the government. No one knew or knows what will happen as the cable box mates with the computer; the phone with the cable wire; the networks with the studios; the studios with cable; the computer with the studios or with telephone, publishing, or electronics companies. To do any of this, Diller understood, required a convergence of three distinct forces: the emergence of "enabling technologies"; alliances among business adversaries; and government approval.

A report issued around this time by the Wall Street firm of Goldman, Sachs—"Communacopia: A Digital Communication Bounty"—reasoned to a similar conclusion: "As a result of rapid technological developments . . . we believe a true revolution in the delivery of entertainment, information, transactional, and telecommunication services may be at hand. Through a confluence of interests, this revolution could bring together a broad cross-section of industries that heretofore have considered themselves unrelated."

In the spring of 1992, the borders between the rivals' domains were blurring. Each rival dreamed of becoming a vertically integrated giant—able to control every step in the process, from the idea to the manufacture of it to its distribution—only to find that such integration was too expensive or too complicated without partners. So Apple Computer already had a joint venture with Sharp Electronics, Pacific Bell, Random House, Motorola, Bellcore, and SkyTel, to provide software and communications through the handheld personal digital assistant, which Apple is calling the Newton. Toshiba, a Japanese electronics conglomerate that excels in appliances and PCs, owned a minority stake in Time Warner. Sony made consumer electronics and owned a film studio and a record company, and so did Matsushita. Overseas, Hollywood studios and TV networks had been seeking local partners, in the hope of avoiding protectionist barriers and making the entertainment product more palatable to local tastes. Cable companies like TCI and Comcast had pushed into the telephone business, and the seven Baby Bells had petitioned the government to relax restrictions so that they, too, could provide cable and other information services in their own regions. (In February 1993, Southwestern Bell pushed into the cable business in another way, by buying two cable systems near Washington, D.C.) AT&T, having been barred from providing wired domestic phone service, had gone into the cellular-phone business and was already poised to make the chips that might convert analog to digital signals.

The first of Diller's field visits was to San Mateo, California, to see an electronics company called 3DO. The date was May 13, 1992. 3DO has developed what it calls a universal box—a device that will make home-entertainment and computing equipment compatible, and able to communicate with each other. This software company was already crowding Nintendo with such popular computer games as "John Madden Football," and now it had invented a box it hoped would be the enabler or control switch to operate all home-entertainment devices.

Diller was impressed with the games and technology, but not with the company's programming sense. He was given a preview of a Sherlock Holmes mystery game, and recalls asking, "How long does it take to play this?"

"Eighteen to twenty-two hours," responded the technician.

"Really? How do you know where you are in the process?" asked Diller.

"Huh? You just know when you're finished," the technician said.

"Have you ever thought of putting in act breaks?"

"That's a great idea! I'll have to work on this a lot longer," said the technician.

Diller remembers he walked away thinking, *He's making this game for true believers, people who already know how to play these things.* "That said to me that most of these people are talking to each other." In fact, the more places he visited, the more certain he would become of this.

Diller traveled on to Fremont, and there he toured the NeXT Computer plant and had dinner at the Palo Alto home of its founder, Steven P. Jobs, who was seeking to re-create the success he had when he co-founded Apple in 1975. After studying NeXT's brilliant software and graphics ("It's the most magical computer," Diller says), he told Jobs, "You've made these things too hard. It shouldn't be this hard."

"No," Jobs answered. "It's like learning to drive. It takes two months."

"No, it takes very little time to drive," Diller said. "A computer is not that—it's hard. Why make it harder?"

This exchange reinforced Diller's conviction that the technocrats were too insular, as did his visit to MIT, seven days later. While the Media Lab impressed him as a sort of Disney World of the future, with its hundreds of gadgets and technologies, he was unimpressed with the practicality of what he saw. But he did see something useful. "What I learned there is how digital can and needs to work," he says. Digital technology meant that television and radio would no longer be bounded by the narrowness of the highway over which they transmitted pictures or sound; instead, the signals would be converted into tiny numbers. In a universe exploding with channel choices, up to ten digitalized channels might be able to travel along a bandwidth formerly reserved for a single analog signal. Eventually, as digital compression technology was perfected, the picture quality would improve.

Throughout the summer, Diller talked on the telephone and met several times in New York with Brian Roberts, of Comcast. Roberts had recently been elevated to president by his father, Ralph J. Roberts, who was the chairman, and he was launched on an odyssey similar to Diller's. The two men enjoyed comparing notes. In a long telephone conversation on the morning of June 12, they swapped information about their latest technological sightings. Roberts remembers telling Diller that services like Prodigy, which allowed computer users to do their banking from home and to call up the news and weather, were maddeningly slow and dull, especially compared with what was coming. Too many offerings were not user-friendly. Roberts knew, as Diller now did, that the "techies" were often brilliant but in some ways dumb.

Roberts talked about Comcast. The company's income, principally from cable subscriptions and cellular telephones—two of the country's fasting-growing businesses—was up 11 percent in 1991, he said. He reviewed Comcast's varied holdings. He talked about QVC, of which Comcast and John Malone, of TCI, owned a controlling majority of the stock. He talked about how Comcast, jointly with TCI and others, had opened cable and telephone beachheads in England and was looking elsewhere overseas, and about Comcast's pay-per-view services and how digital compression and fiber-optic highways would permit hundreds—thousands—of viewer choices. Roberts said that the next step for Comcast and the $20 billion-a-year cable industry was to target the $12 billion video-rental business. "If I get a quarter of the video business, that's huge," he said. Comcast also had its eye on the $80 billion American telephone business. Annually, AT&T and MCI and other long-distance carriers pay roughly $25 billion to local carriers, including the seven Baby Bells, for access to customers, Roberts said. Once the cable companies installed fiber-optic wire, they could handle long-distance interconnections, not to mention local telephone calls. "They can pay us ten percent less. We'll do it," Roberts said. "That's what we're doing in England, where some twenty percent of the accessible homes are switching to our cable telephone."

But the most important piece of the puzzle for Comcast, Roberts emphasized to Diller, was also its most glaring weakness: programming. Unlike TCI or Time Warner, Comcast had not invested in programming. It had the hardware; now it needed software. Join us as a partner, he told Diller, and together we can make science fiction real. As the cable and the computer box and the telephone become linked, we can be pioneers. With Barry Diller, he said, cable can make programs that people want to see.

Diller wasn't ready to make choices, but he does remember thinking, This is one of the people I'm definitely going to talk with. First, however, he would be visiting Microsoft, in July, and he was looking forward to it. In many ways, Microsoft was a model of what Diller was seeking, for it created the software that enabled powerful personal computers to function. At Microsoft's corporate campus in Redmond, Washington, he was taken on a tour by Rob Glaser, the vice president for multimedia and consumer systems, and was shown what Microsoft was currently producing and what was to come. Then he had lunch with the company's chairman and founder, Bill Gates.

Diller assaulted his hosts with questions. He was particularly fascinated by Microsoft's attempts to devise cable boxes that could decode compressed digital signals, retrieve vast amounts of information, and allow viewers to interact with their TV sets. Gates has described such a system as "information at your fingertips"—a world in which, with a few clicks of a mouse, a customer can summon any movie, any program, any sporting event, any weather or news report, from a video warehouse with nearly unlimited storage capacity. "Our vision is to facilitate, to play the same enabling role that we did with home computers," Glaser says. "Today, people don't think of their TV set as having an operating system."

Microsoft's vision is not limited to TV sets. Diller learned—as Brian Roberts had on a visit to Apple Computer—how the personal digital assistant might work. With a handheld device, doctors would be able to swap X rays instantly for a second opinion, or review a patient's full medical history without rifling through a file, or monitor patients at home, or, with a pen, jot a prescription on the screen and send it to a pharmacy. Using the same built-in codes that identify shoppers, citizens would be able to vote from home. With interactive remote-control devices, children could take part in customized tutorials and quizzes, or play chess, instead of sitting passively before TV sets.

Many people have visited Microsoft's campus, including Rupert Murdoch; Martin Davis, the chairman of Paramount Communications; and John Malone, of TCI. "But," Glaser recalls, "of all the folks that have come by, Diller was the most engaged in the stuff we are doing." Diller recalls, "Everything blew my mind. I knew little. And each thing I saw made me think in ways I had not thought before." What he saw was that "a communications enabler could be inside the TV or beside it." He thought that a software company like Microsoft might provide the operating system for anything electronic in the home, just as it did for PCs.

Also in July, Diller accompanied Von Furstenberg to QVC's headquarters. He was captivated by the reach of QVC; through cable, this live, twenty-four-hour home-shopping network and its separate fashion channel reached a potential forty-five million homes. He was staggered by what he saw in the studio: instead of the usual army of producers, camera operators, production assistants, and high-priced anchors, a single producer and a group of product coordinators, accompanied by five robotic cameras, with a director and two engineers in the control room, above, followed the host—described as an "explainer"—as he or she displayed merchandise or interviewed celebrities and designers who marketed products carrying their signatures.

Diller liked the pleasant efficiency of the trained telephone agents there. Each agent completes a transaction in about two minutes, and there is a backup system that can answer up to a thousand calls automatically. Each transaction is recorded in an IBM mainframe computer, and then dispatched to one of three shipping sites, which fill an average of a hundred thousand orders a day. Even more, Diller liked the way QVC made its money: it marked up each item it sold by up to 100 percent; and because it purchased in huge quantities, it still undersold stores.

Von Furstenberg was already preparing to do business with QVC, and Diller would negotiate the deal. She would design what she called Silk Assets—dresses, skirts, pants, and a blouse—under her own label and exclusively for QVC. She would design the clothes and select a manufacturer, but she wanted no responsibility for dealing with distributing or storing the inventory. That task would fall to QVC.

Before the summer ended, Diller made another visit to QVC, to complete the negotiations, and to look around again. But he still wasn't ready to jump. He had more to see; for one thing, he was to pay a visit in early August to John Malone, Comcast's major partner in QVC. Unbeknownst to Roberts, Malone had pursued Diller over the telephone from the day he received word that Diller was leaving Fox. Diller and Malone had known each other for a number of years. Diller didn't have to be told that Malone was easily one of the most powerful figures in television, and not just because his company controlled one of every five cable connections in the United States (about ten million in all), or because it had a bigger cash flow than ABC, CBS, and NBC combined. Malone's influence extended into programming, since TCI had an interest in the Discovery Channel and the Learning Channel, and owned 22 percent of the Turner Broadcasting System, including CNN. Liberty Media Corporation, a cable-operations-and-programming company that Malone had spun off (although he is the chairman), partly to silence congressional complaints that TCI was a monopoly, also had a stake in a number of other cable-programming ventures. These included QVC, the Family Channel, Black Entertainment Television, American Movie Classics, and Court TV. It also owned or had an interest in twelve regional sports networks and in the Prime Network, a national sports-programming service. Another Malone subsidiary, Netlink, USA, was the largest provider of cable programming to owners of satellite dishes. And Malone had plans, which would not be announced for several months, to introduce digital compression, which he said would deliver about five hundred channels, beginning in 1994.

In preparing for those channels, TCI has already become the world's largest industrial consumer of fiber-optic cable. It is also the largest cable operator in England. And it has partnerships in various future-oriented endeavors with AT&T, U S West, McCaw Cellular, the Fox network, and Digital Equipment Corporation, among others. In addition, Malone, who has a Ph.D. in engineering, serves as chairman of CableLabs, a cable-funded communications research center in Boulder, Colorado.

"He's one of the great visionaries of our time," Martin Davis, of

Paramount, says of Malone. A cable-programming CEO who believes Malone squeezes cable networks for discount rates in exchange for the right to appear on his cable box says, "He's an evil
genius." Washington had its own opinions about Malone. Last year,
when Congress was able to override a veto by President Bush and
thus impose new regulations on cable, Malone was cited by fellow
cable operators as the sort of person the legislation was aimed at.
"John Malone is a monopolist bent on dominating the television
marketplace," Senator Albert Gore declared.

To better understand Malone, it helps to see him out West, at his office in Englewood, Colorado, just outside Denver. Here he comes to
work wearing a leather jacket, blue knit shirt, checked gray slacks,
and loafers, not the somber suits in which he appears on visits to
New York or on industry panels. He is square-jawed, dark-haired,
and, at six feet, has the build of a lumberjack. His passion for sailing
is reflected in a large model sailboat that rests on a credenza behind
his desk. In New York, Malone often appears brusque. Here he
speaks languidly, so casually that he can sound like Gary Cooper. His
office windows stare out at miles of empty plains, bordered by the
Rockies. This is a place that obviously gives Malone and TCI a sense
of freedom: there is space to roam and few competitors or customers
to encounter. "It's an outsider culture," Robert Thomson, senior
vice president of communications and policy planning, explains. "It's
a culture that doesn't define itself as establishment. It's not Eastern.
It's a Western culture. Like in *Butch Cassidy and the Sundance Kid*—
Who *are* those guys? That's us."
    Although Diller is no cowboy, he does swagger about how he
cares not a whit about "status" or "glamour" and only wants to be an
entrepreneur. Diller may look like a Fortune 500 executive, in his
dark suits and white shirts, but he thinks of himself as a buccaneer,
like Murdoch or Malone. That's one reason that, in early August, he
was eager to fly to Denver and spend a few days visiting Malone.
    "What's going on?" Diller asked.
    Malone explained that instead of sailing in Maine, as he usually
did, he and Brian Roberts and the executive committee of Cable-
Labs had been on the road visiting computer companies, including
IBM, Apple, and Microsoft. Technology was herding industries together, he said. As broad fiber-optic cable highways became available, and digital compression decreased the space needed on those
highways, the cable industry would look for technical help from
computer and software companies, and help from television pro-

grammers like Diller. The computer companies, he believed, "were stalled a little bit—they have penetrated the workplace but not the home." That's why they looked to cable. The seven communications powers circled one other warily. "It's all a blur," Malone said. He predicted a media landscape where cable companies operated in the phone-company business, and vice versa, and where network broadcasters were bought by cable, and vice versa. "It really all comes down to government regulations." One day soon, he said, technology would make possible a digital compression box "more powerful than any PC," and will transform the TV set into "an input/output device." He went on, "The TV remote control will become like a computer mouse. You've got to personalize television. If there are five hundred channels, you can't just give the consumer a scroll. That's the world of the future." So the one-word answer to Diller's question—"What's going on?"—was "Invention."

What particularly impressed Diller was that Malone, like Brian Roberts and others in cable, seemed to invest time in thinking and learning about technology. In October, Malone and Roberts and the CableLabs group planned to tour Europe, meeting with such electronic corporate giants as Holland's Philips N.V. and France's Thomson S.A., looking at technical advances and exploring possible partnerships. A year before, the same contingent traveled to Japan, where they met with the heads of Sony, Matsushita, Toshiba, and Pioneer.

Malone asked Diller what he was thinking of doing, and Diller said he had narrowed his search to four options: he might buy into cable; he might run a movie company; he might team up with a computer-software company; or he might acquire a television network. They spoke about QVC, and Malone sketched a vision of how technology could customize home shopping, allowing viewers to treat their remote-control devices like personal robots, for ordering merchandise, paying bills, collecting information on airline departures. And, he said, such a system would make a bundle of money. Malone told Diller that he hoped he might join forces with cable.

Diller still had a number of other visits planned, including trips that week to two telephone companies. He came away from those visits convinced that the telephone companies were strong in the laboratory and in cash but weak in one crucial area: although they had a wire reaching into every home, those wires might have to be replaced to carry high-quality video signals. To wire even half the homes in the nation with fiber-optic cable would cost the phone companies perhaps $150 billion and could take fifteen years to complete. And Diller suspected that the phone companies, after years of being operated as government-regulated monopolies, did not have the kind of entrepreneurial culture typified by Malone.

It was hard for Diller to abandon the idea of owning a network like NBC, even though that option seemed less and less attractive. Everything he was learning made it clear that more viewer choices would continue to drain customers from the networks. He believed that viewers no longer needed a network middleman—that technology would enable people to do their own programming. The networks were reliant on a single source of revenue—advertising—at a time when cable enjoyed two sources, ads and subscription fees. He thought that the networks, with about four thousand employees each, were freighted with too much overhead. Yet Diller was tempted. He thought that he could raise the money, and he hoped he could prevail upon General Electric, which owns NBC, to sell it to him for a lot less than the $4 billion GE was reported to want. He had spent too many years fathering TV programs—pioneering made-for-TV movies and the miniseries form for ABC, and promoting shows with an "attitude," like *The Simpsons*, on the upstart Fox network—not to believe in himself. Besides, he had a vision of making the network a healthy business again. There was the possibility of a twenty-four-hour worldwide news service, going toe to toe with CNN. There was also the possibility, he thought, of dumping the two-hundred-plus stations that constitute NBC's distribution system and replacing them with cable affiliates. This would save the $100 million NBC pays to its stations in compensation for carrying network programs. Instead of paying stations, grateful cable affiliates would pay subscription fees to NBC. Thus, NBC—if the federal government allowed it—would no longer be in the free-TV business. And Diller, always a proponent of free TV, would be the architect of its demise. There are NBC executives who have thought about such a revenue-rich plan, but they backed off, fearful that the government would never permit them to use the public airwaves in this way.

Diller had other ideas for NBC. Instead of airing the *Nightly News* at 6:30 P.M., as is done in most places, he wanted to move the half-hour newscast to 10:00 P.M., and combine local and network news within a half-hour wheel. To do this, Diller says, he would try to induce local stations to give up their lucrative late-night newscasts—in exchange for sharing the more bountiful revenues reaped at 10:00 P.M., when the audiences are larger. Then he would shift *The Tonight Show* to 10:30 and *Late Night with David Letterman* to 11:30, swelling network revenues because these shows would now reach more people.

Diller's conversations with NBC stretched into the fall. Owning a network "would be fun," Diller said recently. "But, even as I say it, I bore myself. In the end, I thought, it only involves ego." (NBC executives would later say that Diller's ego remained inflated; they insist NBC never had serious conversations with him about selling the network.) By the end of the summer, Diller had decided, more or less, where his future lay. He had come to believe that cable had the pole position in the race to control access to the home. Unlike the networks, it had its own distribution system, through its coaxial cable. Unlike the telephone companies, cable companies had wire already in place that could handle the video needs of the near future. It was possible, he knew, that a fourth option—direct-broadcast satellites sending signals to home satellite dishes—would win favor. However, he guessed that this technology was too expensive for consumers and was years away, and in any case, cable operators like Malone had hedged by investing in this technology as well.

The future, Diller concluded, would be "led by the cable systems." Sixty-three percent of all American homes were already wired for cable. Cable companies could charge for much of their programming, by pay per view, thereby perhaps alleviating government concerns about steep cable prices. In the words of TCI's chief operating officer, Brendan Clouston, "Cable will go à la carte. You pay for what you watch. Like your phone bill." And Diller agreed with Brian Roberts that some of that $80 billion in telephone bills could be siphoned off by cable. Finally, Diller came to believe that the cable people treated technology not as an adversary but as an ally. "They're livelier than most of the competition in their thinking," Diller says. "Talk to the cable people, as opposed to senior people in the news-gathering business, or at the TV networks, or in the studios. Just line them up, and you find that people in the leadership of cable are students of technology and spend vast amounts of time and capital thinking issues through."

Diller knew that he could bring several things to the table, starting with a flair for showmanship: only more programming could allow cable not only to fill those five hundred or more channels but also to spruce up the presentation of home shopping, or of information and interactive games. Another asset was Diller's relationships in Washington, particularly with the incoming Clinton administration. At a time when government would have more say—in whether to impose new strictures on unpopular cable systems that in recent years had raised rates twice as fast as inflation, in regulating how the Baby Bells would participate in the video-information-and-entertainment business, in whether to lift regulations inhibiting networks from plunging into the production and syndication business, in whether to invest in and own the fiber-optic cables—the cable industry needed political friends. "We bet on Bush," Brendan Clouston would later say. "We lost."

By the end of the summer, Diller and Brian Roberts had spoken so frequently that they had become friends. Roberts proposed that Diller should become cable's programmer. But Diller wasn't interested in being in the service business; he wanted ownership. The cost of buying a cable system was too high. If Comcast was serious about Diller, Roberts realized late in August, he would have to give up some ownership to make him a partner. But in what?

Diller had arranged to be in Philadelphia, where Comcast has its headquarters, on September 28, and to meet with Brian and Ralph Roberts. They spent the entire morning in a suite at the Four Seasons Hotel, looking for an idea they could all agree on. At one point, Diller mentioned his two visits to QVC, noting what an incredible operation it was. He said that he was thinking of producing an on-air segment for Diane Von Furstenberg's Silk Assets. Brian Roberts pointed out that QVC had an operating cash flow of $130 million in 1991 and was nearly debt free. He said that each cable operator received 5 percent of QVC's sales in its territory, which gave the operators an incentive to promote QVC and open more channels for it. Diller perked up.

For the remainder of the morning, the three men spoke of nothing but QVC. An idea popped into Ralph Roberts's head—how to make Diller a partner—and he sketched it out. He recounted how, five years earlier, he had helped the entrepreneur Joseph M. Segel start QVC. Segel had launched eighteen businesses, he said, and he enjoyed start-up situations. Now Segel was hoping to leave QVC at the end of the year. "There's an opportunity, if it is something of interest," Ralph Roberts remembers telling Diller. QVC, he thought, might be the means by which to marry Diller not just to home shopping but also to the cable industry's appetite for more programming. John Malone was already a part owner, and if Comcast and Malone were allied, as they usually were, they controlled a majority of the stock. Brian Roberts pointed out as an added feature that much of the cable industry had a piece of QVC, for in order to get home shopping launched on their systems most of the other cable companies had been offered stock in QVC. And there was always the prospect that QVC's main competitor, the Home Shopping Network, which Malone would soon own part of, could be merged with QVC.

"That was it," Diller recalled later. "After that meeting, I thought I was really onto something. Once I left that day, I thought it would be QVC."

Diller and the Robertses talked again at the Atlantic City Cable Show, in mid-October. By then, Malone was involved and enthusiastic. It could be a three-way partnership, with QVC as the vehicle for Diller's programming prowess. "Having a couple of good partners is as good as doing your own thing," the elder Roberts told Diller. Partners would provide deeper pockets and guarantee distribution of whatever programs or products Diller produced.

Diller made one more visit to QVC, at nine in the morning on Saturday, November 7, for Diane Von Furstenberg's first sale. Surrounded by her silk clothing, she sat beside the host on a small stage in a large brick cavern. Diller stood behind one of eighty telephone operators there and watched the toll-free calls pour in.

Diller was awed. At one point, he looked up and saw Von Furstenberg chatting with a woman from Brewster, New York. At another point, he looked down and noticed a colored bar on a computer screen surging, to register an increased number of calls. In less than two hours, the computer showed, Von Furstenberg had sold twenty-nine thousand items to nineteen thousand customers, for a total of $1.2 million. "This was the clincher," Brian Roberts says. "It was the ultimate Nielsen rating. The phones light up. You don't wait till you come into the office tomorrow to find out how you did." And Diller says, "It was the closest link I've ever seen between action and reaction."

Diller and Roberts flew to Colorado on a Saturday to spend a day with Malone, and a commitment to pursue a deal was forged between Comcast and Malone's Liberty Media. The deal ultimately called for Diller to invest $25.2 million to acquire 840,000 shares (roughly 3 percent) of QVC stock; he would also have an option to buy 6 million more shares, at an average price of just over thirty dollars a share. Diller would become the chief executive officer of QVC; when he exercised his options and his partners sold each other shares, he would own a third of the controlling interest in the company. Peter Barton, the CEO of Malone's Liberty Media, was thrilled. "I think Barry Diller's going to be punching a hole in the sky," he said. The lawyers for the three partners were instructed to move quickly.

Diller sees QVC as much more than a shopping network, and so does John Malone. "The shopping business itself can become a big business," Malone explains. "How big is the shopping-catalogue business? QVC does a billion dollars a year, and it's just scratching the surface." He believes QVC can be as huge as Wal-Mart. He makes it clear, however, that home shopping is not the ultimate aim of QVC. "It's about whatever we can cook up," he says. "It's a vehicle. It's a platform for Barry. We just look at Barry as firepower. We're delighted that he's wearing one of our uniforms."

After the December 9, 1992, announcement, Diller began spending about three days a week on QVC's corporate campus, in West Chester, Pennsylvania. He has a corner office on the second floor of a two-story red-brick building which faces a ridge of evergreens; they are so close that Diller can see none of the rolling countryside. Along the wall to the left of his desk are nine TV sets. In rural Pennsylvania, Diller is no less concerned with details than he was in Hollywood. He interrupts conversations to pick up his glasses and stare at QVC or the Home Shopping Network. "Why does Home Shopping sell pillows all the time?" he asks. "Maybe we should be selling pillows?" A moment later, he spots a QVC game wheel, much like the one on *Wheel of Fortune*, and says he intends to get rid of it, because it looks cheap. He often refers to QVC as "them" and says, "They've had such explosive growth in their business that they're reluctant to change."

Diller wants to persuade QVC to think of itself as more than just a channel. QVC can also be a shopping catalogue, a brand name, like L. L. Bean—a catalogue that Diller can program pictures for. He thinks that QVC's full shopping potential won't be tapped until it becomes truly interactive. As Diller envisions it, the customer will say, "I want a raincoat. Instantly! I want an umbrella," and QVC will figure out which are the cheapest ones, and deliver them to the customer's door. He predicts, "Three years from now you'll say, 'I want shoes.' You'll press a button and see yourself in various shoes on the screen." From their homes, he says, consumers will be able to roam the aisles of Bloomingdale's; avoid the last-minute Christmas rush by calling up a selection of gifts for the "special person," choosing one, and having it delivered the next day; find a hotel in the Caribbean, inspect its rooms and amenities on the TV screen, and then press a button to make a reservation.

Diller foresees selling QVC and other packages or services to disparate customers, including Time Warner, which recently unveiled a digital system that it plans to test in Orlando, Florida, this year—a system that Gerald Levin, Time Warner's chairman, has hailed as "the electronic superhighway of the twenty-first century." If it is successful, says Peter Price, president of Liberty Cable in New York, which is allied with New York Telephone (not John Malone), this "video dial tone" will allow customers to make their own decisions about what to watch and when.

The direction the video business is taking is toward lessening the power of the middleman. Networks and independent stations are middlemen, in that they schedule programs, which someone else usually owns, on certain days and at certain hours, or give the viewers the news they—not always the viewers—deem important. Consumers watching what they want when they want will gain a sense of participation, of empowerment. To this end, Diller envisions QVC providing viewers with news stories that present more historical sweep and context, and also with instant news. "Information services are something I plan to have a real role in," he says.

Diller also says that in several years he expects to be "in the storytelling form as well." He is sure that there will be some interactive element. He is uncertain at the moment whether his entertainment programs will be distributed over QVC channels or sold as packages. "That's around a dark corner," he says.

After ten months of peeking around dark corners, Barry Diller chose to enter the cable industry, but he has no illusions that the future of the TV box is clear. If customers want to discard the middleman, will the losers be the networks and pay-per-view cable channels, who mostly rent rather than produce their products, or will they be any channel, including QVC's? Will AT&T or the other telephone companies make a deal with the studios to bypass both the networks and cable, using the telephone wire as a distribution system? Perhaps talent agencies, which are already exploring this idea with telephone companies, will decide they can sell their talent packages directly to a distributor, thus bypassing the studios as middlemen.

Asked to guess the winners and losers among the seven vying industries, John Malone thinks aloud: "It depends on how you define yourself. If you define yourself as in the transmission business, you lose. . . . You've got to think of yourself as a supplier." The profits of the TV networks dwindled because each acted to protect a single channel, rather than to own or sell to other channels as well.

How advertising will mesh with this video-on-demand future is also not entirely clear. What is clear is that all parties welcome advertising dollars. While it's certain that government will establish the rules, it's uncertain what those rules will be. While it's known that customers want the freedom to choose, it's unknown what they will choose.

Surely it's unknown where technology will lead. Five years ago, the broadcast industry was terrified that it would be decimated by Japanese- and European-produced high-definition television. Playing referee, the Federal Communications Commission stepped in and ordered a time-out so that it could study the matter. Soon after, American engineers discovered digital compression, decimating the European and Japanese companies. Tom Super, who supervises NYNEX's laboratory in White Plains, New York, takes a visitor on a tour of his facility, displaying an impressive array of the latest interactive and compression technology. Back in his office, he tells the visitor that everything he just saw didn't exist twelve to fifteen months ago.

The economic and social consequences of the technology revolution are also unclear, of course, and so is the larger philosophical question of whether Diller's efforts—and those of technology in general—may further weaken our already fragile sense of community. Neil Postman, who has written books describing television as a narcotic, has now written a book about technology, entitled *Technopoly: The Surrender of Culture to Technology*. Although Postman speaks specifically of computers, he is making a broader point, and it is one that could be extended to QVC:

> Now comes the computer, carrying anew the banner of private learning and individual problem-solving. Will the widespread use of computers in the classroom defeat once and for all the claims of communal speech? Will the computer raise egocentrism to the status of a virtue?

In his Waldorf Towers suite, I asked Diller whether by creating narrowcasting channels and catering to individual cravings through video on demand he was further weakening the bonds of community and shared experience that, whatever the many vices of broadcasting, were, at least, a virtue.

"It's an interesting question, as a question," Diller says. "We don't know enough yet. We don't know yet what 'good' is in a more fractionated world of communication. I'm not interested in narrowcasting—that's not the direction I'm going in. As to what its value will be, later gets to judge."

POSTSCRIPT:

The situation would change with astonishing speed. A year later, John Malone tried to merge with a telephone company. Two years later, after trying and failing to buy CBS, the kind of television network he had disparaged as a relic, Diller left QVC, much richer but unemployed. By 1996, the computer and the Internet—not the cable-TV box—were proclaimed the new medium for interactivity; the notion that studios or cable programmers or editors were superfluous middlemen seemed downright silly. And Diller, in partnership again with Malone, was offstage, stealthily building a new television network and what he hoped would be a home for interactive communications.

2

THE COWBOY

John Malone's Cable Kingdom

(*The New Yorker*, February 7, 1994)

There is no television set in John C. Malone's office. This seems odd, since Malone, the president and chief executive officer of Tele-Communications, Inc., is the most influential man in television. After twenty years at the helm of TCI, Malone now controls nearly one of every four cable boxes in the country, and through TCI and Liberty Media, a programming company launched by TCI, he has an interest in twenty-nine cable services, including CNN, TNT, QVC, Court TV, Black Entertainment Television, American Movie Classics, and the Discovery Channel. Malone and his company have important business links to such corporate giants as Rupert Murdoch's News Corporation, Ted Turner's Turner Broadcasting System, Bill Gates's Microsoft, and AT&T; they have branched out into the telephone business here and abroad, and, in addition, they are readying two broadcast satellites. In all, since Malone started at TCI, in 1973, the company has acquired, invested in, or become a partner of about six hundred and fifty companies.

For all his previous success, Malone's most important deal was announced at a press conference in Manhattan on October 13, 1993, when he and Raymond W. Smith, the chairman of Bell Atlantic Corporation, said that their companies would merge, in a stock-for-stock transaction valued at $33 billion. It was the biggest merger ever in the communications industry, giving birth to a behemoth that will reach about one-third of American homes, will be able to call upon $60 billion in assets, and will rank as the sixth-largest American company. The merger increased Malone's worth—now some $1.1 billion—by about $300 million, and made him and TCI's founder and chairman, Bob Magness, major shareholders in the new company. (Malone will own 1.5 percent of the stock, Magness 2.5 percent.) This merger shifts the fundamental balance of power in the communications industry. Only a year ago, Malone boasted of how the cable industry would best the telephone companies. "He was captain of the cable team," says Brian Roberts, who is the president of the country's fourth-largest cable system, the Comcast Corporation, and has been a partner of Malone's in various cable ventures. "But now he is straddling two industries—perhaps as captain of both teams!"

One cable programmer, in discussing the power that Malone has, notes that if TCI's cable boxes aren't available, what's left is too small an audience to support a profitable service. A cable executive who does business with TCI explains the situation this way: "Imagine that you have all these publishing houses but only one book chain in the United States. You can write the best book, but what happens if you can't get into the bookstore? John Malone is the bookstore."

Malone transacts his business from an eleventh-floor corner office in a corporate park in suburban Denver. A block of gray granite serves as his desk; cactus plants surround an L-shaped couch of gray leather. The two exterior walls are windows, and from them one sees nothing but the snow-capped Rockies. Except for a six-foot-high replica of an 1854 America's Cup sailboat that stands on a stark black marble credenza behind his desk, and a smaller carved replica of *Ragtime*, a sixty-four-foot motorized commuter boat built in 1929 to go from Long Island to Manhattan, which Malone restored and now uses in Maine, the office seems impersonal. There are no mementos or plaques. The only touch of whimsy is a large fake-fur gorilla dressed in a white T-shirt and a red CNN tie, which sits on a chair facing the door; Malone instructs entering visitors, "Say hello to Ray Smith."

Malone is a private person—so private that he rarely asks even close business associates personal questions, and expects none in return. Not a single photograph is visible in his office—not even of his wife of thirty years, Leslie, nor of his daughter, Tracy, and her one-year-old daughter, nor of his son, Evan, a senior at the University of Pennsylvania. There is no photo of the man who has been his mentor, Bob Magness. An unframed picture of his wife is kept in a drawer; it shows a plain, trim, jeans-clad fifty-year-old woman with dirty-blond hair worn up in a bun. Hidden in cabinets are Malone's industry plaques and trophies.

When Malone appears on industry panels or testifies before congressional committees, he seems a frosty figure dressed in boxy suits and wearing the strained expression of a scientist or an investment banker. In Colorado, he seems much more relaxed. He is fifty-two years old, about six feet tall, and square-shouldered. Smooth skin, a square jaw, a long, thin, straight nose, a tight smile, and strong arms give him the look of a retired fighter pilot. Sports shirts, slacks, and loafers are what Malone wears to work. He drives a gray 1989 BMW or, on snowy days, a 1990 Chevy pickup with a plow. Of late, an arthritic condition has caused him to walk with a slight limp, which makes him appear almost vulnerable. He speaks in a monotone, but there is about him an unmistakable intensity. His eyes are hazel, and they fasten on visitors. "He's so focused," says Diane M. Grimshaw, who was Malone's secretary and administrative assistant for eighteen years and in 1992 became TCI's director of facilities. "I used to say that I could walk through his office stark naked and he wouldn't notice."

Malone does not indulge in bombast or theatrics. When he is talking, his hands usually remain jammed in his trouser pockets or wrapped around a coffee mug. He is an intense listener. On the phone, when Ted Turner or anyone else calls him, his conversations are unhurried, almost leisurely. He proffers advice, he asks for time to think, he calculates. He is a man of science: he has a Ph.D from John Hopkins University in operations research, which centers on the construction of mathematical models. In his business life, he strives for pure logic, unalloyed by emotion, unswayed by friendship or sentiment. He hates political noise and what he sees as Washington's aversion to facts. Of Turner, whom he counts as a friend and business partner, he says, "He's a very sound businessman. The only problem is that sometimes he lets emotions get the better of him."

The telephone does not ring in Malone's office—except when his wife calls on a line reserved for her. All other calls are screened by a lone secretary. For a man driven to succeed, Malone has an unusually brief office day—five and a half hours. He arrives at about nine, leaves at eleven-thirty to drive home for lunch with his wife, returns just before two, and leaves three hours later, sometimes to meet

Leslie at a local health club. To accommodate his wife, who will not fly, he bought a customized recreational vehicle two years ago—"a land yacht," he calls it, noting that it is the same size as a Greyhound bus. Twice a year (in late January and in early June), they pack their seven dogs into the vehicle, fill the gas tank, and drive nearly forty hours to Boothbay, Maine, where they have a vacation home. They spend as much as five months a year there.

The time away from the office, Malone explains, is both penance and reward. The reward is that Malone loves the sea and his wife. The penance is paid to his wife, who still feels cheated because her husband was off building an empire while she reared two kids. When speaking of Leslie, Malone sheds his reserve and becomes almost confessional. "She always felt she came second," he says. "She's always felt I haven't honored the you-come-first, the-kids-come-second, and career-comes-third commitment, and, I think, with some validity."

Why are the daily lunches, the coming home before dark, and the five months in Maine not sufficient? "It ain't enough," he responds. "The model I think that Les has, and to some degree I also have, is the model of daddy's home every night at five and when he's with the kids he's not thinking about anything else. And when you go on vacation, nobody calls you and you don't have to call anybody. Your focus is entirely on each other, and we've never really had that." She wants him to devote more time to the family, including Tracy, who recently moved with her husband and infant daughter back to Denver. "I'm not a psychologist or psychiatrist," he adds, "but you look back and say that the balance that we have today would have been an acceptable balance had I been able to strike this balance twenty years ago. . . . I have to overcompensate at this point." Leslie Malone is an intensely private woman; few in the cable industry, or even at Malone's company, have ever met her.

One reason Malone rarely appears at social functions is that his wife prefers not to. Another is that Malone and his TCI executives see themselves as outsiders who have successfully bucked the networks and the big boys of New York and Hollywood by helping pioneer CNN and Black Entertainment Television. To do so, they've had to be tough. And though to many Malone remains the Wicked Witch of the West (Al Gore, as a senator, referred to him as Darth Vader), admiring associates see him as both a genius and a liberated, balanced man. His mind is like a Rubik's Cube, according to Peter Barton, Liberty Media's president. "Take all these colors and six sides," Barton says. "Every square moves. It's a lot to keep track of.

Now think of a whole bunch of disparate concepts as a whole bunch of unrelated colors. Now remember what's on each square. And move the cube around until the concepts fit together. That requires an extraordinary amount of memory and processing capability. John does this."

Michael Fuchs, the chairman and CEO of HBO, says that along with the investor Warren Buffett and Bill Gates, the chairman of Microsoft, Malone is one of the best businessmen in the country, and adds, "The guy is brilliant." Fuchs, who serves with Malone on the board of Turner Broadcasting and is a sometime adversary, says, "Malone knows everyone in the industry but is a personal friend of no one. It's all business." Always, Malone is guarded, acutely aware that as cable and telephone and film and electronics and publishing and broadcasting companies continue to vie, an ally in one business or region becomes a competitor in another. "Being a CEO is lonely," he says, unapologetically.

Malone has always prided himself on flying solo. One of his most vivid childhood memories is of walking or driving past cemeteries and noticing the "awful" sameness of the gravestones. "Nothing different," he recalls. "Nothing distinctive. Nothing unique. So I've always wanted to be different, always wanted to be unique." He and an older sister grew up in an eighteenth-century Colonial house in a blue-collar neighborhood in Milford, Connecticut. For several years, his father, who was an engineer with General Electric, commuted to a GE plant in Syracuse, New York, on Sunday night and did not return home until the next Saturday. Malone worshiped his father and refers to him affectionately as "an intellectual with white socks." His mother was a world-class swimmer; she earned a master's degree in education from Temple University and interrupted a teaching career to bring up her children. But he remembers that she was a remote figure, who did not hug her daughter or her son, because "she had some fundamental insecurities." Malone thinks of his father as a strict Calvinist, and he also recalls that he was tight with a dime and insistent on two things: a rigorous work ethic, and a conviction that President Franklin D. Roosevelt was a socialist, or worse.

"I lived *American Graffiti,*" Malone says, referring to the George Lucas movie about teenagers in the early 1960s. He recalls souping up cars and "cruising the boulevard." In school, he had a nearly photographic memory and could recite entire passages from textbooks. When the Soviet Union launched Sputnik I, in 1957, and America was frantic to reclaim the lead in space, outstanding science students were in demand. John Malone was awarded a scholarship to Hopkins Grammar School, in New Haven.

Malone commuted daily to Hopkins, where he shone. But he was torn between his blue-collar friends and his intellect. His friends often taunted him, and he got into numerous fistfights. "I was neither a preppie nor a town kid," he says. "So I put a lot of energy into athletics." He got letters in fencing, track, and soccer. "It was raw drive, not skill," he says.

In the summer of 1958, when Malone was about to enter Yale, he was at the beach with friends one day when his sister approached and asked, "Want to meet a nice girl?" The girl was Leslie Ann Evans, a fifteen-year-old neighbor. Malone recalls strolling along the beach with Leslie, and making a silent vow that they would marry. While she finished high school and worked as a secretary with the Milford Police Department, he attended Yale. Science courses enthralled him, but social sciences did not. Thinking of himself as an "individualistic" conservative, he felt his teachers and most students were "socialistic." He argued with his teachers, and he says that his grades suffered as a result. "But when I got over into engineering, of course, the answers were either right or wrong," he says. "So as soon as I got into engineering school my grades took off."

Malone graduated with a B.S. in electrical engineering, in 1963; he and Leslie were married later that year, and he went to work for AT&T's Bell Labs. Bell helped finance two master's degrees that Malone received—from Johns Hopkins, in industrial management, in 1964, and from New York University, in electrical engineering, in 1965—as well as his doctorate from Johns Hopkins, in 1967. A rising star, Malone may have been smarter and better paid than his peers at Bell, but he felt like one of many tombstones. "The odds were that you were going to spend your whole life writing memos and making basically no impact on anything," he recalls. He hated the timidity, the bureaucracy. Looking around, he says, he saw "guys who had been me, ten years earlier, who had basically given up." He was determined to be different.

In 1968 Malone joined the management-consulting firm of McKinsey & Company, seeing it as "a place to retool"—as an opportunity to go from science to business. The Malones moved into a house in Weston, Connecticut. Leslie Malone expected that her husband would be home as often as he had been while he was working for Bell Labs, and could join in bringing up their daughter and son. He wasn't going to be like his dad, with a home in one state and a job in another. On his first day at McKinsey, Malone found a note on his desk telling him that a client expected him in Montreal for lunch; he stayed in Montreal six weeks. From then on, since his clients included IBM, GE, and other lofty corporations, Malone was often traveling. Without manuals to follow, he remembers, he was on his own, inventing his own answers. He learned as well what he did not like. "My years at McKinsey gave me tremendous disrespect for large corporations," he says. "It seemed to me, almost certainly, in most cases, there were enormous inefficiencies. And the guys who climbed the ladder to the top had become specialists in climbing ladders and really lost touch with the business."

While Malone was at McKinsey, he saw the marriages of colleagues dissolve. He told his wife that he would find a job requiring less travel. In 1970, he joined the General Instrument Corporation and was soon named president of its cable-TV division, Jerrold Electronics. Over the next several years, Jerrold became a dominant force in cable. Malone got to know most of the pioneers in the infant cable industry, including Steve Ross, of Warner Communications, and Bob Magness. A part-time cattle rancher and cottonseed salesman who wanted to improve television reception in rural areas, Magness owned a struggling cable system, based in Denver, that reached into twenty-three states, and also a new microwave broadcast company. In 1972, Ross tried to hire Malone to head Warner's cable operations; Magness offered him the presidency of four-year-old TCI.

Leslie Malone was not happy living in the East. She had a daughter who rarely saw her dad. She wanted to have family dinners at home, and to live someplace where they could keep horses. As for her husband, he recalls, "I wanted to get out of New York. I liked Steve [Ross] an awful lot. I found him to be hypnotic. But I wanted to spend time with my family. I had made this commitment to my wife that we were going to have a more normal life." Besides, he and Magness shared a dream. They envisioned loosening the stranglehold that, in Malone's words, "three channels, controlled by three guys in New York," maintained. On a typical evening in the mid-seventies, nine out of ten viewers watched CBS or NBC or ABC, and the only national cable service was HBO, which was then relatively small. Malone took a 50 percent cut in pay and accepted sixty thousand dollars a year as chief executive officer of TCI. Magness would be his partner, the company chairman. Malone's base would be Denver, which was the capital of the embryonic cable industry.

Tim Wirth, a former Democratic senator from Colorado, recalls attending a cable meeting at a Denver hotel not long after Malone arrived in town. Wirth says, "I remember Malone's determination"—to change cable regulations. "You could see his crooked little smile. He looked at you with that smile and you could tell that behind it was a tank." Malone had reason to feel determined. Soon after he joined TCI, in late 1972, the stock market fell, and interest rates climbed; then, within months, oil prices soared. TCI had $130 million in debt and $19 million in revenues, Malone recalls.

Despite his good intentions, Malone was rarely home. "Of course, I might as well have stayed in the East," he says. "I was living where the banks were—in New York and Boston—for the next five years." His mission was not to expand the business but to save it from bankruptcy. "We looked catastrophe in the face every day," Magness recalls. When he and Malone traveled to make their pitch to creditors, the two shared a motel room. Hourly employees were put on a four-day workweek.

"We had to be very tough," Malone says. The backs-to-the-wall circumstances of their early years, combined with their sense of mission, and with Malone's Calvinist proclivities and Magness's cowboy nature, helped to shape a corporate culture that has by now become either entrepreneurial or ruthless, depending on one's vantage point. It was Malone and Magness against the world. "He was my board of directors," Malone says of the chairman, whom he still calls "a mentor, a father figure, a friend, a partner."

TCI achieved a breakthrough in 1977, when Malone persuaded seven insurance companies to refinance his bank debt with longer-term capital. Money was thus loosened enough to begin making modest acquisitions. More money poured in after HBO and Ted Turner made their cable-programming breakthroughs, in the late seventies, and cable-company stocks soared.

Malone had put out TCI's business fires, but he had problems at home. As Malone remembers it, Leslie told him, in the early seventies, "I thought I was marrying an engineer at Bell Labs. You were going to be home every night. We were going to have a normal middle-class life and raise the kids and take the two-week vacation and so on. And now you're off on this ego trip of building a career. You're off running companies, things that I never bought into. If that's something you gotta go do, so do it. But I'm not coming along. Goodbye, I'm leaving."

Malone had reassured her that Denver would be different, but so far it had been worse. Now he promised to change. They bought a 1904 Stanford White house on sixty-three acres in Boothbay. Then

TCI acquired its first jet, and the plane allowed Malone to shorten two-day business trips to one day. In Denver, he and Leslie started going to a health club together. Diane Grimshaw, who was then Malone's assistant, says of them, "A lot of John's motivation is his wife." David E. Rapley, who runs a successful engineering firm in the Denver area and, with his wife, Sandra, is among John and Leslie Malone's few intimate friends, says, "It's difficult for him to talk about his innermost feelings. Yet he has a real soft center. He's a very analytical thinker. He's an engineer, trained in logical thinking. Facts. He's not as comfortable with emotions." Rapley says that by contrast, "Leslie is a touchy-feely person." Yet Rapley sees Malone as totally dependent on his wife. "If, God forbid, anything happened to Leslie, it would be hard for John to make a life. It would be easier for Leslie to carry on without John. She's had to make her own life. She paints. She buys antiques." She also studied interior design. And she reportedly has more gaiety.

Grimshaw, looking back on her years as Malone's assistant, remembers, "I really noticed a difference in John." He began to listen more, to open up a bit. He began to pick up Tracy and Evan after school when Leslie was busy with her own projects. He began to buy her presents, including a red Jaguar.

Malone could afford the Jaguar and the jet plane because TCI was now thriving. A dollar invested in TCI in 1975 was worth eight hundred dollars by 1989. In 1992, the company's cash flow was nearly $1.6 billion. TCI and Liberty Media are today the largest investor in cable-programming networks. (By the beginning of 1994, the number of cable networks had mushroomed to ninety-two.) Malone serves as chairman of CableLabs, in Boulder, Colorado, which is the cable industry's laboratory to decode the future. Like many of his Japanese counterparts, and unlike the CEOs of most American communications companies, Malone actually knows how to assemble, say, a radio. Because he can ask technical questions and understand the answers, and because he has experience with customers and economics, Malone may see the future a bit more clearly than his competitors do. Consequently, he has been at the forefront of technological advances, among them digital-signal compression, which will multiply by up to ten the number of signals in existing copper wires; the ordering of set-top boxes from General Instruments to convert these digital signals into the analog signals viewed on a normal TV set; experiments with McCaw Cellular Communications to test cellular networks; the examining of video on demand in the Denver area with AT&T and U S West as partners; and the test-marketing of home-energy information services with Microsoft and the Pacific Gas & Electric Company. It was Malone who announced, in December of 1992, the advent of something that is now part of the language (if not yet a reality): five hundred channel choices.

By the early nineties, Senator Howard Metzenbaum of Ohio, among other political figures, was regularly assailing Malone as a monopolist. In October 1993, Malone was invited to testify before Metzenbaum's Subcommittee on Antitrust, Monopolies, and Business Rights. When he declined, saying that he would be traveling overseas, Republican and Democratic senators alike charged that he was arrogant. Some insisted that he appear. When Malone did appear, in December, he suggested that he was being treated like a criminal. "If I relied solely on recent press stories to learn about TCI and the role that it plays in the entertainment-and-information industry, I would think that I was dealing with an entity combining the power of the pre-divestiture AT&T, IBM when it was the only name in computers, and the U.S. Postal Service before anyone ever heard of UPS or Federal Express," he testified. The questioning of the witness, which attracted a throng of reporters, was surprisingly respectful—a spirit enhanced by Malone's adopting an agreeable pose. "By letting Bell Atlantic acquire TCI, you kind of get rid of me," he told them.

Although both men would probably be uncomfortable acknowledging it, liberals like Metzenbaum and conservatives like Malone have some common ground. In many ways, Malone and TCI do and have done exactly what many liberal critics denounce American corporations for not doing. TCI thinks long-term, and not just about the next quarter. It did not fatten its dividends by skimping on research investments. Although TCI enjoys astonishing—some would say excessive—profit margins, of more than forty cents on each dollar of revenue, it witholds dividends and instead invests in research and new ventures. "The nub of the problems of American business is shortsightedness," Malone says. "That's a big difference between here and Germany or Japan, where there's a much longer view and a much longer cycle." Malone, like Metzenbaum, was no fan of the frenzy of hostile takeovers in the eighties, which, in the name of slashing corporate waste, also slashed research-and-development funds and choked companies with short-term debt. This is one reason Malone holds most investment bankers in contempt, and often serves as his own banker, as he did in the Bell Atlantic merger. Malone believes that bankers are preoccupied with fees. He also believes that most businessmen, like most politicians, are conformists, eager to follow the herd. "I have an investment mentality rather than a control mentality," Malone says. "Time Warner has a control mentality."

Malone can be both predictable and surprising. His political heroes are Ronald Reagan and Dwight Eisenhower, since he shares their belief that big government and big taxes sap initiative. Yet he calls himself "a libertarian," and disagrees with Reagan-style conservatives who want the government to tell a woman that she may not have an abortion. But he is not libertarian in his views on violence and sex on the small or large screen, and he has kept the Playboy Channel off most of his cable systems. On theoretical grounds, he doesn't want the government to prevent people from buying guns but he acknowledges the carnage on city streets, so "as a realist" he supports stringent gun controls.

Malone also surprises by saying that the TV networks, which he has spent many years assaulting, are still attractive properties. He is counseling his friend Ted Turner, who would like to buy ABC or NBC. "My guess is that CBS is not long for this world," Malone says. "It will probably merge with Disney within twelve months." Barry Diller, the chairman of QVC Network, might want to acquire NBC, but, Malone thinks, government regulations might prevent this. If Turner could acquire a network and combine CNN and a network news operation, he could realize $200 million in savings, Malone guesses. (Whether the endeavor would improve news is another matter.) One impediment, he warned Turner, is that the government might not look kindly on such a deal, since Malone is Turner's largest shareholder and is already being accused of monopolistic practices for his proposed merger with Bell Atlantic.

Despite Malone's self-image as a man of logic, his response to certain matters seems to be emotional. He repeats the word "socialistic" like a mantra, saying that he picked it up from Rush Limbaugh, whom he applauds for a willingness "to say politically incorrect things." In fact, he picked it up from his father. He uses the term "socialistic" to berate liberals, and even moderates, who support government social-welfare programs. He wants NBC, CBS, and ABC to be uprooted from New York, so they will be less elitist. "There's a huge diversity of values in this country between what people in central Manhattan think of the values of our society and what people in Peoria think the values of our society are," he says. "I think Manhattan thinks that single parents are fine. I think it's a much more socialistic world there, where the government does everything for you." The networks, he thinks, "suffer severely from an inner-city orientation that is reflected in our media very heavily, and it drags our society away from what I call the traditional American values. I would have loved to see Ted Turner take CBS to Atlanta and leave most of the overhead behind. I think you'd see a lot less violence, frankly." Not only does Malone ignore that most network programming comes out of Los Angeles, he also speaks of Manhattan and New York City's four other boroughs as separate places, as many tourists do. John Malone's New York is a stereotype, where the only good schools are private, and anyone with money or fame is menaced.

Yet Malone's own life defies stereotype. Preferring privacy to convenience, the Malones live without servants—no cook, no driver. Their boat has no crew. "We're not socialistic," he says, this time meaning not elitist. "In Maine, we exist. We go out to the movies, or rent them. We go out to dinner by boat in Maine." When they visit their house in Boothbay in winter, they watch a lot of movies and ice-skate on their one-acre pond. In the summer, Malone is up at sunrise each morning and walks the dogs, feeds the fish in their pond, checks on his apple orchard and vegetable gardens, spends an hour in his office reading his e-mail and going through material sent by his office, then goes downstairs and makes tea for two. Often, he rows two hours to the boatyard. When they spend the day on the boat, Leslie will pack a picnic lunch and the two of them will be off. Leslie has a studio, and sometimes she will spend the day painting while he putters about in his garden. Sometimes they will visit antique stores. Occasionally, they will have nonfamily guests—like the Rapleys from Denver. Maybe once a week, the TCI plane will ferry him to a meeting in New York or elsewhere. He has two faxes, three phone lines, and a video telephone for conferences, but he says he tries to restrict his business to an hour a day. No one from the office calls him there, says his secretary, a cheerful red-haired woman named Martha (Marty) Flessner.

In Colorado, the Malones live in a rural area, on a hundred acres bordered by a security fence. Their driveway winds for three-eighths of a mile, leading to a comfortable but not opulent house and a barn with three horses. Though the house is only twenty minutes from the TCI offices, few who work for TCI have ever visited it, for Leslie Malone insists that her husband's personal and business lives be kept separate. Bob Magness says "She's a nice lady," but admits, "We don't see them a lot socially." Ray Smith, of Bell Atlantic, who has been negotiating the merger with Malone since June 1993, and who will become the merged company's chairman, says he has never met Leslie Malone.

The Malones are close to one Denver couple—Sandra and David Rapley. She is an accountant who first met Leslie at an aerobics class about seven years ago, and he is an engineer and the son of an Oxfordshire cobbler. The couples often work out together and meet two or three times a week for a movie or a low-fat dinner. Dressed in jeans and sweaters, the foursome spent New Year's together at the Rapley home, eating burgers and leftover pecan pie, Rapley recalls, before watching a tape of the original *Citizen Kane* that Ted Turner had sent Malone as a gift. At the movie's conclusion, Leslie Malone declared, "That's you. You identify with that guy!"

"It's true," Malone recalls saying. "I do identify with him."

Malone may identify with a tycoon, but as a chief executive officer he has a management style that is curiously indirect. He relies on Brendan R. Clouston, the chief operating officer of TCI, to run cable systems, and on Peter Barton, the president at Liberty Media, to run programming operations. Malone himself concentrates on strategy, deal making, technology, and cable programming. "Brendan is basically the president of the company, though we don't call him that yet," Malone says. His spending so much time in Maine, and declining to attend most operations meetings or give direct orders, "lets the organization grow up," he says. "I've sort of worked my way out of a job."

He has not worked his way out of the job fast enough to suit quite a few people in the cable industry and elsewhere, however. The criticism of Malone and TCI amounts essentially to three charges: that Malone is a ruthless bully; that Malone can't be trusted; and that Malone is a monopolist. Accounts of TCI's coarse tactics are legendary. In 1973, when TCI and city officials in Vail, Colorado, could not agree on cable rates, TCI pulled all its programming for an entire weekend, substituting on TV screens the names and home phone numbers of city officials. The city surrendered. Eight years later, in Jefferson City, Missouri, TCI threatened to let the screen go blank if the city didn't renew its franchise. Again, city officials succumbed. A cable competitor filed an antitrust lawsuit against TCI, and a jury fined Malone's company $36 million. "Keep in mind that in those days there were no rules or laws defining what a cable franchise was," Malone says of both instances. As for Jefferson City, Malone told Congress in December that the TCI employee responsible "was discharged once TCI learned of his unauthorized actions." He noted that in 1988 Jefferson City renewed TCI's "nonexclusive franchise" for eleven years.

Within the cable industry, there is a widespread belief that TCI treats its cable-box customers arrogantly, and this has blackened the eye of the entire cable industry. Malone acknowledged that TCI's service was inferior when he testified at Senator Gore's November 1989 cable-oversight hearings. The truth is that the flip side of TCI's entrepreneurial culture—of fighting against the odds to build a successful company, of pouring every spare dollar not into dividends but into new investments, of acting as if surrounded by adversaries—is that TCI has sometimes been guilty of roughhouse tactics.

The second charge—that Malone can't be trusted—was made by Martin Davis, the chief executive officer of Paramount, in a recent *Vanity Fair* account of the battle for Paramount between Viacom and QVC. It is a charge that has also been made by some members of the cable industry. "His word is useless," the CEO of one major cable rival says. "He's the best businessman I ever met. On the other hand, in terms of character, trust, or honesty, he's one of the most dangerous men I ever met." Another cable-programming executive, who, like most Malone critics, asks to remain anonymous—in his case, because his service is one of the twenty-nine partly owned by Malone's companies—says that even after you think you've reached an agreement with Malone and his people, they often reopen negotiations. "On Monday, he can say to me, 'I want to invest in this, it's great,' " this executive says. "On Wednesday, he or Peter Barton will say, 'I don't think it makes sense.' " TCI, another cable operator says, has this philosophy: "I want mine. And I want what's not yours."

Ted Turner dismisses the criticism that Malone is a greedy partner. "He's a wonderful partner, and I say that after seven years or so of having him on my board with veto power over what we do," Turner says. "He's been exemplary as a director and as a partner. He's as honest as the day is long. He's fun to be with." Malone's frequent changes of mind, his defenders explain, stem from his restless intellect. Another Malone cable partner sees Malone's mind in a somewhat different perspective, defending him as personally honorable but so intellectually adept as to be frightening. "This is a man who in a two-day period can have four hundred different transactions going on," says QVC's Barry Diller, who is a sometimes wary but always respectful partner. "As someone once said, 'He has a frictionless mind.' The next idea has no historical relevance. He forgets what he said eight minutes before. But if you make an agreement, you can count on it completely."

The third—and potentially most damaging—charge made against Malone and TCI is that they are a monopoly and use muscle, just as the robber barons of the nineteenth century did, to gain advantage. The claim received new prominence last September, when Viacom, which competes with TCI in various programming ventures, filed a lawsuit charging that Malone had violated the Clayton and Sherman Antitrust Acts, by maneuvering to deny competitors access to TCI's markets. The heart of Viacom's suit is contained in this passage:

> Malone's scheme to monopolize cable television begins with his empire of cable television systems. Malone-controlled defendants Tele-Communications, Inc. ("TCI") and Liberty Media Corporation ("Liberty") have amassed local cable monopolies controlling approximately one in four of all cable households in the United States. . . . Defendants' monopoly power as cable operators, together with their expanding interests in all other aspects of the cable industry, gives them unparalleled power to dictate terms to cable television programmers, such as Viacom. Without access to Malone's systems, cable network programmers cannot achieve the "critical mass" of viewers needed to attract national advertising or a sufficient number of subscribers required to make the network viable. . . . As a result of Malone's unique control over this life-line, he can—and does—extract unfair and anticompetitive terms and conditions from cable programmers, including Viacom.

Among the coercive tactics alleged to have been employed by Malone, perhaps the most serious concerns his relationship with the Learning Channel. Back in 1990, this fledging cable-programming service was put up for sale. The Lifetime Television Network, which is owned by Viacom, the Hearst Corporation, and Capital Cities/ ABC, stepped forward, ultimately offering $39 million. Sumner Redstone, the chairman of Viacom, charges that after Lifetime's owners negotiated a sale agreement they went to see Malone and he told them he planned to drop the Learning Channel from many of TCI's cable systems. As the Viacom lawsuit says, "Lifetime then withdrew its bid because of his threat." Four months later, the Discovery Channel, which is partly owned by TCI, bid $30 million and ended up buying the Learning Channel.

One executive who was not a principal in the negotiations but is close to the Hearst Corporation agrees with Redstone's claim that Malone used inappropriate strong-arm tactics. Herbert A. Granath, who oversees all cable investments for Capital Cities/ABC, and who attended a meeting of the three owners of Lifetime with Malone, offers an account of the meeting that, while critical, is closer to Malone's version. "We flew out to see John and said we were willing to turn it into a viable channel and invest more," Granath recalls. "But we wouldn't do that unless John agreed to keep it on the air a couple of years. He said he could not make that commitment." The reason Malone gave Granath was that many of his cable systems had already been told they "were free to take on Mind Extension University," Granath says, "And some had. That could be true." Granath says of Malone, "I never found him untrustworthy. He tells you what he's going to do, and he does it. He's just very tough."

Being tough is not a felony, says Malone, aware that all business executives seek an advantage on behalf of their shareholders. "I play hardball," Malone admits. "But I'm very direct. And I win, frequently." What happened with the Learning Channel, he says, is that it was both a financially and a programmatically weak service at the time, and TCI, to protect itself against the possible bankruptcy of this programming service, sent a standard notification to the Learning Channel's principal owners that TCI reserved the right to stop carrying the channel. Many of TCI's local franchises never carried the Learning Channel, he says, and the Learning Channel was airing a lot of infomercials. "We contracted to distribute Mind Extension University, which we don't own a piece of," Malone says. When the three partners from Lifetime visited, he says, he told them his complaints about the quality of programming on the Learning Channel and reviewed the number of local operators that had already made other choices. He also says he offered to reconsider, and to pay more if they made the programming "first class." But they decided not to go forward. When Discovery bid, he says, it was "the only bidder." Since he owns a piece of Discovery, he is boosting the value of a TCI investment.

There has been no shortage of allegations made against Malone and TCI. But they all dance around the same basic question: Is TCI a monopoly? Here, too, there seems to be no definitive factual answer. Redstone says yes. Others say maybe. Malone says no. People who say TCI is a de facto monopoly, even though it controls almost one in four cable boxes, predicate this claim on their belief that no programming service could survive without access to Malone's subscribers.

One cable programmer explains: Nearly a third of all the cable systems in America can accept only about fifteen channels, which eliminates newer programming services. This leaves a target audience of about 66 percent of cable boxes. Subtract TCI's boxes, and what's left is too small an audience to support a profitable service.

Malone rejects the charge that TCI is a monopoly. "TCI is a large company, but we are certainly not a giant, whether compared to other telecommunications and technology firms or American companies as a whole," he testified before Congress in December. "TCI's 1992 revenues (including Liberty) were a little over three billion seven hundred million dollars. By comparison, AT&T and IBM each had 1992 revenues of about sixty-five billion." TCI, he said, was about a quarter the size of Time Warner, the nation's second-largest cable-system owner. It was about a third the size of each of the seven Baby Bell telephone companies. Besides, he operates under Federal Communications Commission guidelines. The FCC has ruled that ownership of more than 30 percent of the nation's cable boxes would threaten to become a monopoly, and the commission restricts ownership to that limit—a restriction that Malone says is reasonable.

Malone asks, "If we're cutting such a hard deal, how come Sumner Redstone is making so much money?" The real money in cable, he says, is being made by programmers like Redstone and by HBO, ESPN, USA, and, of course, Malone's Liberty Media. The chief reason cable-subscription rates have risen sharply in recent years is that programmers have boosted rates between 30 and 50 percent annually. "No one in Congress has sat down and looked at it," Malone says. The debate, he contends, is all about emotion, not facts. Malone is not wailing about his own cash flow, which is robust, partly because TCI and Liberty have bought minority interests in so many programming networks. He is simply arguing that the balance of power belongs to programmers, because even if he wanted to, he could not deny his subscribers such popular fare as HBO, MTV, Nickelodeon, and CNN. "What Redstone wants is to raise prices on his product, and I won't resist," he says.

Where does the truth lie? TCI is not a monopoly in the traditional sense. After all, TCI controls only about one-quarter of all the cable boxes. Nor, even though Malone has fought with both Redstone and Time Warner, can he easily deny a place on his shelves to such brand names as MTV or HBO, which are owned by Redstone's Viacom and Time Warner, respectively. Malone bridles at being called "evil" for doing what he believes that most business colleagues would do. "People ascribe venality to someone who is only acting in his own interest," former Senator Tim Wirth says. Some part of the criticism of Malone is envy.

Some part is not. Malone's company does, as he acknowledged, play hardball. It did remove HBO from some cable systems in Texas, because "we needed to get their attention," a TCI executive concedes. This executive says that TCI was losing subscribers, and felt it needed to renegotiate its agreement with HBO. It is also true that with the number of channels still limited, Malone does possess life-and-death power over new programmers that have yet to establish themselves. And he has not been timid about asserting his power, sometimes ruthlessly.

By the early nineties, Malone could not shake another truth: He had become, for the cable industry, the Wicked Witch of the West. For one thing, he was mysterious. He rarely granted long interviews. He didn't linger at industry conferences. TCI didn't advertise, or pour large sums into political campaigns, or retain high-priced public-relations handlers or lobbyists. Malone didn't court members of Congress or the executive branch. He remained a loner. He maintained a naïve faith in the power of his argument, in logic, even though he had no faith in the logic of Washington or the media.

The attacks mounted. There were anonymous physical threats, Malone says. He hired security guards for Leslie and for his children. His hair, which had been almost black, turned gray. And Leslie became more agitated. She couldn't sleep, he says, and worried about the children, and bridled when he was called Darth Vader. She pressed him to leave TCI, to strike a better balance between family and business. Last fall, when Malone couldn't shake a persistent flu, his doctor told him he was suffering from stress.

Whenever Malone thought about Washington, he felt even more stress. A Democratic government was in the White House. Malone's old nemesis Al Gore was now vice president. And Malone worried about what the government would permit. Would it allow him to run both a cable system (TCI) and a programming company (Liberty Media), or would it compel him to divest? Would it allow him to make more deals? Portrayed by his foes as all-powerful, Malone actually felt powerless to stop the freight train he saw barreling down the tracks toward him and TCI. The uncertainty made him more tentative.

Malone's confidence that cable could be king was also shaken. The telephone Goliath loomed larger in his mind. He believed that the Clinton-Gore administration would unshackle the telephone companies, allowing the seven Baby Bells, AT&T, and other long-distance carriers to carry video on their wires, thus challenging cable, without at the same time unshackling cable companies so that they could enter the phone business. He knew that balance sheets were not cable's friend. "The cash flow of the entire cable industry is one and a half times the size of one Baby Bell," he has said more than once. Huge amounts of capital would be required to build an information superhighway—to lay the fiber-optic wire, to invest in programming to fill five hundred or more channels, to order the high-speed switches and set-top boxes that would let viewers call up a movie or a sporting event instantly on their video jukeboxes, or interact on their computer network. The cable industry alone didn't have the capital. Nor did it have the capital to fulfill Malone's longtime dream of opening a second front by challenging the Baby Bells in the telephone business. Critics said that Malone was too big; he thought TCI was too small. "It would be a seven- or eight-front battle for us," Malone told Metzenbaum's subcommittee in December.

"Eighteen months ago, I started telling Bob Magness and the board that I felt we had to do a strategic deal with a phone company or we were going to have serious problems long-term," Malone says. To go into the phone business alone, he felt, would risk too much of their capital. His first preference, he says, was a "strategic partnership" with AT&T, in which TCI could draw on AT&T's capital and technology—could remain independent while having "a big brother." There were many meetings and discussions between Malone and AT&T. Malone sat on the board of McCaw, the largest cellular-phone company, and was aware that that company was about to merge with AT&T. Together, the three of them could compete against the Baby Bells—or cable, for that matter—in providing wired or wireless services. But such a joint venture would require a nod from government, and that was uncertain. Besides, Malone says, AT&T feared that if government deregulated the phone business too fast, the Baby Bells would challenge AT&T's long-distance business before AT&T was ready to challenge their local phone business.

Malone looked to other partners. "I made a deliberate effort to get to know the phone company CEOs," he says. There were advantages and disadvantages to each company. U S West, for example, would have been "the easiest way to go," says Malone. TCI and U S West were already partners in the cable/telephone business in

England, and they knew each other well because they served the same western states. But because their service area overlapped, he feared that the government would require them to divest assets. And U S West had invested $2.5 billion in another strategic partnership, with Time Warner, which was both Malone's largest cable competitor and his partner in several ventures, including Turner Broadcasting. Malone had a preferred partner. "Emotionally, I'd love to do a deal with MCI," he says, "because they're our kind of company. Very entrepreneurial; very much the Don Quixote approach to life." But, he adds, "it would be a real go-it-alone, declare-war-on-the-rest-of-the-world approach. It would be hard to do to AT&T, when they've been so good to us." Perhaps.

Craig McCaw, the chairman and CEO of McCaw Cellular, suggested that Malone get better acquainted with Ray Smith, of Bell Atlantic, whose headquarters are in Arlington, Virginia. Malone had already met Smith while appearing on an industry panel with him, and Malone had visited Bell Atlantic two years before and dined with Smith. In May of 1991, they had become partners in a satellite programming service in New Zealand. Malone admired Smith's courage in challenging, in federal court, government prohibitions that kept telephone companies out of the video business. Smith claimed that the prohibitions were tantamount to a denial of free speech, and in May 1993 he won his case. Bell Atlantic's balance sheet also impressed Malone, for in 1992 the company generated revenues of more than $12 billion, had relatively little debt, and produced net income of $1.3 billion. Although he was a CEO with more than thirty-three years in a government-regulated monopoly, where customers and profits were assured, Smith had managed to retain an entrepreneurial attitude, which surprised Malone. He thought that Bell Atlantic was running the kind of slim headquarters organization he himself favored. "It was clear to both of us that we saw the world similarly," Malone says.

There was a bonus with Smith, and it was one that mattered to Malone: Smith was a skilled politician. He was on a first-name basis with leading Washington figures. He liked dealing with the press. He was used to having a public-relations employee sit in during press interviews, which Malone, who rarely gave interviews, found strange. One cannot be greeted by Ray Smith—as he comes out of his office to grasp a hand and flash a bright smile, as he offers guests a drink—without realizing he has the personality of an accomplished politician. Smith counts among his friends Washington officials of both parties. Ray Smith, Malone remembers thinking, could deal with *them*.

Unbeknownst to Malone at the time, Smith, too, was searching for a partner. He had four reasons to target TCI as his top choice, he told me in his roomy Arlington office. First, he thought TCI "had national coverage"; second, it owned cable boxes in key major markets. "Then there was John Malone," he went on. "John's understanding of the business, and his skill at handling programming—that was the third aspect of it." The fourth: TCI had "very good cash flows." Smith phoned Malone and asked if he might pay him a visit.

They met in TCI's boardroom on June 16, 1993. Instead of getting right to the point, which was that he wanted to propose a partnership between the two companies, Smith recalls, he started by talking about their common vision—about how, as he is fond of saying, the information superhighway means that "your computer will speak, your TV will listen, and your telephone will show you pictures." Smith spoke of how he was leading a cultural revolution at Bell Atlantic, encouraging managers to be more nimble—to be more like TCI. And in the future they, like TCI, would plow profits back into expanding the business rather than pouring profits into shareholder dividends. He spoke of how each of them imagined cable and telephone slugging it out, and joked about their sparring sessions on various panels. Then he got serious: "Do you think you'll be able to do this yourself?"

"Of course," Malone said.

"But then he quickly retreated," Smith recalls. "He made it clear he could learn everything, but it would be a lot faster if he had someone who understood switching, regulation, operating systems, and service requirements—the stuff that we do very well." Smith then suggested a strategic partnership.

"The chemistry with Ray was terrific," Malone says. "The other guys are old school. There is no way companies like U S West would walk away from dividends. Ray was clearly a cat of a different color. Very entrepreneurial. He's not the stiff three-piece suit."

He also plays hardball, just as Malone does. Although Smith was equally impressed with Malone, he says, he wanted to leave Malone with one other message, which was: "If we did not succeed in partnering with TCI, we were going to partner with someone else." The meeting ended with Smith saying, "I'll write you a letter tomorrow."

Within days, Malone received a four-page letter. It asked that he consider three options: Bell Atlantic could make a small minority investment in TCI, much as U S West did when it invested in Time Warner; or it could take a larger minority stake, supplying more capital but without gaining control; or, finally, the two companies could merge. Smith also sent a confidentiality agreement, which both men would sign.

Malone was interested. It was the right thing for TCI to take option number three and merge, he felt. And it was the right thing for Leslie and John Malone. He didn't want to run all or part of a large organization. Earlier in the year, when IBM had been scouting for a new CEO, its search committee visited Malone and asked if he might consider becoming CEO of the computer giant. "I'm at the point in my life where I don't want to run a large organization," he recalls telling the search committee. "I'm not good at it. I'm a builder. I like to invent." Leslie Malone desperately wanted her husband out of the public eye. He says that at one point he imagined taking Liberty Media and riding off into the sunset. "Les was real happy with that," he says. "She likes Peter." Peter Barton, an ebullient man of forty-two, is one of the few TCI employees who have been inside either of the Malone homes. Malone envisioned staying on as the chairman of Liberty Media, with Barton running it.

Still, though Malone liked that scheme, he was uncertain what to do about the rest of TCI. As it happened, two previously scheduled business conferences—a meeting of the National Cable Television Association in Lenox, Massachusetts, on July 8, followed by an annual CEO retreat in Sun Valley, Idaho—helped crystallize Malone's thoughts. The twelve-seat TCI Canadair Challenger jet picked him up in Maine and stopped in Lenox, where he stayed just half a day. Ranking cable executives were there with their spouses. "I'm feeling like a left thumb, wondering why I am there," Malone recalls. At five in the afternoon, he headed for Sun Valley, pulled out a yellow legal pad, and begin making notes. At the top of page one he wrote, "John and Leslie's goals and objectives," and he listed six: (1) to reduce stress; (2) to have more fun; (3) to insure a safe and liquid personal investment portfolio; (4) to generate predictable income to fund their lifestyle; (5) to reduce government and legal and media exposure by taking himself out of the public eye; (6) to honor commitments and moral obligations to his family and business associates.

These goals, in turn, led to a set of conclusions, which Malone also jotted down: (1) to retire from TCI; (2) to reduce outside public board memberships from eleven to four; (3) to remain chairman and controlling shareholder in Liberty and stay as chairman of CableLabs; (4) to remain on the Turner board, because it was fun, and because it was an important programming investment for TCI and would allow him "to maintain my relationships with the old gang"; (5) to "get the government off our back"; (6) to generate predictable income by deferring TCI compensation payments, which, with stock dividends, should produce sufficient cash to maintain their lifestyle; (7) to say nothing publicly about the contemplated change until Bob Magness was comfortable.

Magness, who is sixty-nine and spends less and less time in the office, had been kept abreast of Malone's search for a partner. To bring Magness and Smith together, a dinner was arranged in New York on July 12. Sharon and Bob Magness, Malone, Brendan Clouston, and TCI's outside attorney, Jerome Kern, of Baker & Botts, would greet Smith and a few of his key executives at an apartment that TCI maintains in Manhattan. The purpose of the evening was to see whether Magness was as comfortable with Smith and Bell Atlantic as Malone was. Smith acted like the suitor he was. Before going to dinner, he talked about the benefits of joining the two companies into one giant, and he made it clear, in his gentle way, that if they were to merge, he wanted Liberty to be part of the deal. Smith knew he needed programs to fill his cornucopia of digitalized channels. He wanted Liberty, he said. He wanted TCI people. He wanted the same kind of entrepreneurial culture that Magness and Malone had helped inspire. Most of all, he wanted Malone, his brain, and his cachet on Wall Street.

They walked to a restaurant. When the two groups parted, after nearly five hours, Bob Magness was sold on Smith and Bell Atlantic. "I was very comfortable with them," Magness recalls. Malone was comfortable with Bell Atlantic, yet uncomfortable with the thought that he might not be able to go off into the sunset with Liberty. "They were getting into a business they didn't know well," he later said of Bell Atlantic, "and they were afraid I'd outsmart them and they would buy obsolete stuff and I'd keep the good stuff. This is where my reputation hurt me."

A few days after the dinner, Malone telephoned Smith. "I agree with you: we should merge the two companies," Smith recalls Malone saying. "And that means Liberty and TCI?"

"Right."

Smith wasn't sure what role Malone wanted to play in the merged company. He was sure that he, not Malone, would be the CEO. "I felt that I was the better person to run the combined company," he says. "Having the larger company. Having the experience in regulation." And he knew that Malone had no interest in being CEO. So he wasn't surprised when Malone said on the phone, "You should run the whole thing."

With the Bell Atlantic deal proceeding nicely, Malone knew that it would complicate, if not sabotage, an earlier deal he had made to own 22 percent of QVC and to be partners with Barry Diller and Ralph and Brian Roberts, the father and son who are, respectively, chairman and president of Comcast. When they announced their partnership, in December of 1992, they had said that Diller would do more than build a home-shopping service. He would become cable television's programming maestro. Diller and Malone and the Robertses together would bend the future to cable's will. Now three impediments had presented themselves. First, Malone had been told in midsummer that Diller wanted to make an offer to acquire Paramount, and Malone now says that that offer conflicted with discussions he had been holding which were meant to facilitate a link between Turner Broadcasting and Paramount. Second, if the merger with Bell Atlantic went forward, Malone would be unilaterally altering the nature of his partnership with Diller and the Robertses. And, finally, he would be aligning TCI with a phone company that wanted to enter the cable business and would be poised to challenge Comcast, since Comcast had cable boxes throughout much of Bell Atlantic's region.

Mindful of his confidentiality agreement, and of the possibility of subsequent shareholder suits, Malone said nothing to Diller or to the Robertses. When Diller proposed to the QVC board members that they make an unsolicited bid for Paramount, Malone recused himself, he has stated in a sworn deposition, telling Diller and the board that "it was inappropriate for me to put myself in a position where TCI's discussions, ongoing discussions, with Paramount would be in conflict with something that QVC and the QVC board might want to do." Malone says that over the previous three years he had had discussions with Martin Davis, of Paramount, about merging Paramount with Turner Broadcasting. This idea sank over "social issues," Malone says. "Ted's attitude was, 'Maybe I can let Marty stick around a couple of years before I get rid of him.' And Marty wanted Ted to sell advertising!" As late as last April, Malone now says, he received a proposal from Davis to link but not merge Paramount, Turner, and Liberty. Each would own a piece of the others but retain its own independence. Davis and Turner would continue to run their own empires, and Liberty would own more than 20 percent of Paramount to go along with its nearly 25 percent ownership of Turner.

Malone says that he still had hopes of pulling off such a deal when Diller proposed going after Paramount in the summer of 1993. Since Malone continually explores possible deals, he was still talking to Davis through the summer. Malone said in his deposition that Davis had repeatedly asked whether Diller was coming after him and demanded that Malone put Diller "on a leash."

Because Malone played his cards close to the vest, there have been conflicting interpretations of his behavior. Friends maintain that he was trying to strike a delicate balance between his partners and his own shareholders, since he honestly believed that his earlier proposal for Paramount was better than Diller's. He did recuse himself—that is, until Davis chose Viacom as a partner, in September 1993. Detractors say that Malone's behavior in regard to Paramount is one more example of why he can't be trusted, for he didn't tell his partners, Diller and the Robertses, of his discussions with Davis, and he didn't tell Davis of his discussions with Diller. Whether one believes Malone was honorable or duplicitous, this much is certain: even when Malone recused himself, he had a strategy. He was guessing that the threat from Diller might drive Davis into his own arms. To some observers, it seemed to be another case of "I want mine. And I want what's not yours."

Unlike the discussions with Turner, or the subsequent negotiations between Sumner Redstone and Martin Davis (as recounted in voluminous depositions), the so-called social issues between Malone and Smith—Who would be CEO? Which company name would be listed first? What titles would be conferred?—were quickly resolved that summer. They agreed that Smith would run the company, that the TCI name would not be part of the new entity, and that Malone would serve as a kind of consultant, based in Denver, and would shepherd strategy, programming, and technology. "Conceptually, we were aligned," Malone says. "The real issue came down to price."

On August 30, Smith and his team flew to Portland, Maine, to meet Malone and his team and take Malone's boat out to sea. Once on board, they wrestled much of the day over price. One of the issues was Malone and Magness's insistence that they be paid a 10 percent premium for their Class B stock. Malone maintained that this was not unusual. "Bob feels strongly that he is giving up control and ought to get a premium," he told Smith. Bell Atlantic was asking

Malone and Magness to agree not to sell their stock for five years, and was asking Malone to sign a non-compete clause preventing him from working elsewhere in the industry for a number of years. Malone and Magness wanted something in return.

No definitive agreement was reached. But in coming weeks Smith embraced the 10 percent premium as fair, and, since this had always been seen as a stock-for-stock swap, they agreed on a method for determining the dollar value of TCI's shares. However, as is often true with Malone, until a letter of intent was signed he kept tinkering. On the morning of October 12, the day before the marriage was publicly announced, both parties met at the Manhattan law office of Skadden, Arps, Slate, Meagher & Flom, which was representing Bell Atlantic. They haggled over price—how to value each of the components of the deal, including Liberty Media's programming investments.

"I was ready to walk away," Smith says. When they finally settled these matters and were in separate rooms drafting statements to be included in a press release distributed the next day, they had forgotten one important detail. Smith remembers that Malone entered the room to say, "My guys tell me I have to have a title."

"What do you want?" Smith asked.

"They say 'vice chairman.' "

"OK," Smith said. To Smith, this was the ultimate demonstration that "neither of us has position ego." He says, "We don't care about where we sit or about titles." Malone doesn't care about titles, others say: he cares about control. At TCI, Magness had the title, and Malone had control. And Smith did, after all, insist that he be CEO of the combined entity.

Though Senator Metzenbaum and others reacted with alarm to the prospect of Malone at the nerve center of a behemoth that would reach into about a third of American homes, the reaction from the Clinton administration was muted. The emerging consensus in Washington is that as long as the wires are open to any programmer, the merger encourages competition by provoking warfare between cable and telephone companies, and thus inviting two or more wires into most homes. The Clinton administration also had a political concern. They were getting tagged as proregulation, probureaucracy, and the charge stung.

So what does Al Gore say about the man he once compared to Darth Vader? "I find myself engaging in fewer ad hominems as vice president than I did as chairman of a Senate subcommittee," Gore told me, his hands folded in his lap as he sat beside the fireplace in his White House office. "I think he's"—he giggled nervously—"Luke Skywalker!"

In the winter on 1994, Malone and Smith remain uncertain about what the federal regulators will do. They know they face a wait of about a year before the federal regulatory bodies rule on the merger. When Malone spoke to TCI employees at a videotaped question-and-answer session right after the October merger announcement, he himself did not sound like an employee. "Keep in mind, for those of you who feel bought or traded, that the two largest shareholders of the new Bell Atlantic are going to be the two largest sharehold-ers of Liberty and TCI—that's Bob Magness and me," he said. "In a tug of war, I'd bet on our managers." And he made it clear: "TCI did not sell out. We got married."

In January, when Malone spoke to his managers at a Denver hotel, he guessed that there was "a seventy percent likelihood that the deal happens." Why so low a percentage? Because, he has said since, the government might object to the marriage, or insist that the new company shed assets that Bell Atlantic would not want to shed. Per-haps, he says now, the complexity of the details to be settled between the two parties—as of last week, a definitive merger agreement had yet to be signed—could prompt one or both to back away. If the deal should collapse—and Malone does not believe that this will hap-pen—he tells his managers, he would "give up on the idea of a sim-ilar merger." Instead, TCI would go back to square one and look for a strategic partner, with Bell Atlantic being his first choice. If that happened, Malone might slip off with Peter Barton and Liberty Media.

And what of Malone's former partners, Comcast and QVC? They were unhappy, because the proposed merger with Bell At-lantic meant that the Federal Trade Commission would require Liberty to divest itself of its stake in QVC—if, that is, the home-shopping service should capture Paramount. To stay in the bidding contest with Viacom, Diller and Comcast scurried to find new in-vestors. If Diller loses Paramount, Malone will remain as their partner—unless the federal government should veto that arrange-ment. No matter what happens, Malone's present, or former, part-ners are apprehensive. Overnight, Comcast went from being an ally of Malone's to being a competitor. "I'm not saying John shouldn't have done the deal," a cable rival says. "But there was no emotion. This guy is a machine."

Comcast's president, Brian Roberts, who brokered the original partnership between Diller and Malone, still professes admiration for Malone's abilities, but he does not camouflage some hurt. "I was disappointed on several levels," he says. "John was not up-front that he was talking to a phone company and might sell his company. Yet we were in the middle of a ten-billion-dollar deal for Paramount that relied heavily on him. You can argue that Bell Atlantic was a thirty-three-billion-dollar deal and he didn't want to risk it by telling us. But he could have indicated that something was up without revealing any details. Deep in my heart, I say he didn't trust us— or perhaps anyone—enough to take a chance." Malone and Smith maintain that if Malone had told Diller and Roberts earlier than October 12, which is when Malone called them, it would have violated Securities and Exchange Commission regulations. "We didn't make the deal till the last minute," Smith says. "John had to swallow the discomfort. We had agreed to that. I had a partner who was totally shocked"—a Mexican cellular-telephone company, in which Bell Atlantic had announced just two days before that it had acquired a billion-dollar stake.

The fraying partnership between Malone, Roberts, and Diller may be a harbinger of the mercenary nature of business alliances in an era when communications companies are scrambling to converge. Malone says that he regrets that his relationship with Brian and Ralph Roberts has been altered. "We have great relations with Comcast," he says. "We have been partners in many deals together." Of Diller, Malone says, "He's probably the most professional businessman I ever met. He blends sensitivity and toughness in a way that you rarely find. I regret that I have not been able to be of counsel to Barry to the degree I had hoped."

But Malone is not looking back. In the future, he says, business relationships will get more complicated. "Virtually everybody who is not on your team or in your company will be both a friend and a competitor," he observes. "We're partners with Time Warner on a whole bunch of stuff. Yet we are also competitors." He will seek to invest not in any one Hollywood studio but in several, he says, so as not to be just a renter of software. "If you have cross-investment, it increases the likelihood that your purposes are aligned," he explains. He does not want to own a studio outright. "It's like racing horses or owning a ball team," Malone says of the movie business. "There's too much ego in it to be a business."

Today, TCI has an equity stake in Turner Broadcasting, which has bought two small, independent Hollywood studios and the MGM library; it has a joint pay-per-view venture with Rupert Murdoch's

Twentieth Century Fox; Malone retains a relationship with Diller if QVC should fail to take over Paramount; and he has been talking to both Sony and MCA about an equity investment. Malone's game plan has never wavered. What he wants, he says, "is a piece of everyone's business."

Malone believes that companies have no permanent allies, only permanent interests. Such an approach both enhances Malone's flexibility and constricts it. His alliances rest more on business logic than on personal loyalty and shared passions. There is no fellow businessman, Malone says, with whom he'd rather kick off his shoes and strategize than Rupert Murdoch. Yet so aware is he that Murdoch is also a competitor—his Fox network competes with Malone's cable channels; his BSkyB satellite distribution system in Europe requires TCI to pay a toll to distribute its programs, just as Malone's cable systems require Murdoch to do in the United States—that, Malone says, he is always a bit guarded with his friend. Actually, they are not friends. In their embattled world, no one can be simply a friend or a foe. In Malone's universe, everything is business.

Because Malone's intelligence and cunning are widely admired, speculation in the business and media world posits that he must know something denied to everyone else. Maybe he has figured out that the superhighway and its five hundred or more channels will not be successful. Maybe he has discovered that the cable industry is a dinosaur. Maybe he made a deal with the government. Maybe he is keeping Liberty or other assets.

But it may be, ironically, that this time, Malone is simply following his heart, not his head. Maybe, for once, he's following human, not business, logic, and stepping down, as the Wall Street mutual-fund legend Peter Lynch did in 1990, in order to spend time with his family. Perhaps we should take Malone at his word when he says that although he had multiple motives for merging with Bell Atlantic, his overriding motive had nothing to do with business. Mind you, he's not about to retire, he warns, and then explains, "Ray can make the speeches; he's better at it than I am, and he likes to do it. I can spend a lot more time with Les and the kids, and still be sufficiently involved to keep my intellect and creative side occupied. The Promised Land."

What, he was asked, would he say to someone who thought this tough-guy Malone sounded henpecked? Malone did not shift uncomfortably in his office chair or move his hands or act in any way as if he were under assault. "Call it what you will," he responds quietly. "It's something I feel." The feeling, he continues, is that he has not honored "the deal I made when I married her." And there's a selfish side too, he says. "When I'm in Maine, I'm a different person. I smile. I sleep well. I'm more fun."

Whatever Malone's personal motives, it's commonly assumed that Ray Smith will not be lonely running the merged TCI–Bell Atlantic, for many people believe the fish has swallowed the whale. Soon after the merger, Brian Roberts marveled of Malone, "Think how smart he is. Here's a guy whose company the federal government was thinking of breaking up on one day because he was too powerful. And the next day he has tripled the size of his business and the government appears to like it." Since Malone will be fifty-three in March, people can't believe that he is truly stepping down. Tim Wirth says of Malone's biggest deal, "I wonder who this guy from Bell Atlantic is. He's in for a ride."

POSTSCRIPT:

The merger between TCI and Bell Atlantic collapsed later in 1994, as Malone warned it might. When the federal government froze cable-rate increases, cash flows and stock prices shrank. With TCI less valued, Bell Atlantic wanted to reduce the purchase price; Malone refused. Nor did the two corporate cultures mesh. Bell Atlantic, for example, decided not to slim the steady diet of dividends it fed shareholders, and Malone insisted that it do so.

By the end of 1994, Malone's cable-distribution system looked more vulnerable than it had a year earlier, and the telephone companies looked more potent than they would only a year later. While Malone continues to go home for lunch and to spend large stretches of time in Maine and on his boat, he no longer talks about his impending retirement.

By early 1997, Malone was no longer the most influential figure in television. His company was choking on debt; his stock price plunged. Malone was compelled to curb R&D spending and to become more involved in running the day-to-day business than he—or Leslie Malone—wished. Once again there was talk of finding a merger partner. For the first time, it was not unusual to hear it said that Malone was not a visionary but a charlatan. His lower standing mirrored the view of the moment on Wall Street and in the media toward the cable industry. Cable is in trouble, it is now commonly said. Maybe cable's woes are permanent. Maybe the stock market and the media are just demonstrating that the only thing more fickle than technology is the opinion of the herd. Perhaps we are only in the fourth inning of a nine-inning contest. Or perhaps the game is over for John Malone.

If the game is over for Malone, one reason is that cable has become a commodity, its value cheapened by direct-broadcast satellites and newer wired and wireless routes to the home. As technology opened new paths to consumers, the need for programming expanded. Which is why the 1993–94 battle for Paramount was so brutal.

3

THE WAR FOR
PARAMOUNT
(AND SOFTWARE)

(*The New Yorker*, October 4, 1993)

Although Sumner Redstone hadn't played poker in twenty years, he knew he couldn't lose this hand. He had five aces—the supreme hand in a seven-card game called day baseball. The Viacom chairman was traveling with three colleagues from the West Coast to Minneapolis, and this was the final game before their rented G-3 jet landed. Redstone did not want just to win; he wanted to empty his friends' pockets.

"Explain to me again the rules of this game?" he asked the other players, who sat in leather armchairs around a table in the rear of the plane. Nines and threes are wild and the best hand is five aces, Redstone's colleagues told him. After the cards were dealt, Redstone hesitated, munched some nuts, stared at his hand, and lamented, "I don't know whether I should stay in." The other players—Frank Biondi, the chief executive officer of Viacom International, Inc.; George S. Smith, its chief financial officer; and Julian Markby, then an investment banker with Drexel Burnham Lambert, all of whom, along with Redstone, were on a five-day tour to sell five hundred million dollars' worth of Viacom bonds—paid little heed to Redstone's plight. Each felt that he held the winning hand. Biondi had five kings. Smith had five jacks. Markby had a full house.

While Redstone seemed to dither, the others pushed their remaining capital—matchsticks—into the pot. Then Markby showed his hand. Smith topped Markby's full house, and then Biondi revealed his five kings, certain that the pot was his. With a perfectly blank expression, Redstone slapped his aces on the table one by one, and when he was finished he jumped up and whooped, "I wiped you out! I wiped you all out!"

That poker hand reveals the essential Sumner Redstone, Biondi says, a man who is "focused" and absolutely determined to win. Redstone himself says, "I have an extremely obsessive drive to win."

In 1954, seven years after graduating from Harvard Law School, Redstone joined his father in a family-owned movie-theater business in Massachusetts, and parlayed some shrewd investments into a company called National Amusements. In 1979, after he was badly burned in a Boston hotel fire, Redstone was told that he might never walk again; he spent five months in a hospital, and, though his right hand remains gnarled, he says he plays a better game of tennis today then he did then. He displayed similar tenacity when, in 1987, everyone insisted that he would never acquire Viacom, a diversified cable-and-broadcast-programming company; he persisted, sweetening his bid three times, and finally succeeded.

Today Redstone, who is seventy years old, hopes to gain control of Paramount, perhaps the last available major studio. On September 12, 1993, he announced that Viacom would acquire Paramount Communications for $8.2 billion, or sixty-nine dollars a share. Redstone thought he held the winning hand. Yet, eight days later, Barry Diller, who two decades earlier had been the chairman of Paramount Pictures, announced that the QVC Network, his home-shopping company—backed by QVC's principal partners, John Malone, CEO of Tele-Communications, Inc. (TCI), and chairman of the Liberty Media Corporation, and Brian and Ralph Roberts, president and chairman of Comcast Corporation—was offering eighty dollars a share. At stake is something much larger than a single studio: the purchase by either bidder would create a global giant involved in film, cable, broadcast television, publishing, sports, and the budding world of interactive electronics.

The battle for Paramount could trigger a bidding war reminiscent of 1980s dealmaking. In part, the contest pits software providers, who want to circumvent the middleman, against gatekeepers, who own a distribution system—the cable wire—and want to own the software as well. If Redstone wins this contest, he will own a global programming colossus that includes MTV, Nickelodeon, VH1, Showtime, the 890 or so films in Paramount's library, network-TV programming produced by Paramount, and also such syndicated successes as *Cheers, Roseanne,* and *The Cosby Show,* not to mention one of the world's largest publishing enterprises, Simon & Schuster, and the New York Knickerbockers and Rangers. Redstone believes that he does not need to own the highway that distributes his products if he owns brand-name products. By contrast, if Diller and his cable partners prevail, they will own both the cable wire that goes into one-third of all cabled homes and the software required to program their channels both here and abroad. There is a human subplot here as well. Barry Diller's close friend Diane Von Furstenberg says, "Barry's been dreaming of this for a long time"; Paramount was "the first mountain he climbed" when he was a young Hollywood executive, she adds. Now he sees an opportunity not just to reclaim that mountain but also to oust the man who shoved him from the top in 1984—Martin Davis, the chairman and CEO of Paramount.

As early as last spring, Martin Davis had reason to be concerned. Since assuming control of the company, in 1983—at the time, it was known as Gulf + Western—he had succeeded in sharpening its focus as an entertainment company: he had shed non-entertainment and non-publishing assets; he had reduced the company's debt, renamed it Paramount Communications, and amassed more than a billion dollars in cash. Yet he knew that Paramount was vulnerable to a takeover. He knew that all the channels that were going to be available someday would need programs. He knew that over the past five years, since 1988, his studio had lost market share, and that even with a new team running the studio the numbers were not good. Paramount's quarterly earnings for the period ending October 31, 1993, might alarm Wall Street. After a string of flops, the studio was expected to miss its private budget targets by up to eighty million dollars, according to two sources. Such swings are not uncommon in the movie business, but Davis knew that investors might panic, and that Paramount's stock price might fall sharply. And he knew, too, that investors were impatient, because Paramount had not made a big transaction, even though he had been promising one.

Over the past four years, Davis had had a series of exploratory conversations about a merger of some sort, but nothing had come of these discussions; the companies that he had been talking to turned out to be not for sale. The ever-wary Davis kept his eye on John Malone, the one contemporary he feared, and regularly invited him to dinner. Davis could not have been comforted to see Malone and Diller become partners last winter, but he was sure that his own relationship with Malone was secure. "He's one of the great visionaries of our time," he told me then, insisting that I attribute the quote to him by name.

Davis grew anxious when Malone, in a telephone conversation late last winter, seemed to confuse something that Davis had said about him with an unattributed quote from a cable-company CEO in the same article. The CEO had complained that Malone was "an evil genius." Davis reassured Malone that he had not said this. He went so far as to order a paperweight with his compliment embossed on one side and the "evil genius" line on the other and sent it to Malone. "I never heard from him," Davis says. His suspicious nature was now on full alert. He worried that his old nemesis, Barry Diller, was up to something. This suspicion hardened into certainty when Davis invited Diller to lunch last summer and asked whether QVC hoped to acquire Paramount. "I got a vacant answer," Davis recalls.

Davis knew he had to do something. "We felt that the world was changing," Stanley R. Jaffe, Paramount's president, who is second in command to Davis, says. "It was becoming a world of a few giants in media and communications. We worried that competitors like News Corporation"—controlled by Rupert Murdoch—"and TCI were getting too strong a hold on the gate." Among possible merger partners, Davis felt comfortable with Sumner Redstone, whom he had known for forty years. The relationship had solidified in 1965 when Herb Siegel's Chris-Craft made a hostile tender offer for Paramount and Davis was put in charge of marshaling forces to stave him off. As a theater owner, Redstone was a principal Paramount customer, so Davis recruited him. "I confided in Sumner. I trusted him," Davis remembers. "He became chairman of the Paramount Shareholders Protection Committee."

By the spring of 1993, Davis had relented on something he had long insisted on: that in a merger Paramount must be the buyer, not the seller. Redstone would be the buyer now. Neither man needed anyone to tell him how snugly the assets fitted: Viacom's growing cable channels—particularly MTV and Nickelodeon—could make movies and have them distributed by Paramount's worldwide distribution system; Paramount's plethora of television programs could appear on the companies' combined twelve broadcast stations, or on Viacom's 1.1 million cable boxes; MTV and

Nickelodeon's characters and merchandise could appear in Paramount's five theme parks; Paramount's book division was a natural match for the interactive technology that both companies had invested in. Together, the two companies would rank among the world's largest communications companies, rivaled only by Time Warner, Murdoch's News Corporation, Capital Cities/ABC, the Walt Disney Company, and Bertelsmann. Diller and Malone would need Paramount/Viacom to provide programming for their cable channels.

Discussions between Davis and Redstone grew more intense in June, when Davis asked his investment banker, Felix Rohatyn, of Lazard Frères, to begin serious negotiations with Redstone's banker, Robert Greenhill, who had just left Morgan Stanley and become president of Smith Barney Shearson, a firm better known for brokerage than for dealmaking. Redstone would say, "This has nothing to do with money, this has nothing to do with glory. . . . This is an act of destiny."

At various points over the next two months, the deal seemed destined, only to collapse. Negotiators haggled over price and over the amount of cash versus the amount of stock the merger would involve. Redstone and Davis squabbled over what one banker called the "social issues": who would be CEO; what role the number two man at each company—Stanley Jaffe at Paramount, Frank Biondi at Viacom—would play; whether the name "Paramount" would precede the name "Viacom" in the new company. "I never thought it would happen," Redstone recalls. Several times last summer, he broke off negotiations.

But by Tuesday, September 7, 1993, after meeting much of the day in a conference room at Smith Barney, the negotiators had reached a tentative agreement on the financial outlines of a deal: Viacom would pay Paramount shareholders $69.14 a share in stock and cash, depending on the price of Viacom's stock, which had risen (partly because Redstone was buying it); Sumner Redstone would become chairman of the combined company and own nearly 40 percent of its common stock and 70 percent of its voting stock; and Martin Davis would become the CEO of the new company. Biondi's and Jaffe's roles would be defined after the merger was completed.

Rohatyn asked that his team members be left alone in the room so they could call Davis.

Davis said into the phone, "I want to see him face to face."

"Hiya, boss" is how Redstone greeted Davis when he picked up the telephone. It was an appropriately soothing opener, for, though Davis would be the CEO, Redstone would sign his checks. Rohatyn, who has done many deals in his sixty-five years, observes, "Most deals are fifty percent emotion and fifty percent economics." With Davis, emotion probably counted even more.

Redstone and Davis dined that night in a suite at the Hotel Carlyle. They talked about how, together, they could build a global communications giant, how they would be partners. They also talked about how they were determined to block what they saw as the monopolistic designs of Malone and Diller. Redstone, who calls himself "an old-fashioned liberal Democrat," and who worked as a young lawyer for Harry Truman and was a supporter of Bill Clinton, accused Malone of seeking a cable monopoly, because he was both a gatekeeper and an owner of much of the programming that he allowed to pass through his gate. That night at the Carlyle, Redstone and Davis shook hands on a deal.

At the press conference several days later, in the Viacom cafeteria, Redstone dominated the podium. "Software is the name of this game," he declared.

Reporters asked: What about a potential hostile bid from Barry Diller, John Malone, and Comcast?

Redstone said he had known Diller a long time and did not believe he would attack this deal. Of Malone, Redstone said, "I respect him, but I don't fear him." Besides, he added, the political environment was adverse to Malone. Privately, both Redstone's and Davis's investment bankers saw the possibility of at least one bid. "We recognize that we're opening the door to other bidders. No question," Steven Rattner, a partner at Lazard Frères, told me. The biggest threat, he predicted, was likely to come from Diller and his allies. "They want a studio in order to claim its library and its capacity to produce product," Rattner said. "They know this is the last studio play—until the Japanese get tired of their toys."

At the press conference, however, Redstone and Davis behaved as if this deal were destined to forever redefine "synergy." In separate turns at the microphone, both practiced the new old math, asserting that one Viacom plus one Paramount equaled not two but four, or even six or eight. "This is Time Warner without the debt," Redstone declared.

Diller did not agree with this math. "I am not a big believer in synergy," he told me on the eve of the Redstone-Davis announcement. "More money has been thrown away on empty synergistic arguments. It's mostly noise." Nevertheless, he obviously contemplated a QVC bid for Paramount. "There are circumstances that might allow us to make a move," Diller said, elliptically.

One of Diller's two main partners in QVC—Brian Roberts, of Comcast—was eager to pounce. After word of the impending marriage between Redstone and Davis leaked, Roberts said, "When the new highway technology comes down, a way for a cable company like ours to play an even greater role is to own programming."

For several days after his merger announcement, Redstone basked in his new celebrity. When he arrived in the lobby on the twenty-eighth floor of Viacom, at 1515 Broadway, for a picture-taking session with a *Newsweek* photographer, he could not contain his glee. "See that national press!" he exclaimed to a press assistant. He agreeably posed before a panel of twelve TV screens, which displayed MTV and other channels of his. Obviously, one part of Redstone is ham. "This guy really wants to be a media mogul," says a business associate.

But there is also a modest side to Redstone. He wears off-the-rack suits, draped loosely on a six-foot-plus frame. His black loafers are scuffed, his shirt cuffs don't show, because he wears short sleeves, and his ties, which are wide, can also be garish. He has lived in the same house, in Newton, Massachusetts, for the past thirty-five years, and owns neither a private limousine nor a corporate jet. "That's not a pretension I ever wanted," he says, adding, "It was also the wrong message to send in this company." His company is perhaps the number one disseminator of pop culture, yet Redstone's tastes run to Glenn Miller, Tony Bennett, and Artie Shaw. He admits that when he attends the annual MTV awards, he often sneaks out early.

While Redstone can be a demanding, even a rough, boss, his style of management tends to be collegial. He likes to say he treats people as "partners." He is in the office most days, attends the regular meetings that Biondi conducts with other Viacom executives, participates in strategy and deal conversations. In day-to-day management, Biondi sets the corporate style, and the style is one of trust. Family pictures dominate Biondi's office. He does not micromanage. He listens. He analyzes.

The management style practiced by Martin Davis and Stanley Jaffe at Paramount contrasts sharply with Viacom's. While Redstone is somewhat disheveled, Davis always looks as if he has just come from the barber. Not a strand of parted silvery hair is out of place. He wears monogrammed white shirts, crisp and fastened at the sleeves by cuff links, and tailored pin-striped suits, the pants held by sedate suspenders. Yet there is about Davis a jumpiness, a sense that he is about to pounce. Unlike Redstone or Biondi, Davis regularly converses only with Jaffe and two or three other executives. "Marty does not talk to a lot of people," says Jaffe, who is not only his deputy but his friend and admirer. Jaffe's style is more hands on, but, like Davis, he is known to holler at underlings. While Jaffe is respected as a talented producer, associates snicker at his imperial manner. The most recent Paramount proxy statement reveals that Paramount constructed an elaborate screening room in Jaffe's Westchester County home, at a cost of $1.5 million.

To employees, Davis and Jaffe are imposing figures, and are often viewed as mirror images of each other. "Marty Davis would enjoy pulling the wings off flies," one business rival says. Davis concedes that he is tough but maintains that he isn't nasty. "Some people misinterpret me," he says. "If I have something to say, I'm direct."

When Barry Diller ran Paramount in the early eighties, he found Davis too direct. Two days after the merger with Viacom was proclaimed, Davis sat in his office, high above Central Park, and said that even if Diller and his allies made a bid and offered more dollars per share than Viacom, he was prepared to recommend that his board reject the offer. Ironically, he echoed the line taken by executives at Time and Warner when they opposed Davis's hostile 1989 bid for Time. Now Davis declared, "Nobody can offer this combination. A few more dollars can't do it, because it can't equal the value we're creating. You can't compete with this future." The offer, he added, would have "to be so extraordinary that Sumner and I would look at each other and say, 'It has to be.' " He would not define "extraordinary."

Although observers offered general praise for the way the holdings of Viacom and Paramount meshed, in the aftermath of the merger announcement people began to wonder how other elements would mesh, starting with the contrasting management styles. "These people can't get along," says a Hollywood studio head who knows them all. "On a scale of one to ten, the possibility of the four of them surviving together is zero." Redstone challenges this view, saying of Davis, "He'll have a good time with this, because I've never acted like a boss in my life." Davis will be a partner, as Biondi has been at Viacom, he says. "If there was something important that we disagreed on, it wouldn't happen." For his part, Davis says he thinks he will get along better with Redstone than he did with the man whose protégé he was, Charles Bluhdorn: "Bluhdorn was domineering, unlike Redstone. Bluhdorn had limited interests. Redstone is a fantastic visionary. I fit better with him."

As for the potentially awkward partnership of Biondi and Jaffe, each of whom slides down the corporate ladder, Davis says, "Is it difficult for both of them? Of course." He praises each for not making demands, for accepting that their new job descriptions are vague, yet for putting the good of the company first. Then he adds: "Let me be candid: If I can step down and give up a chairmanship, I expect the same of everyone. On that, I'll be tough."

The reaction from the Diller camp was harsh, and concentrated on another perceived weakness of the merger: the absence of a dominant creative figure, like Diller or Disney's Michael Eisner. "You have two perfectly adequate pieces of concrete placed next to each other," Diller says. "There's no energizing force. No one who understands programming. Synergy is different from energy."

"That's a stupid response," Davis says. "Look at MTV. It's creative. I'd match Jaffe's creativity to Diller's. Diller was brilliant in creating a fourth network, but I haven't seen anything happen after that. I respect Eisner, but he relied on people he inherited; Jeffrey Katzenberg has built the Disney studio."

Another criticism, one heard within the halls of Viacom and Paramount as well as outside, is that while the two would constitute a powerhouse software company, they remained relatively weak on the distribution side. "There is no link to the superhighway here," says a top executive in one of the two companies. With only one million cable-system subscribers, their cable-distribution system is relatively small—so small that Redstone recently explored selling the franchise, and may still do so. They have no major link to the telephone wire, or the cellular phone, or the direct-broadcast satellite system, or a television network. By contrast, a linkup with Diller would marry Paramount to the cable highway. Redstone disputes this, saying of cable, "It is not the same business it was before regulation."

Redstone believes that the cable sun is setting. Whether he succeeds or stumbles in his bid for Paramount, this core assumption provides a much-overlooked motive for his proposed merger. When Diller joined up with Malone and Comcast to run QVC, in an attempt to become cable's principal programmer, he bet that cable would dominate electronic access into homes. Redstone and Davis are making a different bet. They see that the public is angry about rising cable prices. They see that Congress and the Clinton-Gore administration are inclined to treat cable the way Washington once treated the three TV networks: as a monopoly. Already, the courts are relaxing strictures that prevented the telephone companies' entry into the video business. There is growing support within Congress to compel cable companies to choose between owning the hardware (wire) and owning the software (programs), just as TV networks like CBS were once compelled to choose between owning a distribution system and owning a studio. Meanwhile, by rapidly paving the way for other highways into the home—including the telephone wire, the satellite dish, and wireless radio frequencies—technology has spawned competition, not retarded it, while weakening smaller competitors.

Against this backdrop, one can better understand the business motives of Redstone and Davis versus Diller and cable. Redstone is betting that with an ample supply of programming for the much-heralded five-hundred-plus-channel future, software manufacturers like Viacom and Paramount will be able to treat the cable gatekeeper like a superfluous middleman. "Software will get distributed because the public will demand it," he declares. "People want their MTV." He is suggesting that MTV, Nickelodeon, and TV shows like *Cheers* are brand names, and distributors must carry these products on their shelves. And if they don't, "there are other ways to distribute programming," he says. "As competition develops, the life of the programmer improves."

With other means of distributing products, and with more channel choices, the demand for products outpaces the supply, says Robert Iger, the president of Capital Cities/ABC. "Shelf space is now so enormous that the ownership of shelf space is less important." A brand name can guarantee its own distribution. "There's been a little bit of a power shift," Steven Rattner says. "Programming will be king."

Accessing entertainment-industry mergers, Rattner says communications companies have three options. The first is to be a studio and produce products. The second is to be a wholesale distributor of products, as MTV, CBS, and HBO are. The third is to be a hardware-delivery system, whether that hardware is a cable wire or a Walkman. MCA's merger with Matsushita and Columbia's with Sony were meant to create synergies between Japanese-produced consumer-electronics hardware and studio software—between options one and three. The union between Viacom and Paramount would strengthen the joint company in options one and two, but not appreciably strengthen it in option three.

It was with options one, two, and three in mind that on September 20 QVC bid $9.5 billion to acquire Paramount. This bid was about eleven dollars a share more than Viacom's accepted offer. QVC would pay about thirty dollars a share in cash and the rest in QVC stock, and that was about twenty-one dollars in cash more than Viacom bid. "Sumner Redstone can't compete with this offer," a member of Diller's group said. "His package is worth about sixty dollars a share. To compete, he has to come up with thirty-five or forty percent more cash. And in a bidding contest Diller can call on backers other than just Malone and Comcast"— including Ted Turner, the chairman of the Turner Broadcasting System. Of course, Redstone can also come up with another partner, or sell an asset like his cable system. But by the end of last week one of Davis's advisers was gloomy: "Diller is Secretariat in the 1973 Belmont Stakes. He's lapping the field." To slow him down, Viacom filed an antitrust suit in federal court in Manhattan last week, charging that QVC's bid represented an attempt by John Malone to extend a monopoly.

In one sense, Redstone's suit is on target. Behind the Diller-Malone-Comcast bid lurks an ambitious attempt to forge a perfectly vertically integrated company, controlling everything from the idea through the production of that idea to the dissemination of it. Look at it this way, Diller said late last week: "If you were at Paramount and you could buy QVC, you'd get a company that is on the frontier of interactivity, a company that is countercyclical to the vagaries of the TV and movie business. Along with it, you get a management with a history with Paramount and maybe an ability to operate it. Plus you'd get an alliance with Liberty cable"— John Malone—"and Comcast. If you could do all that, the market and the public would say, 'Wow!' That's what will happen if we complete this."

With the stakes so high, the battle between Redstone and Diller might attract even more players. For the moment, however, Barry Diller—not Sumner Redstone—seems to have all the aces.

POSTSCRIPT:

In the end—after a lawsuit, which Diller won, and after more rounds of bidding and vituperative personal attacks—Redstone won the bidding for Paramount. Those who said the personnel could not mesh were proven correct, for even before the merger was complete Davis turned on Jaffe and ousted him. And when the merger was consummated,

Redstone showed Davis the door, selecting Biondi to be CEO of the new Viacom.

These machinations and maneuvers to control "software" were often undertaken without a clue as to what to do with the software once it was acquired, as we shall see.

4

WHAT WON'T YOU DO?

(*The New Yorker*, May 17, 1993)

The producer Lawrence Gordon was convinced that his movie *The Warriors* would be a major success. The research told him. Word of mouth told him. The year was 1979, before Gordon had produced such hits as *Die Hard, Predator, Field of Dreams,* and *48 Hours,* and he thought that *The Warriors*—a movie about street gangs—was his ticket to producer Heaven. When the movie opened, long lines at theaters across the country seemed to confirm his hopes. Gordon, who grew up poor in the Mississippi Delta, thought he would finally be fabulously rich. But his euphoria was short-lived. When teen-agers left those theaters, violence erupted. In the first week, three killings were linked to the movie. "People went out and pretended they were warriors," Gordon says. He and Paramount recalled the film.

It was a humbling experience. Now fifty-six and with a graying beard, Gordon still produces action-adventure movies. But when he speaks of the impact of on-screen violence he does not sound, as some others in Hollywood do, like a cigarette manufacturer insisting that there is no conclusive proof that smoking causes cancer. "I'd be lying if I said that people don't imitate what they see on the screen," he says. "I would be a moron to say that they don't, because look how dress styles change. We have people who want to look like Julia Roberts and Michelle Pfeiffer and Madonna. Of course we imitate. It would be impossible for me to think they would imitate our dress, our music, our look, but not imitate any of our violence or our other actions."

Compared with our current movies—or TV fare, music, books, or advertisements—Gordon's *The Warriors* was tame. The average American child, watching around three hours of television a day, has by seventh grade witnessed eight thousand murders and more than a hundred thousand other acts of violence, according to the American Psychological Association. What we see or hear does affect our behavior, even if it doesn't transform most Americans into killers or rapists. How could it not? More than three thousand studies arrive at a similar conclusion: what appears on the screen influences the behavior or attitudes of viewers, particularly young viewers.

In 1991, the Motion Picture Association of America rated only 16 percent of American movies as fit for kids under thirteen. Yet a PG film is more than three times as likely as an R-rated film to gross over a hundred million dollars at the domestic box office. Twenty-two of the forty-six films that crossed this threshold between 1984 and 1991 were PG rated. Yet the percentage of films with a PG rating has dropped, according to the media research firm Paul Kagan Associates. Which leaves this question: If PG movies generally do better at the domestic box office, why are there so many bloody action-adventure films?

The simple answer is profits—particularly overseas profits. For a hit movie, the studio makes more money from foreign rights, video sales, and rentals than it does from domestic movie theaters. Larry Gordon's *Die Hard 2*, for instance, is expected to gross nearly five hundred million dollars, only a third of that total from domestic theaters. "The action genre travels well around the world," Gordon says. "Everyone understands an action movie. If I tell you a joke, you may not get it, but if a bullet goes through the window we all know how to hit the floor, no matter the language."

Few deny that violence and sex are often sensationalized to excite rather than inform customers. The studios play a game of Can You Top This? Gordon adds, "We're aware that people expect something. You need excitement." Much of this competition is like a mindless arms race, says Hollywood producer Norman Lear, creator of such enduring TV series as *All in the Family.* "I just brought home *Paths of Glory,* and there's not a single 'fuck' or a 'shit' or even a 'damn' in it."

Of course, another reason for the genre is that America, the country most strongly addicted to the moving image, is among the most violent of Western societies. On average, a violent crime is committed here every seventeen seconds. The entertainment industry alone cannot be blamed for this, any more than guns alone, and not the people who pull their triggers, can be blamed for gun-related deaths. But the connections are inescapable. If there were fewer guns, fewer people would be shot to death; if there were fewer violent images, fewer people might think that violence is a viable option. Little wonder, then, that the American Psychological Association issued a report last year entitled "Big World, Small Screen," which concluded: "Accumulated research clearly demonstrates a correlation between viewing violence and aggressive behavior—that is, heavy viewers behave more aggressively than light viewers. Children and adults who watch a large number of aggressive programs also tend to hold attitudes and values that favor the use of aggression to solve conflicts."

The people who help shape our culture are often questioned about their success but rarely about their values, except by the likes of former vice president Dan Quayle, whom they smugly flick aside like a flea. Quayle's motives were dismissed as political, but "family values"—as any desperately poor teenage mother trying to raise a couple of kids while battling the influence of the streets and the TV screen knows—are not some right-wing confection. Bill Clinton, in an interview with *TV Guide* which was published just after he was elected president, echoed Quayle, saying that he was "mortified" by much of what is created in Hollywood, and he urged the industry to lead in "deglamorizing mindless sex and violence." But the entertainment industry as a whole has probably given more thought to the pollution of rivers than it has to the pollution of minds. "They don't even think about what they put in movies," a key figure at one of the six major studios says of his colleagues. "The same people who are so enlightened and socially responsible don't even think about it."

Don't they? And, if they do, what limits do they put on themselves? I interviewed a cross section of the executives and artists who decide what we watch on the large or small screen, surprising each of them by asking, "What won't you do?"

In the office where three of Rupert Murdoch's secretaries sit, just outside his own office on the Twentieth Century Fox lot, four clocks show the time in Los Angeles, London, Sydney, and Hong Kong, representing four of the continents on which Murdoch's worldwide media empire operates. No matter how busy he is with his television, movie, cable, newspaper, magazine, and book-publishing enterprises, Murdoch seems to suspend the clocks and the phones when you enter his office. Wearing a white shirt and tie, he greets a visitor and welcomes him with undivided, courteous attention.

What wouldn't you do?

Murdoch broods a long while before answering the question, his eyes closed. "You wouldn't do anything that you couldn't live with, that would be against your principles," he says. "It's a very difficult question if you're a man of conscience. If you thought that you were doing something that was having a malevolent effect, as you saw it, on society, you would not do it. We would never do violence such as you see in a Nintendo game. When I see kids playing Nintendo, and they're able to actually get their character on the screen to bite his opponent in the face, that's pretty sick violence. And you watch kids doing this to each other and they're yelling and laughing for hours on end. Is it all fantasy, and is it all harmless fantasy? I don't know. There has been violence in movies that we put out. Some of it I dislike. . . . But is violence justified? Is the violence of *Lethal Weapon* OK? I think so. If it involves personal cruelty, sadism—obviously, you would never do that. The trouble is, of course, that you run a studio, and how free are you to make these rules? The creative people give you a script and are given last cut on a movie. The next thing, you have a thirty-million-dollar movie in the can which you may disapprove of."

Murdoch's *Sun*, in London, publishes a photograph of a barebreasted woman on page three every day, and he regards that as harmless fun, but he is critical of the sexually prurient Sharon Stone/Michael Douglas movie *Basic Instinct* and says, "I wouldn't have made that picture. The violence, the homosexuality, the varied aspects that were added just for shock effect—it was a film of no redeeming moral values."

Does he believe that screen violence has an impact on the audience?

"I would tend to believe that," he answers. "There are different orders of this, and there's a question of what age people see movies and are affected by them. I've seen a lot of very violent movies in my lifetime, and I've never been violent."

Murdoch assails Hollywood liberal groupthink and its knee-jerk reaction against "family values" and against fellow conservative Dan

Quayle. He sees Hollywood as a town populated by too many insecure, eager-to-please people, who probably spend too much time
consulting their psychiatrists. "This town has a very monolithic view
of life," he says. "You mention things like family values, and they're
terribly suspicious that you're talking sort of religious rules. It really
is very hard to have a discussion here with people of a different opinion about things, the way you can in New York." In Hollywood, he
says, "certain things are accepted as absolute givens—abortion, gay
rights."

How does he reconcile the two Murdochs—the citizen who em-
braces "family values" with the publisher and programmer who
sometimes undermines those values?

"Without being specific or apologizing for anything—I'm sure
I've made lots of mistakes in sixty-two years—I'm not going to
spend my life looking back," he says. On the other hand, referring
to Gerald M. Levin, the CEO of Time Warner, who defended
gangsta-rap recording artists, he says, "I'm not going to do a Jerry
Levin and say, 'Hey, everything's fine under the First Amendment.
We publish everything and anything by everybody.' We don't. We
reserve the right to edit. I think you should not give offense to people's religious beliefs. For instance, I hope that our people"—at
HarperCollins, the Murdoch-owned book publisher—"would never
have published the Salman Rushdie book. It clearly went out of its
way to give great offense to a lot of people. Now, obviously, I'm not
supporting anyone saying, 'Let's kill him for it,' but I think it went
to the point of being an abuse of free speech."

Most senior people in Hollywood, he believes, do wrestle with
these value issues, though there are too many, he says, who are "de-
luded" and think "you can do almost anything under the name of
art." He cites director Oliver Stone's most recent movie: "I thought
that *JFK* was a fundamentally dishonest movie, and I hope it was a
movie we wouldn't make. Yet I went to see *Malcolm X* expecting to
dislike it and I loved it. I didn't know much about the subject. And I
realized this wasn't necessarily perfect biography. But it was a very
interesting point of view, with interesting insight into some of the
black experience."

During the course of our interview, Murdoch says he is annoyed
that a "generally flattering" article in *The Economist* suggested that
his London newspapers have contributed to a "coarsening of British
public life." I ask: Don't you think a tabloid TV show like Fox's
much-imitated *A Current Affair*—some of whose recent segments

have been headlined "Hollywood Sex," "Topless Haircut," "Killer Doctor," "Sexy Calendars," "Super Bowl Hookers," "Felony Nannies," "Teacher Pervert," and "Sex Addiction"—have had a coarsening influence on American life?

He replies, "If you want me to get up and defend every film, every program, I won't do it." He acknowledges that sometimes *A Current Affair* "got out of control. 'Coarsening'? I don't know. If you were to say that there have been occasions when *A Current Affair* has treated some subjects sleazily in the past, I'd have to say yes. To say that *Hard Copy* is sleazier is no defense." He says he wants *A Current Affair* to be "a popular edition of *Nightline*," and adds "It is changing. That's what it was intended to be in the first place." But it is changing very slowly. If one judges by the titles of the 108 half-hour segments of *A Current Affair* aired between October 1992 and February 1993, sex or violence dominated more than 90 percent of these segments.

Until March of 1993, the executive producer of *A Current Affair* was a former NBC News producer, John Terenzio, a short, affable man who worked out of a crowded Fox office on East Sixty-seventh Street in Manhattan. Terenzio also says that Fox is planning to clean up the show, so as to attract quality advertisers and respond to research suggesting that viewers were tiring of gratuitous sex and violence.

So what wouldn't Terenzio do?

He has been accosted by so many sleazy people who wanted money or publicity or retribution that Terenzio seems to have thought more about the question than Murdoch and many of the other people I would interview. He ticked off a list: He wouldn't pay a convicted felon when there was a victim of the crime, or in the case of a homicide. He wouldn't invade the privacy of a child. He wouldn't invade the privacy of an adult by identifying someone as being HIV-positive. He wouldn't out a homosexual. He wouldn't, he says, "do anything that makes fun of the average guy—like the story I killed about the controversy in a little town in Kentucky over the fact that a guy had followed the local sheriff around and got a picture of him necking with a woman other than his wife.

"I like to see the program as a guerrilla army," he explains. "Guerrilla armies are predicated on being armies of the people. They are predicated on having an irreverence towards authority. They are predicated on never being part of the establishment. I like to say it will never be a member of the country club. All of that is good. What I think happened was, the irreverence toward authority creeped into irreverence toward the average guy."

Terenzio describes other changes the show has made consistent with Murdoch's vision of a tabloid *Nightline*. The focus-group research they did late last fall conveyed the strong impression, he says, "that viewer patterns were changing and becoming more conservative. . . . family values is a real phenomenon. And even in the focus groups we're doing right now for other shows, they're indicating a very strong drift away from the freak show, the transvestites, toward 'What can this magazine show do for me?' " Instead of a story ending with a tragic death, says Terenzio, they now try to find, for instance, one about a mother of a victim who launches a one-woman crusade and brings the killer to justice.

"I think there's a big pullback on sex and violence in general and toward a more meaningful, fulfilling content," he says. "I'm not saying that they have got to laugh, because the world is so depressing. I'm saying the viewer wants to feel up."

But is "up" what Terenzio was giving viewers? In the executive producer's small office a monthly schedule of shows is posted on a board. The shows' titles holler of scandal, sex, mayhem, and murder.

Terenzio concedes this is true, saying, "With a show like this, we've got a moneymaker here. It's a slow transition." He admits that the previous night's program "was a sexy show," with three screaming segments: "Swaggart Scandal," "Dead Soap Star," and "Carnival." There was a recent segment about the dark-haired woman who slept with Fergie's father, and a segment on the thirtieth anniversary of the opening of the first Playboy Club. "We have a one-on-one interview with Hugh Hefner," boasts Terenzio.

And no pictures? Terenzio smiles. Of course there are pictures.

What else won't you do?

"This program did not pay a hundred thousand dollars for an interview with Amy Fisher, although we were offered it first," says Terenzio. "We pay for a lot of exclusivity, but not for convicted felons."

Does being the parent of two young children affect the programs he selects?

"I've never put anything on the program I couldn't defend journalistically, and I've put lots on the program I'm really proud of," he says, before adding, "There's lots I've been uncomfortable with. And when your wife and your friends question it repeatedly, it takes its toll. Unquestionably."

Did he let his kids watch daddy's show?

"That's where I'm going," says Terenzio. But he's not there yet.

Oliver Stone, the director of *Wall Street* and *JFK*, is a First Amendment absolutist. Stone is currently editing a movie he just filmed for Warner Brothers, *Heaven and Earth*, and is working out of a makeshift Santa Monica office that looks like a warehouse. He is an intense man whose thinning hair lifts from his head as if electrically charged. He wrestles only briefly with the question "What won't you do?"

"Off the top of my head, I'd pretty much do anything," he says. "I don't view ethics from the outside, only from the inside. What you would find shocking I probably would not. For me, it's a question of taste."

For example?

"Lurid, sexual, kinky stuff which I might like privately I don't necessarily want to do publicly. As Oscar Wilde said, 'I just don't subscribe to your bourgeois morality. It bores me.' You can do anything as long as you do it well. I think Hitler would make a great movie."

Does he believe, as Bill Clinton does, that Hollywood is too preoccupied with violence and sex?

"That's an old issue," Stone says. "I don't believe that government has the right to legislate art or censor it." A movie is "a limited art form that sells for three to seven dollars and fifty cents a ticket, and it's a person's choice whether to buy it or not. It's like buying a book. Buying *Ulysses* once made you commit an illegal action, made you subject to fine and imprisonment. So where is this going to go? Is Tipper Gore going to be our cultural commissar? I resent that. Bill Clinton is talking through his asshole. He's just catering to the body politic. Nobody's forcing anybody to go see *Bad Lieutenant*. But thank God that Abel Ferrara made it. It was an act of liberation. As is Madonna's *Sex* book. She has a perfect right. And if Ice-T wants to say what he wants to say about cops, he's got a right."

Since few say that Madonna or Ice-T don't have that right, why does Stone equate criticism with censorship?

"It depends on what form the criticism takes," he replies. "Aesthetic criticism is fine. If you're saying, 'I don't like the subject matter and I don't think you have the right to say that,' you're engaging in a form of criticism that borders on censorship."

Does Stone reject the argument that there's too much violence in movies?

"Yes and no," he says. "Yes, there's too much violence when the violence is badly done. I go back to my aesthetic defense. If it's badly done it becomes obscene. It's not real. If it's well done, it has impact, it has a dramatic point, then it has meaning. It's valid."

Some people—Rupert Murdoch among them—accuse Stone of dishonesty for promulgating conspiracy theories as facts. How does Stone justify putting words in the mouths of famous men, as he does in *JFK*?

"It comes out of context," Stone says. "If you examine the movie, you'll see that nothing is factually put in. It's surmised. Donald Sutherland describes a scenario: 'This could have happened. And that is possible.' And he lays out a paradigm of possibilities. And you choose. In fact, Lyndon Johnson was quoted in Stanley Karnow's not necessarily great book as saying, 'Just let me get elected and then you can have your war.' " The quotes from Chief Justice Earl Warren, Stone adds, were taken from a transcript of an interview Warren had had with Jack Ruby. "I didn't put anything in Warren's mouth."

But the movie does put words in people's mouths, I assert.

"In the suppositions, I put things in Oswald's mouth," Stone says.

The movie leaves the clear implication that Lyndon Johnson was part of a conspiracy to murder John F. Kennedy. Does Stone think that questioning this implication is "bourgeois morality"?

"In a sense, it is," he says. "It's a description of what Barthes called 'context.' It is a form of censorship. I'm not saying it's necessarily an evil one. Understand what I'm saying, because I don't want to be misquoted on that."

This is being tape-recorded, I respond.

"Then you better check it," says the director, who seems to make no connection between his desire to be quoted correctly and his own flexible use of the words of others. Somehow, Stone sees his own practice as artistic freedom, yet if a journalist doctored quotes it would be an outrage. But Stone, at least, has a rationale, however odd it may be. "It is a restriction and it is a form of censorship to demand of history a fact-only basis, because history is subject to interpretation, and reinterpretation," he asserts. "And the facts are often in dispute."

Stone believes that the movie-ratings system—a voluntary system—amounts to a form of censorship. "There is a natural law of ethics that operates as a money law," he says. "There is an economic ethics where if you don't make an R rating you get the NC-17, and you get kind of frozen out of the theater business. They're all afraid of that. What is necessary therefore to turn that around is to have a very strong and striking and original NC-17 movie that cleans up and does a tremendous amount of business and excites the world and becomes the next *Clockwork Orange*, or whatever, that breaks the barriers. . . . I think *Midnight Cowboy* got an X in 1969, and it was a powerful and original movie."

Does Stone oppose any and all ratings systems for movies?

"Oh, yeah," he says. "I think the ratings system is a consumer label that is put on the package to deal with an age-old fear from the nineteen-twenties, of Hollywood being satanic and taking over the minds of the young."

The man who perhaps more than any other influences the minds of the young is Michael Eisner, the chairman of Disney, and he agonizes very little about morality or violence in films. While he has not made many violent movies, Eisner says of studios that do, "It's not a moral issue. I'm glad they do it. It brings people to the movie theaters."

Eisner bristles when asked: What wouldn't you do?

"I believe there is nothing you should not be allowed to do," he says. "I don't believe, strongly, that government has any right to be involved in anything, or almost anything," related to entertainment. Seated behind a vast blond-wood desk and beside a giant model of Mickey Mouse in an office whose picture window helps convey the feeling of a luxurious chalet, Eisner makes clear that he's "not interested" in "gratuitously violent movies," and he rarely made violent or even action-adventure movies when he and Barry Diller ran Paramount Studios.

What does Eisner say to Bill Clinton and others who urge Hollywood to tone down the sex and violence, and who quote from more than three thousand studies showing that television affects the behavior and attitudes of viewers—particularly young viewers?

"There are studies that show the opposite, too," he says. "That it's a release from built-in tension. I do not think the president of the United States has an obligation to encourage censorship. There's nothing wrong with him expressing his opinion. I don't disagree with that."

Why is Eisner raising the specter of censorship? Even when Dan Quayle criticized *Murphy Brown* last May, he was not advocating censorship.

"The majority of people don't want the government to tell the writers of *Murphy Brown* what to write," Eisner says. "I'll tell you what I am offended by. I'm offended by those who get on a platform and berate Hollywood for violence in the movies, on the one hand, and ignore the proliferation of handguns—something that they could do something about—on the other. That hypocrisy really annoys me."

What about the assertion that people in Hollywood don't think about the social consequences of what they do?

"What is Hollywood?" Eisner responds heatedly. "I personally think that I'm very responsible. And I think our company is very responsible."

Does Eisner agree with a producer who said he would not allow his ten-year-old to see any of the movies he made?

"I would never make a movie that I would not allow my ten-year-old to go to," he says. "I find that disingenuous. Now, maybe ten is not the line. Maybe it should be twelve or thirteen."

Would he encourage a child of ten—or thirteen—to see Disney's violent and scary *The Hand That Rocks the Cradle*?

"To me, *The Hand That Rocks the Cradle* was a complete fantasy," Eisner says. "It was a fantasy. It ended up being pro-social, in that there was a whole re-look at the question of leaving your children with people who haven't given you decent references. That was a silly movie. A fun movie."

Does Eisner see a distinction between real and cartoon violence?

"I don't think that anybody thinks that movies like *The Terminator* are real," he says. "I'm not sure that they don't relieve pressure more than they create it. I don't know the answer to that. I don't want to sit in judgment. I don't think about it that much." Eisner does get upset, however, "when Hollywood is lumped together in a homogenized group. I object when they depict everybody as kind of striving for that last dollar. It's very easy to hide your own lack of action by blaming some film producer who's made some action movie that doesn't fulfill all the moral criteria that a perfect society would dictate. That's an easy shot."

The producer who says he wouldn't let his ten-year-old see any of his movies is Martin Bregman, who has produced such films as *Sea of Love, Scarface, Serpico,* and *The Seduction of Joe Tynan.* Bregman works in Manhattan, in an office overflowing with movie scripts. He speaks in the accents of the streets of New York, on which he was raised. Like Larry Gordon, he's an independent filmmaker.

What won't you do?

Bregman glances at a table beside his desk, where there are photographs of his family, and says, "I have a ten-year-old daughter.

She's never seen any of my movies. I have no complaint with the R rating. I don't let my kids see any R-rated film."

Asked for an example of a film he wouldn't make, he says, "A film that extolls the virtue of crime. I don't want to knock any studios, but there are things that would be personally offensive to me. I wouldn't do a film that denigrates any particular group. I wouldn't do what I wouldn't do in my life."

Are there movies he's made, scenes he's shot, that he regrets?

"The whole movie? No," he says. "I don't think there is a movie that I have been involved in that I wouldn't redo a small piece or some sections of it. But I'm not unusual."

What about *Scarface*, which received an R rating because of its violence?

"*Scarface* was always conceived as an opera," Bregman says. "Always conceived as larger than life." Directed by Brian de Palma and scripted by Oliver Stone, the film is not particularly violent, certainly when compared with contemporary films like *Lethal Weapon*. The gory scene in *Scarface* where a drug dealer gets tortured with a chain saw is imagined, not seen, as blood splatters a bathroom. Since this movie is about pathological drug dealers, Bregman assumes that for an audience to feel the characters they must feel their madness.

Still, Bregman, like others in the movie business, is torn by contradictory impulses. As a moviemaker, he believes that the voluntary ratings system is a form of censorship. "I don't like someone telling me what I can say," he says. As a parent, however, he welcomes it. "If there were no ratings, I'd first have to go see each movie before allowing my daughter to go," he says. On the one hand, Bregman continues, he thinks the influence of film violence is overstated and believes the entertainment industry is blamed when responsibility properly rests with the family. On the other hand—he lifts a copy of a glossy international film magazine thick with advertisements, points to the full-page ads that beckon viewers with violence and sex, and says, "There are a lot of irresponsible people out there. Warner Brothers, Disney, they all make them. They make these films to make money."

There are those who say that Columbia Pictures' *Lethal Weapon* is too violent and might pollute the minds of kids. Mark Canton, who has been the chairman of Columbia since 1991, is not among them. In Canton's grand office, he is surrounded by bouquets of fresh pink and white and yellow flowers, framed photos of Redford and Stallone, and posters of movies he's made. He was born in Bayside,

Queens, forty years ago, and New York still flavors his speech. He is wearing a cream-colored double-breasted suit, which he keeps buttoned.

Canton speaks grandly of movies, the way a De Mille or Selznick might. "Unless we continually reshape the future, continually reinvent and create original stories and new things, the audience isn't going to show up," he says. ". . . What wouldn't I do? So far I haven't found it out. All I know is that in my heart I wouldn't do what I've already done—do it again, the same way. And that even goes for a sequel. That I wouldn't do."

You won't do sequels?

"Oh, I will do sequels. But I won't do sequels that are the same as the movie that made you want to make a sequel." He mentions the most successful sequel in years, *Lethal Weapon 2*, which he bought when he was at Warner Brothers. "The original movie grossed sixty-three million in theaters," he says. "*Lethal Weapon 2* grossed a hundred and forty-eight million. . . . The message was bigger. The likability was bigger. The entertainment factor was bigger and better."

He speaks of his movies as if they were art, with sweeping narratives and profound messages. "It is true that I would not want to make movies that are socially irresponsible and that could cause real harm," he says. "I approach each day by thinking that the art form, that the opportunities that lie within the process of making a movie are almost limitless. . . . Often, movies anticipate what society is about rather than merely reflecting what it's about. You end up on both sides of that equation. I think that by and large there are several messages in the first *Lethal Weapon*. Dick"—Richard Donner, the director—"had a little thing about 'Don't eat tuna.' And he got that in. And there was also a message about condoms, which he got in. There were lots of messages within the drama, and that is part of the reason people felt it was very accessible to the real world. So I think if you can have stories, and I think *Lethal* is one of them, in which the bond, the relationship between the cops and Danny's family, and Mel as a loner, was such that you have an emotional connection—when you have that, you succeed. When you have movies that are violence for violence's sake, you don't succeed."

So what won't Canton do?

"I would not consciously involve us in any motion picture that I really felt was without any logical component to the story, any redeeming overt value."

So what movies does he think went over the edge?

Canton declines to name any. Instead, he retreats to the high ground, saying that he senses—and that Columbia's research confirms this—that "PG movies, by far, have become more popular." Perhaps this is why the new Arnold Schwarzenegger feature, *Last Action Hero*, is more like a James Bond movie, without blood, without graphic violence or language. "I believe there is starting to be a turn toward the family movie," Canton says, calling it a "new mood" in Hollywood. Yet a moment later he interjects, "What I don't like, what I won't allow myself to be, is censored by the critics—the Michael Medveds."

Hollywood has a "bias for the bizarre," Medved wrote in his 1992 book *Hollywood vs. America*, which catalogues one gory detail after another in mass-marketed films, videos, and records—episodes in which performers drink urine, rip toenails out with pliers, and torture women. Why is it "censorship" if a movie critic and author like Medved urges—as he does, and as Canton seemed to be doing at one point in our conversation—that moviemakers think more about the consequences of what they put on the screen?

Canton replies by flashing his liberal badge. "We were the people in the sixties who were advocating peace," he says. "We were out there. We worked at the social issues. And we are the responsible citizens and leaders now. I believe we know how to manage ourselves." He becomes heated. "I think of a guy like Michael Medved," he says, "and I watch this guy in Texas on television who's decided that he's Jesus Christ . . . and ten to fifteen people are dead around him."

What does cult leader David Koresh, who killed four federal agents when they stormed his Waco, Texas, compound, have to do with Michael Medved?

Canton slows his motor, and says, "It's dangerous for anyone to be so set in their ways that they feel that anything that's outside of something that has specific rules is irresponsible and unsafe."

A small studio that has managed itself well, both making money and producing or distributing quality movies, is Miramax Films, which was recently acquired by Disney. Miramax's output includes *The Crying Game, Cinema Paradiso, My Left Foot, The Grifters, Mr. and Mrs. Bridge, Passion Fish, Reservoir Dogs,* and *Enchanted April.* Miramax is the creation of two brothers from Queens, Harvey and Robert Weinstein, who, as boys, went to see every Fellini and Visconti film and argued for hours afterward about what the filmmaker meant. Miramax's offices in Tribeca, in lower Manhattan, help explain why they make movies so cheaply. There is no aquarium-size fish tank, such as one finds at Warner Brothers, no sleek blond-wood walls and desks and reflecting pool, as at Disney, no stained-glass windows and fresh flowers, as at Columbia. Miramax's offices look like those of a college newspaper, with metal desks jammed together and mountains of papers strewn about. And Harvey Weinstein looks like a delicatessen owner—unshaven, smoking constantly, his belly bulging under a short-sleeved green knit shirt.

What won't you do?

"I wouldn't put violence for violence's sake on a movie screen," he says. But, he adds, the subject is more complicated than that. For instance, his wife found *Reservoir Dogs* excessively violent and walked out when Miramax screened it. "It was too real," he says. "It wasn't cartoon violence." Still, Weinstein feels that the movie accurately depicts the banality of the lives of six small-time hoods with hair-trigger tempers. To his wife, it was gruesome; he thought it was art. "At the end of the day, it's back to your own personal taste," he says. "Is Clint Eastwood's *Unforgiven* art, or is it gratuitous violence? I thought it was a great film. Yet I know a lot of women who felt it was gratuitous violence."

What else wouldn't he do?

"I don't know," he says. "I don't want to do things to inspire criminal activity. On a prejudice level, I don't want to do anything to encourage racism of any sort." He is not troubled by the ratings system. "I understand the need for it. Children are involved. You have to let children and parents know." The censorship Weinstein worries most about is not the ratings system, or politicians who criticize Hollywood, or the political correctness that Murdoch complains about. Rather, it is the homogenization of "a company town where ideas tend to get mingled" and where high-overhead studios search frantically for formulas or "the idea or flavor of the month" and end up putting material through the same Cuisinart, thus failing to do what he says a good movie must: "It takes you to a place you've never been to before."

As the mother of a six-year-old boy, Deborah Winger gets livid when she hears that Oliver Stone calls movie ratings a form of censorship. "That's bullshit!" the actress exclaims in a telephone interview. "He gets to do whatever he wants. We just want to let people know by some code what it is, so they can decide whether to take their six-year-old or not. Where's his conscience?"

Unlike Stone, Winger finds "gratuitous violence" and kinky sex in too many movies. Like Murdoch, she has not always lived up to her professed standards. But her attitude changed when her son, Noah, was born. "I don't have any nannies, or anything," says Winger, who left Los Angeles to bring up her son alone in upstate New York. "So when I go out to the movies I usually have to go see something he can see, unless I have a night out. But when I want to see something that's going to entertain me as well, I find it sort of startling." She scans the small boxes in movie ads, searching for PG ratings. She worries about violence. She worries about foul language. She worries about explicit sex. As an actress in search of interesting roles, she saw these ratings as a threat. Now that she is a parent, she sees them as a guide.

Sometimes she won't let Noah see films that carry a PG-13 label. "I wouldn't let him see *Home Alone*," she says. "I hate films where the parents are idiots." She also hated the joy that the son, played by MacCauley Culkin, took in committing acts of violence. "I took Noah to see *A River Runs Through It*, which was great," she says. "But, my God, getting him through the coming attractions—I had to throw a body block!"

What won't Debra Winger do?

"Gratuitous violence, to me, is not entertaining," she answers. "A lot of people can file it as pure entertainment. I have not been without violent moments in my films, but they're limited, and they're very specific. And, even then, it's my least favorite thing. I mean, somebody blowing away fifteen people in the first reel!"

Besides gratuitously violent movies, what else won't Deborah Winger do?

"Gratuitous sex, though that may be an oxymoron! I think 'gratuitous' is the key word. I like things that are integral."

Does she have a theory as to why Hollywood makes so many movies with gratuitous sex or violence?

"It's cheap, and it appeals to the lowest, the basest, part of a human being. . . . I'm not putting down Arnold Schwarzenegger, but let him say one stupid line and it's a formula now. I'm not saying every film has to be an art film. But when you look at Errol Flynn in the original *Robin Hood*, the dialogue is fantastic. It's witty. It's really action-packed. It's adventurous. It's romantic. And it moves."

Why do so many Hollywood figures get defensive about what they view as censorship?

"They're all killers," she says. "They go from being devoted to their families to being killers. They're cutthroat. When I talk to them about their kids, this is where I find a big defensiveness comes up. They're doing things out in the world that are very, very questionable. And they're all these sort of liberal Democrats. It's all very confusing. It's almost as if they never paused and looked at the whole picture. I see people who go along and their kids are gathered like assets—you know, 'I have three children now.' " In Winger's view, Hollywood itself, with its insularity and its comfortable way of life, shields people from reality, and that may explain why there is a disjunction between the movies made and the individuals who make them. "I'm a strong believer in adversity, and, with the weather out there being what it is, and everybody with three cars, and everything there for you, it's sort of tragic, in a weird way," she says. "There's no feeling of how the world really works."

Steven Seagal writes, produces, and stars in the kind of violent action adventures that Deborah Winger shields her son from. A former martial-arts instructor, Seagal has now starred in five popular films, often playing the role of an avenger who takes the law into his own hands to crush the forces of darkness. The violence can be gruesome. In *Marked for Death*, for example, a film made by Murdoch's studio, Seagal cuts off and holds up the head of a drug-dealing Rastafarian. Seagal is currently casting for a movie he is making for Warner Brothers, and works out of a bungalow on the studio's Burbank lot. For a visitor he affects a solemn mien, moving languidly, speaking in a sweet, almost Michael Jackson–like purr. He is wearing a double-breasted blazer over a dark shirt buttoned at the collar. His dark hair is swept straight back into a short ponytail.

What won't you do?

"The no-no's for me certainly include making pictures that are simply exploitive. I've been forced to make movies that I didn't care for, and tried to turn them into something that they originally weren't. And I'm finally getting the power in my career to make the kind of movies that I want to make." He mentions *Hard to Kill* as a film he "didn't want to make," as if he had been forced to make it. The film was "about nothing," he says, and was "a piece of shit."

Does he worry about the impact of the violence in his films?

"Absolutely," he says. "The only thing I can say is that I get thousands and thousands of letters from all over the world—I guess probably hundreds of thousands." Most of those who write look upon him "as a positive role model," he says. "So I must be doing something in my films to give that impression. I never did violence in any of my pictures that was unjustifiable." He defends the vigilante roles he has played, saying, "The judicial system is very flawed, and it's very seldom that the bad guys get their comeuppance."

Is he concerned that his movies might encourage vigilantism?

"No," he answers, "because I think history has sort of proven that if people were more rebellious in their thought the system would have to change for the better, because it's not working. History has proven that people are so complacent that they are being slowly devastated by urban life the way it is."

Michael Ovitz, who is the chairman of the Creative Artists Agency, one of the three major Hollywood talent agencies, was once one of Seagal's martial-arts students. Though Ovitz doesn't run a studio, or make movies or television shows, or write books and screenplays, he represents many of the most influential people who do, including Seagal. There is about Ovitz none of the flamboyance usually associated with Hollywood—no shirt opened to the navel, no sunglasses, no sneakers or high fives. His manner is earnest, his tie is always fixed tight, his short sandy hair is neatly parted, and a slight gap between his two front teeth lends him a boyish appearance. In his imposing office, divided into a seating area with an oversized L-shaped couch and a work area with an antique desk with fastidiously arranged notepads and piles of paper, visitors are surrounded by the African sculptures and modern art that Ovitz avidly collects.

There is a widespread belief in the Hollywood colony that Ovitz doesn't talk to the press. In fact, he talks to many reporters but usually insists on not being quoted. On my visit, he stares disapprovingly at the tape recorder, saying he prefers not to have his words preserved. Then he relents, and sits quietly, rarely using his arms to gesture.

What wouldn't Michael Ovitz do?

"Let me start by telling you how we operate on a day-to-day basis," he says. "We have meetings every single day. All of the projects—incoming rights materials, ideas, newspaper stories, magazine articles—are reviewed." He mentions the saga of Amy Fisher and Joey Buttafuoco. "In the context of one of those meetings," he says, "it came to my attention that we were offered by the lawyer for one of the principals the rights to put it together as a movie or a television movie. And I declined to get involved. I really thought that it was not something that this company should associate itself with. Now, by the way, three networks did movies on it. And reasonable men differ. I'm not sitting here passing judgment."

Why did he make that choice?

"I was just exceedingly uncomfortable with the whole story and the reality of it—the tabloid reality of it. That's not to say that some of the fiction that we get involved with is better or worse."

What other projects would CAA decline to represent?

"I can't say in a blanket statement like that," Ovitz replies. "I would never comment on creative people's work. It's not our job to do anything but advise and attempt to be almost pre-editors for them. It's not our job to tell creative people what not to do—unless we think it's morally reprehensible. These are not legal, or even ethical, issues. We are the agents, not the principals. We're not a studio."

What about Madonna's *Sex* book, which many critics dismissed as pornographic? Did Ovitz have qualms about representing the book?

"I think she is brilliant," he says. "I have to tell you that. So I have a personal bias. Whether I agree or disagree with the content of what she does is not relevant. That's a personal issue. When she described the book to us, I thought the whole concept was quite well thought out. The idea of a book being sealed, so it took on a certain taboo, if you will—I'm using her words." He says that CAA represented Madonna on the basis of her oral description of the book: "This was her vision. I had no sense at all of what the content of the book was going to be like. It covered a subject that's as old as the hills, just in a different way. This is a woman who consistently reinvents herself every year. That doesn't happen by accident. I think she's really smart. Did I agree with all the pictures in there? It's not relevant. I didn't know what was in her mind. I only had a vague sense of it when she laid it out for us."

Ovitz has three young children. Would he let them read the book?

Ovitz hesitated, seemingly embarrassed by the question. "I'm not going to get into it," he says. "I don't believe that anyone's forced to go see anything. No one was forced to go see Madonna's book, by the way. And it was in a sealed cover. In order to see it, you had to buy it. Or see somebody else's. That's a personal choice. It's like going to see movies. It's self-choice."

Does he believe that violence has an impact on audiences, particularly kids?

"Yes and no," Ovitz replies. He says that seeing a violent film is a question of choice, and the violence in movies is often "not real." He continues, "People aren't really getting killed. When you were a kid people said, 'Let's make believe.' "

Asked again if he thought movie violence had impact, he says, "I absolutely think it has an impact on kids. It becomes a framework on which children build. I remember all the things of my childhood.

They've been my framework for my own value system, and I grew up in the fifties in Los Angeles."

Ovitz believes the movie business corrects its excesses, and that if there are too many violent and too few PG movies, the situation will change. Yet he also believes that what drives the rush for action-adventure movies is the need to top the other guy with slam-dunk special effects, with novelty, with huge hits that can help subsidize other pictures. Then he veers off and says, "I don't believe this has to be a business of hits. I believe it is possible to have a very mixed business. I think Warner Brothers has proven that very nicely. Warners has had its share of hits, but nowhere near what a lot of other companies have. It's hit a lot of singles and doubles. It has a real mix of movies, and it's to be complimented for it. The same company can do *Driving Miss Daisy* and *JFK* and *Batman* and *Lethal Weapon.*"

Of course, *Batman Returns* is a grimly violent film, and the body count in *Lethal Weapon* rivals that of the Vietnam War. But Ovitz still believes that Warners is special, because it has been managed by Robert A. Daly, its chairman, and his president, Terry Semel, for more than a decade, whereas most studios change their management far more often. "In a lot of companies, there's an enormous amount of turmoil and turnover," Ovitz says. "And what that creates is short-term thinking. . . . That instability creates bad product."

Few producers turned out more product or made more money in the eighties than Don Simpson and Jerry Bruckheimer. Together, they produced such box-office blockbusters as *Top Gun, Beverly Hills Cop,* and *Flashdance.* Today they make their home at the Walt Disney Studios, where they have a number of film projects in various stages of development. The partners' personalities differ, as is signaled by their choice of drinks at the bar of the Beverly Wilshire Hotel. Simpson, the more voluble and flamboyant of the duo, with blondish hair that cascades in his face, orders a martini. Bruckheimer, who is rail-thin, wears a short beard, and defers to his partner to answer questions first, orders a Coca-Cola. However, they think as one about the kind of movies they wish to make.

What won't Simpson and Bruckheimer do?

One no-no, says Simpson, is this: "We've never had a principal character in our movies that smokes." Nor, he adds, are they comfortable with "gratuitous violence," because "I believe the excessive amount of violence on television and in movies contributes to the plethora of violence on our streets."

The partners, particularly Simpson, who grew up in Alaska, see themselves as renegades, as not part of the politically correct Hollywood culture. They share Murdoch's concern that self-censorship is the greatest menace. "This town reminds me of the most cliquish high-school environment," says Simpson. "It's people trying to take the appropriate moral high ground: 'How could you not be for this concept. It's the right thing.' This is a town that engages in its own political censorship." Where Simpson and Bruckheimer join the Hollywood chorus—and diverge from Murdoch—is that they fear the heavy hand of the government censor. Asked about President Clinton's critique of violence, Simpson says, "I don't want anyone to tell me how to run my life. . . . We're here to tell the truth."

Who said anything about censorship?

"Clinton is president of the U.S.," Simpson explains. "Our franchise is popcorn. I don't fear censorship from Clinton. I think he's a sane man. But there are people in Washington who don't want to see the word 'fuck' in movies."

Hollywood may traffic in violent movies, but it doesn't traffic in public criticism of fellow members of the colony. Few executives speak on the record of actual movies they wouldn't have made or actual performances they wouldn't have produced. Music executive and movie producer David Geffen (who sold his Geffen Records to MCA in 1990) is an exception. When Geffen is asked, What won't you do? he answers instantly: "Rather than talk about what I wouldn't do in the abstract, I wouldn't put out the Geto Boys record or Andrew Dice Clay." These performers, he says, "celebrated murder" or were "homophobic." (Geffen did produce albums by Guns N' Roses that were widely regarded as containing homophobic material.) Unlike Time Warner, Geffen says, he would not produce an album by Ice-T: "I'm not going to put out a record about killing policemen."

What would Geffen say to those who claim that he threatens artistic freedom?

"They're free to make these records," he says. "And other companies are free to distribute them. I'm not going to do it. I'm not saying they don't have the right to do this. I'm simply saying I have the right not to sell them. It's about responsibility. It's not about artistic freedom or censorship."

Geffen, who dresses Hollywood casual in T-shirts and jeans, who is a devout liberal and a billionaire who regularly jets between New York and Los Angeles on his private Gulfstream, nevertheless enjoys going against the grain of the Hollywood colony. He believes Hollywood is an often silly town, populated by executives who too rarely read or reflect. He dares defend Tipper Gore, who was denounced as a censor, or worse, when she proposed that records be voluntarily labeled. "I agree with that," Geffen says. "Labeling is not an infringement on anybody's rights. It comes off with the plastic cover. This is about giving parents a way to be responsible with their children."

Aside from questions of personal taste, does he believe music or pictures can do harm?

"There's no question they do," he says one Sunday morning over a bowl of hot oatmeal at E.A.T., a food emporium on the Upper East Side of Manhattan. "To this very day, Stanley Kubrick has not allowed *Clockwork Orange* to be distributed in England because he lives there. The normal person will not become violent from a movie. But there are people who are encouraged by things they read, things they hear or see." He is a fervent political supporter of Bill Clinton and agrees with his criticism of Hollywood.

Geffen is quick to emphasize that he does not advocate censorship—that he believes "there are a great many people" in Hollywood "who really do care" and think about quality. But he also believes that the limitations—and the excesses—of the entertainment business spring from the weaknesses of the people in charge. "Too many people who are involved in the world of making movies don't read, don't have a sense of the written word," he says. "They have no sense of story, and so they're not burdened by seeing a movie that has no story. They have one overriding concern: Will it make money?"

The director James L. Brooks is burdened by a quality that seems foreign to, say, Oliver Stone: ambivalence. He has written and directed *Broadcast News* and *Terms of Endearment*. He is executive producer of *The Simpsons* on the Fox network and was a co-writer of such television classics as *Taxi* and *The Mary Tyler Moore Show*. Yet when Brooks is asked "What won't you do?" he answers, "I'm confused. There's almost nothing I can say that I could not contradict. There's no question in my mind that as the parent of two young children, I have to wait a long time to find movies I can take them to. Yet the movie I'm making now—*I'll Do Anything*—I wouldn't take my kids to."

Brooks concurs with the belief of Oliver Stone and Michael Eisner that any strictures would place a ceiling on his imagination. Yet

he applauds the Federal Communications Commission's recent announcement that it will more strictly police the content of Saturday-
morning cartoons. "We wouldn't question it for a second if the
government moved against drunk driving," he says.

But Brooks comes back to the gray areas. A different standard
should apply to comedy, he says, because the implicit message is
"We're kidding." So the no-no's on *The Simpsons*, he says, tend to be
" 'Oh, come on. Not another magic joke!' " In making this popular
show, Brooks says, "The number one rule is 'It's got to be funny.'
One rule that I'm not confused about is that Bart Simpson has to be
a role model. I don't believe that. Once we were told, 'Mary Tyler
Moore has to be a model for feminism.' That's bizarre. You don't
write Bart Simpson or Mary Tyler Moore to be what children want
to be or should be. They're characters."

The creators of *The Simpsons* meet regularly, he says, and if they
err, it is usually "on the side of bad taste. It's the nature of our show.
. . . The kid is brash, the family is dysfunctional. I hope we don't do
that so we lose Marge's good heart. We have a self-righteous good
neighbor who drives Homer crazy. The guy's a good man. Somebody's going to tell us one day in a book what the show's about."

Sometimes, as Larry Gordon learned with *The Warriors*, it's impossible to anticipate audience reaction. Should a filmmaker think
about how the audience may respond? No, says John Landis, who
worries that such thinking contains the seeds of censorship.

At forty-two, Landis has directed thirteen feature-length movies,
many of which have fared better with audiences than with critics, including *National Lampoon's Animal House, The Blues Brothers, An
American Werewolf in London*, and *Three Amigos*. He has also directed
two Michael Jackson videos. In discussing one of these videos—
Black or White, made in 1991—Landis inadvertently provides ammunition to those who claim Hollywood indulges itself too much
and thinks too little of consequences. In making this eleven-minute
video, Landis recalls, Michael Jackson looked at some footage and
said, "It's not dazzling enough."

The last half of the video "was basically an improvisation," Landis says. The video was being shot outdoors in downtown Los Angeles, and Jackson had the idea of letting the music and his mood
take him where they would. The intensity mounted. As Jackson
danced, he noticed a garbage can, and, impulsively, he grabbed it and
heaved it into a store window, shattering the glass. He picked up a
crowbar and smashed up a parked car. The violence was unscripted.

Landis liked it. "Any Saturday-morning cartoon show has more violence than that," he says.

But then Jackson further indulged himself, grabbing his crotch and simulating masturbation. He rubbed or squeezed or pulled down the zipper of his fly a total of thirteen times. The choreographer applauded, mentioning that Madonna and Prince did this as well. Jackson saw this as an act of liberation from accepted norms.

At first, Landis was uneasy. "I pointed out that it's not Michael," he recalls. But Jackson felt that this was "what he wanted to do," Landis says, and he went along. At least, it was a bold attempt at self-expression, he says, adding, "Who's to say that's a bad thing?"

Thousands of parents said it was a bad thing. The public furor prompted Jackson to apologize and Landis to quickly sanitize the video. (Today, Landis notes, MTV airs the original, uncut version of the video.)

What wouldn't John Landis do?

"Off the top of my head, I couldn't give you a strict Islamic code," he says. "It depends on the context." As a general rule, doing a film about a rape was always a no-no—until he saw *The Accused*, with Jodie Foster, and glimpsed that it could be done in a meaningful way.

Landis has no trouble with the movie-rating system, believing that it was established "to prevent government censorship" and pointing out that "the MPAA doesn't say to you, 'You can't do this.' " However, like Oliver Stone and others, he seems to believe that censorship is a real danger to Hollywood. "Right now in America, people are taking *Huckleberry Finn* off bookshelves," he says. When asked about Dan Quayle's criticism of Hollywood's "cultural elite," he says, "The last time I heard that was in Berlin!"

Tom Freston had final say on whether MTV aired the Michael Jackson video. For the past thirteen years, he has run the MTV network, which now reaches fifty-seven million American homes and two hundred and fifteen million worldwide. Another twenty-five million homes are reached though Nickelodeon, the children's channel Freston also oversees. About 90 percent of MTV's audience is between the ages of twelve and thirty-four. Although he is considerably older, at forty-seven, than his core audience, Tom Freston doesn't fit the stereotype of a CEO. With a thatch of brown hair flopping on his forehead and his tie askew, he plops onto a couch as the TV set behind him blares an MTV report of a Calvin Klein fashion show, with bare-chested young men swaying down the fashion runway, thumbs stuck in their Calvin jeans.

What won't you do?

Freston rears his head back to collect his thoughts, then rests his chin on clasped fists and says, "The cardinal sin has always been, I think, in television, to do anything that really affronts anyone's religious beliefs. That has been the cardinal sin. We saw that recently with Sinéad O'Connor"—the singer tore up a picture of the pope on *Saturday Night Live*. "We have standards in place for MTV and Nickelodeon that are all the standard no-no's. You know, gratuitous sex, gratuitous violence."

What's an example of gratuitous violence or sex?

"Gratuitous violence is violence for the sake of violence," says Freston. "There doesn't seem any reason for it to exist other than to shock. The person who inflicts the violence largely goes unpunished, either directly or by inference." He hastens to add that neither MTV nor Nickelodeon is "a green vegetables network," nor do they carry only "goodie-goodie programming."

He describes MTV's screening mechanism, which is a miniature version of the broadcast-standards departments of the big three television networks. Profanity and nudity are not allowed on MTV or most cable networks, he says.

Do MTV's broadcast standards operate by some set formula, or are subjective judgments made on a case-by-case basis?

"It's a bunch of people that sit there and stuff comes in and they'll look at it. They'll say, 'This is questionable. This is no. This is fine.' And if it's questionable they'll generally go over it with the programming person and they have discussions. Because it's usually discussions about gray issues. . . . We did a video once for Cher where she was dancing around some battleship in a rather scanty outfit and we heard loud and clear from our viewers. We had tons of callers who complained about it. It was unsuitable for young children, who happened to be in the room even though it wasn't meant for young children. So basically we wouldn't air something like that until ten P.M. or eleven P.M."

What about the MTV videos with young women who simulate fellatio, or the dance partners who grind and squeeze each other's buttocks?

"There's an inference of sex," says Freston. "You won't see any actual sex." In any case, he carefully adds, "We don't make the videos, so we don't control what is in them."

So he ducks responsibility?

No, Freston says. "We turned down Madonna's *Justify My Love*, because there was nudity in it. We did not run it. She ended up going on *Nightline*, which I find particularly interesting because [it] was rated the highest *Nightline* episode since they went on the air. Better than the Iranian hostage crisis!"

Now that MTV is a brand name, and President Clinton has said it helped him get elected, does this power make Freston think more about the nature of MTV's responsibility?

"We think about it a lot," he says. "We know we have a big responsibility." Yet Freston describes another, sometimes contradictory, responsibility, to his business. Rock and roll, "which is our sort of major currency," he says, is "this rebellious art form that's anti-authority, anti-establishment. . . . You want to be somewhat true to the notion and soul of rock and roll being sort of not an establishment art form. Does that mean you have to be irresponsible? I don't think so. I think, actually, if you listen to a lot of the lyrics of the rock-and-roll music, there are probably more prosocial messages buried in their lyrics about hope and love and transcendence than a lot of other stuff that's being put out."

As chairman and CEO for a dozen years of one of the world's five largest companies, General Electric, Jack Welch is a man of ferocious energy. Uncomfortable in a chair, he paces his office, offering staccato replies to most questions. Yet he slows and settles into a leather chair in his conference room on the fifty-third floor of 30 Rockefeller Center when asked: What wouldn't you allow NBC to do?

"I'm not the right person to ask this," Welch says, noting that GE owns but does not run the network. "I think Bob Wright [president of NBC] is clearly the right person to ask this. The one thing I think you can't do, you can't in any way violate . . . standards of integrity. . . . You can't shade the truth."

Asked for an example, Welch immediately cites *Dateline*, the NBC News magazine show that recently was caught doctoring and sensationalizing the fiery crash of a GM truck. "But that wasn't a Jack Welch or GE decision," he says. "That was an NBC decision to do that. You can't do that."

What of other shows—such as the Amy Fisher/Joey Buttafuoco soap operas that NBC and ABC and CBS rushed to make? "I didn't watch it on any one of the three of them," he says, later adding, "I'm not into programming like [Michael] Eisner is and other people are. That's their business. It's not my business. It's Bob Wright's business.

It's the station general managers' business. They've got to operate within licenses. They've got to operate within a certain set of standards. I presume they do that."

As a parent or a citizen, I ask, does Welch find himself blanching at what he sees on his network, including teasers for a series on teenage sex that WNBC was promoting as we spoke?

"I don't watch those shows," Welch says. "I'm never home at six P.M. for the news." What Welch watches regularly on television, he says, is the morning shows, particularly on CNN, when he exercises, and the weekend news, particularly on Sunday mornings. (This is one reason GE is a big advertiser on NBC's *Meet the Press* and ABC's *This Week with David Brinkley*.) But with Bob Wright and Donald Ohlmeyer, whom he and Wright recently installed as president of NBC Entertainment in California, Welch feels he has delegated well. "I have, in Bob Wright, the CEO of this business," he says. "In this *Dateline* thing, for example, Bob Wright made me feel very proud of him, very proud of his behavior. He looked at it. He said, 'Jack, we've got an issue, we've got to deal with it straight up.' Absolutely. I think Bob Wright has good taste. I think he watches a lot of television. He does have kids. He has wonderful values."

Was there a dichotomy between Jack Welch as citizen, father, friend and Jack Welch as businessman, whose network and stations air programs and promotions that he wouldn't want his own young kids to watch?

"I think Larry Tisch [CBS] and Dan Burke [ABC] have different answers to this," he says. "Tisch and Burke are CEOs of their own networks. It is absolutely their job to be sensitive to taste questions. If I got a number of complaints about this, I would have to look into it. But I delegate to Bob Wright the responsibility. Just like it's not my job to build appliances. This is the job of the CEO of the business."

Since January of this year, the key decision-maker for all entertainment programs that appear on NBC is Don Ohlmeyer, a successful Hollywood producer and before that an Emmy Award–winning producer for NBC and for ABC's *Monday Night Football*. A burly, friendly man who chain-smokes Marlboros and wears a crew-neck sweater and a gold ring on each hand, Ohlmeyer assumed command of a network that has recently fallen into third place in the three-network ratings race. He works out of a makeshift office at NBC's squat red-brick entertainment complex in Burbank.

What won't Ohlmeyer do?

"I wouldn't put on a program I wouldn't want my kids to see. I always think that's a pretty good rule of thumb," he says.

What about NBC's Amy Fisher movie?

"It's how you handle it," he says. (The movie aired before he joined the network.) "Television is a very democratic medium. If people like what you put on, they'll watch. If they don't, they won't. Now, you have to make some determinations. If you put on a live execution, you'd probably do a ninety share. But is that the kind of thing you should put on television? No, I wouldn't put it on! I think what's unfortunate is what's happening to television. If you dropped down from the moon and just watched television from three in the afternoon through access [8:00 P.M.], you would think that everybody in this country was either a cross-dresser, had murdered somebody, had a child that was molested, was a transvestite—it's unbelievable! It's like four and a half hours a day of the *National Enquirer*. The public has a mammoth appetite for it. That doesn't necessarily mean it's the best use for television."

Is Ohlmeyer saying that he considers it his job to resist the public's appetite?

"Yeah, sometimes you have to. But," he adds, softening the declaration, "it's also my obligation to try and give people as much of what they want as I'm comfortable giving them." A network, he says, must offer a varied menu, much as a newspaper does with a front page, editorials, gossip columns, cartoons "and a whole potpourri of things. Not necessarily any one of them defines you. Taken as a whole, it defines you."

So what are his programming rules?

One, he says, is to keep his nine-year-old son in mind, and if he airs something he wouldn't want his son to watch, it should run later in the evening, probably around ten o'clock. Second: "We're also in a business that doesn't want to deliver all sorts of programs that advertisers don't want to be in."

Bottom line, what won't Don Ohlmeyer put on NBC?

"What makes the question difficult to answer is that I don't want to contradict myself in six months," he says. He pauses, then adds, "I would not have put on Geraldo Rivera and Satanism. . . . I'd try not to put on something that was intentionally divisive." Ohlmeyer declines to cite specific examples, except to mention "certain talk shows."

Ohlmeyer has two goals, and they're somewhat contradictory:

"We have to hold on to moments where television is more than a merchandising medium. I'd like to think network TV would put on *The Civil War.* But I can't do that when we're in third place."

Ted Harbert is president of the entertainment division of ABC, which in May 1993 finished the season in second place behind CBS in the network-ratings race. Like Ohlmeyer, he has held his present job only a short while—he was promoted in January 1993 from executive vice president for all prime-time entertainment for ABC. Unlike Ohlmeyer, he has spent his career at one network, having joined ABC one year after he graduated magna cum laude from Boston University seventeen years ago. A handsome man, Harbert works in a Century City office building in Los Angeles, in a lovely corner office dominated by a vast brown-leather couch. There is a single television set. A bookcase fills an entire wall; it contains few books but does display a framed photograph of his daughter, Emily, who is three and one-half years old, and whose pretty face is at Harbert's eye level when he sits in his leather chair.

What won't Harbert do?

After a long pause, Harbert says that he recently refused to approve a made-for-TV movie—"I don't want to name it, because it would be unfair to the producer," he says. But he will say this: It was about nothing but "titillation." He was convinced that it would "get a big number," he says. Still, he refused to make it, because he felt that doing so would be "pandering to the audience."

Why, then, did ABC broadcast an Amy Fisher movie?

"Good question," he says, before conceding that the networks have no reason to be proud of their Amy Fisher movies. "Yet that's not the full analysis," Harbert says. The real question, he says, is this: Why did more than a hundred million people watch these three movies? "Part of me—I'm not sure how big a part of me, probably a small part of me—was hoping that after NBC put on the first one the audience would say, 'OK, I've seen my Amy Fisher story. . . . I'm not gonna watch.' This would be a good message to network television if the audience said, 'One's enough!' "

Why are viewers so eager for these stories?

"My perception is that Americans don't talk to each other very much," Harbert says. "People used to sit on the back fence and talk to each other. They'd sit on the front porch and neighbors would talk. Television has replaced the back fence. Americans love to gossip. It's just something that's part of who we are. We get our gossip from television. . . . Americans now use made-for-TV movies as a way to look in their neighbors' window. That being the case, then what's the programmers' decision about whether or not to do Amy Fisher?"

What's Ted Harbert's decision about what he won't do?

He points to the picture of his daughter, Emily, and mentions that a son, William, was born in April. "Emily's entrance into the world totally changed the way I look at television," he says. "I have a massive problem, a personal problem, with violence now on television. I am working very hard to minimize the amount of violence on our air. Frankly, I think we already do a pretty good job of it."

Why did the arrival of a daughter alter his thinking as a programmer?

"Because when she sits there in front of the TV with me, if a promo comes on that I would never let her sit there and watch, or if something comes on that is violent, or the news comes on and she looks at it and this look of bewilderment comes across her face—'What is that man doing, Daddy?'—I don't have a very good answer."

ABC recently broadcast a movie, *Between Love and Hate*, that ends with a youth firing six bullets into his former lover. Harbert's defense is that a network, like a newspaper, offers choices. "I'm a firm believer that there is adult time, and adults get to watch adult programs," he says. "And adults can handle that kind of television. Children can't. This will sound like a paradox, but I don't believe we have to program the network and absolve the parents of responsibility, as if it were our problem and not the parents' problem. Parents have to be responsible for what their kids watch."

Parents might agree with Harbert, while also stressing the responsibility of Hollywood programmers for what they produce. Many Hollywood figures lead two lives—a truth they avoid by complaining about government censorship, or by shifting responsibility from themselves to parents. "We all know they're good citizens," observes Grant Tinker, the founder of the MTM Enterprises studio and the former chairman of NBC. "They give generously. They're good parents. Then, on the lot, they make creative decisions for the wrong reasons—to save their job. They are schizophrenics."

POSTSCRIPT:

Of all the pieces in this collection, this one is a personal favorite. I arranged the interviews by telling each prospective participant that I was surveying leading figures in the entertainment/information business. I said on the phone that I would ask each participant the same question, but I did not say what this question was, and from the shallow way most responded, my guess is that they were surprised. The question—What won't you do?—provoked an inordinate amount of squirming, as individuals strove to justify what they do. I had a sense that few had given a moment's thought to the disjunction between their business and their personal selves. So they resorted to denouncing critics as proponents of censorship. They were exposed, embarrassed; Michael Ovitz phoned after the piece appeared to complain that he had thought we were speaking off the record—despite the debate we had had in his office about the tape recorder! This was just another reminder of the power of shame—witness the role peer pressure and sneers played in reducing smoking.

But shame often passes quickly, particularly when it conflicts with profits. Don Ohlmeyer may say he programs for his nine-year-old, but he soon scheduled at 8:00 P.M.—a time when his kid could watch—a show, *Friends*, that, however funny and skillfully written and acted, dwells on a single theme he probably doesn't want his preteen exposed to: sex. Similarly, as we will next see in a portrait of software giant Viacom, its chairman, Sumner Redstone, yearns for a less salacious press—yet owns one of the most salacious tabloid TV shows, and pushes for more.

5

PORTRAIT OF A
SOFTWARE GIANT
Viacom

(The New Yorker, January 16, 1995)

Eleven of twelve invited executives took seats at a conference table at the headquarters of Viacom, in Manhattan, one day in November of 1994. Frank J. Biondi, Jr., Viacom's chief executive officer, sat alone at the head of the table. An empty seat to his left was reserved for Sumner M. Redstone, the chairman and majority owner of Viacom, which is the world's second-largest media and entertainment company. Biondi picked up a phone at his elbow and dialed. "Harriet, it's Frank," he said. "Is Elvis in the building?"

As the question suggested, Redstone, who is seventy-one years old, has become a business luminary. In fact, his prominence has all but obscured Biondi. An investment banker who has worked for Viacom describes Biondi as "a battered wife," a smart cipher. Others speak of him as being tough. Those who have dealt with him know that he is intelligent, knowledgeable, and attractively modest, and makes terrific corporate presentations. Yet his role remains somewhat mysterious, because he travels just beneath the public radar.

Biondi has never been the subject of an article in a major publication. Last October, *Vanity Fair* ran a photograph of the media executives who attended the annual Sun Valley conference, held by the investment banker Herbert A. Allen in July. Both Redstone and Biondi were at the conference; Redstone was in the picture, Biondi was not. Nor did Biondi make *Entertainment Weekly*'s 1994 honor roll of the hundred and one most powerful figures in the entertainment business, even though several people who work for him did. The billionaire Redstone, who transformed a small collection of movie theaters into the country's biggest privately owned theater chain, and who bought Viacom in 1987 and Paramount earlier this year, was in second place on that list, after Rupert Murdoch, the chairman and CEO of the News Corporation, which owns, among other properties, Twentieth Century Fox.

Casual observers might mistake Biondi, who is fifty, for an accountant. He has a handsome face and is relatively trim, but he wears boxy navy-blue suits, white shirts, sedate ties, and wire-framed eyeglasses. His hair is short and parted precisely. "I'm CEO, but if you read the press I don't exist," Biondi says in a stoic, matter-of-fact way. "Sumner is the embodiment of this place." Was this fact sometimes frustrating for him? "Sure," he responds. "It's not his fault. It just works out that way."

Redstone bought Viacom, but Biondi is the one who has built it. It is Biondi to whom other Viacom executives come for day-to-day decisions, Biondi who tells them when they have strayed from their budget plan, Biondi who fires people, and Biondi who presides over the monthly meetings of the company's key managers. In the entertainment business, where the clash between the creative people and the business side is often intense, and those who actually make products complain about "the suits" from New York who crunch numbers, Biondi is the proverbial suit.

Last October, at a weekly lunch attended by the top four Viacom officials—Redstone; Biondi; Thomas E. Dooley, the executive vice president who oversees finances, corporate development, and corporate communications; and Philippe P. Dauman, the executive vice president, general counsel, and chief administrative officer—Biondi demonstrated his skill at maneuvering Redstone away from a course of action that he thought to be ill advised. At one point, Redstone said, "You know who's called a couple of times to get together in California is Ovitz." Michael Ovitz, the head of the Creative Artists Agency, had offered to have a party for Redstone and Biondi and their team during a trip to California the week of November 7, for strategy and budget reviews at Paramount. But Jonathan L. Dolgen, whom Biondi and Redstone had recruited in March to serve as chairman of Viacom's Entertainment Group, and to oversee movie and television production at Paramount, had reservations about the idea. He did not think a party was necessary, but if there was to be one, he felt it would be more appropriate if he was the host.

"Jon is dead set against it," Biondi said. "He doesn't want us brokered by Ovitz."

Tom Dooley, who, though only thirty-eight, has worked at Viacom since 1980 and is now the de facto chief operating officer, said the company should defer to Dolgen's wishes.

"I don't liked to be restricted," Redstone said. "Ovitz is a friend."

In California a week later, Redstone complied and had breakfast alone with Ovitz. Biondi had quietly got his way.

The company that Biondi supervises has branches in most areas of entertainment, including movie and television production (Paramount Pictures and Paramount Television); music programming (MTV and VH1); children's television programming (Nickelodeon); pay TV (Showtime, the Movie Channel); books and CD-ROMs (Simon & Schuster). It owns the world's dominant video-store chain and one of the world's busiest music retailers (Blockbuster), and five regional theme parks. It owns twelve television stations and fourteen radio stations. Its cable systems have 1.1 million subscribers. It has a 78 percent interest in the Spelling Entertainment Group, whose head, Aaron Spelling, is the most prolific television producer in Hollywood. It owns 50 percent of the All News Channel and the USA Network and 50 percent of the Comedy Central channel. Viacom is the largest single customer of the Hollywood studios (through Blockbuster, Showtime, and the Movie Channel) and of the record companies (through MTV and Blockbuster). It has joint ventures with several other companies, and this month it is launching, with Chris-Craft, a new TV network. Through Viacom Interactive Media, it is also involved with the development and distribution of interactive-television programming, video games, and on-line computer services. And it owns a library containing a total of fifty thousand hours of TV programs and feature films.

What Viacom hopes to be, apparently, is the Microsoft of the entertainment business—in effect, the essential software provider for every distributor. Viacom executives assume that desirable software can create its own distribution system, but as insurance they are planning to use Blockbuster's distribution system to gain an advantage. Last fall, Blockbuster had nearly four thousand video stores and more than five hundred music stores to provide shelf space for Viacom products. Blockbuster's forceful president and CEO, Steven

R. Berrard, told an October meeting of employees from all the Viacom divisions how they could favor Viacom products. He said that he envisioned giving preferential, eye-level shelf space to movies from Paramount and to CD-ROMs and books from Simon & Schuster. Or Blockbuster, with a database of fifty million customers, could use this database to send a mailing to those who might like a certain Paramount film, and offer them a discount on video rentals if they produced a ticket stub to prove that they had seen the film. Or, to boost its pay-TV arm, Berrard said, Blockbuster might offer ten free video rentals to new Showtime subscribers.

Viacom even seems to make Redstone's old nemesis John Malone, of Tele-Communications, Inc., envious. On October 28, at an outdoor cocktail party at Bear, Stearns' eighth annual media and communications conference, at the Boca Raton Resort & Club, in Florida, the two men ran into each other. With a glass of juice in hand, his cheeks rosy from a day in Florida's sun, his wavy and still blondish hair helping him look younger than seventy-one, Redstone was standing near the swimming pool with his wife, Phyllis, when John Malone appeared, smiling, and asked, "Sumner, when are you going to sell me your cable systems?" Both men laughed, knowing that their agents were already negotiating for Malone to buy Redstone's cable systems. Both knew that the sale would go through— when federal regulatory agencies, which Redstone had done much to steam up with a 1993 lawsuit claiming that Malone was a monopolist, cooled down.

"You're in the right business, Sumner," Malone said, referring to Viacom's concentration on owning content rather than the means of distributing that content. "You can't keep up with technology, it's moving so fast." In the past year, Malone and other distributors have lost clout as telephone companies and direct-broadcast satellite dishes and wireless technologies and Blockbuster and computers have paved new roads to consumers. Programmers like Viacom gain by pitting distributors against one another. "One of your good dreams is to have a bidding war break out for your programming services," Biondi says. "That's why—with more than one wire into the home, and more than three networks, and more business overseas—we decided to concentrate on being a programming business."

Sumner Redstone had reason to smile as he settled into his white leather armchair in the Gulfstream that Viacom acquired from Paramount, which was whizzing him and Biondi back to New York the day after this encounter with Malone. "He was very warm last night," he told Biondi. "There's an element of fear, and doubt, I see."

Redstone and Biondi's sense of satisfaction is justified, according to Peter R. Barton, president of Malone's programming arm, Liberty Media. "They outfoxed and outmaneuvered us several times," says Barton, speaking of the past year. He calls Viacom a Superman of programming and adds, wistfully, "I would love to have their assets."

Frank Biondi grew up in a strict Catholic home in Livingston, New Jersey. His father was a scientist who worked on the ultrasecret Manhattan Project during the war and at the Bell Labs afterward. His Irish Catholic mother quit a job as a receptionist to raise three sons. "She was like Vince Lombardi," her eldest son remembers. Biondi guesses he derived his competitive ferocity from his mother, his mellowness from his father. He was sent to parochial schools, after his father had made sure that their math and science teaching was sufficiently rigorous, and attended a public high school where the chemistry and math instruction was superior. "We had probably fifty magazine subscriptions in our house," recalls Mike Biondi, who is twelve years younger than Frank and is today a managing director of the investment-banking concern of Wasserstein & Perella. A middle brother, Robert, is a custom woodworker and lives in lower Manhattan. Frank Biondi did not date much, or smoke, or roll up his T-shirts and loiter on street corners. He concentrated on two things, getting straight A's, and sports.

At Princeton, he made the baseball team as a freshman and was a smooth center fielder with a rifle arm, a swift base stealer. "He was a very intense guy," remembers Senator Bill Bradley, who played first base on the same team. "He wanted to win." Biondi kept his head down and blocked out the tumult of the mid-sixties. He dated the same girl for four years, studied hard, and was voted the most valuable player on the Princeton team in both his sophomore and senior years. In contrast to those who protested on behalf of civil rights or withdrawing from Vietnam, Frank Biondi was detached. He was, he recalls, "very traditional—'My country, right or wrong.' A traditional Catholic. Very black and white. There wasn't a lot of nuance to my thinking. It wasn't until I got out of graduate school and went to work that I made linkages." After graduating from Princeton, he enrolled at the Harvard Business School, broke up with his college sweetheart, and concentrated on his studies. He received his M.B.A. in 1968.

With no particular career in mind, he worked at various Wall Street brokerage firms over the next four years and set up a financial-consulting firm, which foundered. Rick Reiss, one of two managing partners at Cumberland Associates, who shared a conference room with Biondi at the former brokerage firm of Carter, Berlin, and Weil in 1968, says of his close friend, "I think it must have been tough for him for a while when other guys who should have been carrying his briefcase were jumping ahead." Biondi concedes confusion: "The one thing I did agonize over a lot was what to do with my life. I never had a clear view of what I could do for an occupation." By the mid-seventies, he was infatuated with the baby cable industry, and because he easily processed numbers and facts, Biondi became a human data bank.

In late 1972, he went to work for one of the early cable franchises, TelePrompTer, in New York, as director of business analysis. There he met Carol Oughton, who worked in the franchising department. She was as outgoing as he was private. If Biondi had any interest in Oughton at first, he concealed it well. He was all business for six months, and then one night she was working late and he wandered by. They started talking, and he asked her if she'd like to get a hamburger. "Neither one of us ever went out with anyone else again," she recalls. In November of 1973, Biondi proposed, without fuss or flowers. "It wasn't down on one knee," she recalls. "It was matter-of-fact." They were married at the Episcopal church in Dwight, Illinois, her hometown, in March of 1974.

In Dwight, the Oughtons owned the biggest farm, the biggest bank, and the only private swimming pool in town. Carol's great-grandfather founded the Boy Scouts of America, and her grandfather, who was a prominent doctor, founded a famed sanitorium for alcoholics. Carol was a child of privilege. She had gone to finishing schools and college, lived in Switzerland for a time, and returned home in 1966 to work for a family friend, Republican Charles Percy, in his successful quest for the U.S. Senate. She moved to Washington, became a reporter, was employed at the Federal Communications Commission and at the Ford Foundation, and then joined TelePrompTer.

Perhaps the best-known member of the Oughton family was Carol's sister and best friend, Diana, who became a fugitive member of the radical Weather Underground and was killed in 1970 when the explosives she was making in the basement of her Greenwich Village town house went off. For the next few years, Carol was racked with grief and guilt. But by the time she married Biondi she was able to laugh about the FBI agents who attended their wedding on heightened alert because Patty Hearst had just been kidnapped.

Carol Biondi could not dispel her husband's shyness, but she could temper it. Biondi's brother Michael observes, "Frank's marriage to Carol opened Frank up. He's much more social. He has been much more comfortable with people." She introduced him to Michael Fuchs, a young attorney in the business affairs office of the William Morris talent agency, and the two became close friends. Fuchs also helped alleviate Biondi's shyness. "If Fuchs and I went into a room, he would go and meet everyone," Biondi remembers. When the Biondis bought a house in Riverdale, in 1974, Fuchs came up to play tennis nearly every weekend. He became the godfather of their second daughter, Jane, who was born in 1978. When the Biondis got a black Labrador, in 1981, it was named Mickey, after Uncle Mickey, which is what Jane and her sister, Anne, who is two and a half years older, still calls Fuchs.

Fuchs was responsible for the second chapter of Frank Biondi's business life. In 1973, Biondi had left TelePrompTer and started a consulting firm, focusing on the nascent cable-TV industry. A year later, he joined the Children's Television Workshop, where he was assistant treasurer and associate director of business affairs. In 1976 Fuchs left the William Morris talent agency to join HBO, which was owned by Time Inc. A year and a half later, he had been elevated to head of entertainment programming, and he lured Biondi to direct his program-planning department. This, Biondi realized, was his calling. The competitor in him relished the excitement, the uphill struggle. Like baseball, this was about joining a team, about camaraderie, and about winning.

Within a year, Biondi was promoted to vice president for programming operations. He continued to report to Fuchs, who in turn reported to the president of HBO. The pay-cable service was taking off, partly because of deals that Biondi negotiated with studios like Columbia and Orion, which gave HBO exclusive pay-cable rights to their movies, in return for HBO's underwriting a substantial portion of the production costs. By the early eighties, HBO was spending some $250 million a year to finance Hollywood movies.

Soon HBO was a Goliath. The Hollywood studios that had myopically fought the video store and HBO as a bane rather than a benefit to their business were suddenly feeling vulnerable. They feared that HBO would steal theater audiences and would use their cable distribution to sneak into the studio business. Fuchs and Biondi were a happy, winning team.

In 1982, to cope with expansion, HBO reorganized itself into three basic divisions, and Biondi, Fuchs, and Winston H. (Tony) Cox, who was the senior vice president in charge of sales and marketing, were made coequal executive vice presidents. Fuchs would oversee all programming, Biondi would run operations and finance, and Cox would supervise sales and marketing. For the first time in Biondi's career at HBO, tensions between him and Fuchs surfaced. Looking back, Biondi believes that Fuchs came to resent him. "It was really a good partnership," he says. "The thing that disappoints me most is that I viewed our friendship as one that wasn't based on a business relationship. I worked for him for three years out of six. The relationship changed when I became a reporting equal." Fuchs denies that he resented Biondi's equal status, but acknowledges that there were business differences with his friend, particularly about the deal Biondi negotiated with Columbia's Tri-Star unit to finance "fifty percent of the cost of their movies without a production cap"—if the cost of one or all of their movies doubled, HBO was required to kick in half. Fuchs thought Biondi had been "soft."

In February of 1983, Biondi was promoted to chief executive officer, and the fraying of his friendship with Fuchs accelerated. As he recalls, "Michael was nuclear" about the promotion. Fuchs admits that he had a psychological adjustment to make. "I had been big brother," he says. "Now the roles were reversed. It was hard." What was harder, he says, was the change that came over his friend. Within forty-eight hours of Biondi's promotion, Fuchs says, "it wasn't a partnership anymore."

Biondi enjoyed successes at HBO, but he also incurred losses, and he came into conflict over them with Nicholas B. Nicholas, who had recently replaced Gerald Levin as chief of Time's video operations. "Frank wanted to make HBO into an entrepreneurial company, and not just a pay-TV company," says Edward D. Horowitz, who started at HBO in 1974 and is today a senior vice president at Viacom. He pushed to expand into the direct-broadcast satellite distribution of HBO's programs, which was fiercely resisted by the cable industry. Since expansion involved new expenditures, and since Nicholas wanted to focus more on costs, a natural clash loomed. The conflict was speeded by a downturn in business, which coincided with Biondi's tenure as CEO. With major cities delaying the award of cable franchises, HBO's subscription growth stalled. Worried Time executives began to lean on Biondi.

Then one day in October of 1984, Biondi came home and announced to Carol, "I've been fired. Nicholas told me, 'You're a square peg in a round hole.' " Nicholas, who today praises Biondi for bringing "a laserlike focus" to Viacom, says, "I'm not going to tell you, in retrospect, that it was the right thing to fire him." He says that one reason he made the change was that Biondi and Fuchs and Cox "were fighting like three brothers competing with each other." Even after Nicholas told them to stop, "it didn't stop," he says. "So I changed the team." There may have been another factor: Fuchs was complaining. "I made no secret of the fact that I didn't like the way the company was being run," Fuchs says. Time Inc. issued a press release saying that Fuchs would replace Biondi. The friendship was severed.

Biondi felt betrayed. From this point on, observes Carol Biondi, her husband "was usually tense around Michael." Among those in the industry, their rivalry was well known. "We were once on a panel together and they played 'Rocky' when we came in the room," recalls Fuchs. "It was a major feud, perceived as such."

Paul R. Grand, a lawyer who helped negotiate Biondi's severance and who has remained a close friend, remembers this as a life-shaping experience for Biondi. He learned, Grand says, that a big firm was not family, that talent was no guarantee of success. "I think Frank was devastated. There was a period of time—maybe two weeks, maybe three months—where no offers came in. Frank was really hurting. He looked terrible. We'd sit around his kitchen and my sense was that the devastation was not so much that he was fired as it was the emptiness, as if he were thinking: 'Maybe I'm not what I thought I was?' " The experience "hardened" his brother, Michael Biondi remembers.

In January of 1985, Fay Vincent, who was then running Coca-Cola's Entertainment Business Sector, which included Columbia Pictures and Columbia Television, opened the third chapter of Biondi's business life, hiring him as executive vice president of the Entertainment Business Sector. Over the next two years, the TV division accounted for all the Entertainment Business Sector's profits, partly as a result of two big deals that Biondi made—for Norman Lear and Jerry Perenchio's Embassy Communications, and for Merv Griffin's production company.

In November of 1986, Biondi was promoted to chairman and

CEO of Coca-Cola Television, and soon thereafter the Biondis sold their house in Riverdale, planning to move to Beverly Hills. Before they left, various friends and neighbors wanted to give farewell parties for them. Henry S. Schleiff, a close friend, whom Biondi had recruited, and who ran business affairs at HBO, arranged a party at his apartment, in Manhattan. Carol Biondi thought that Michael Fuchs should be invited, and as a magnanimous gesture, Fuchs said that HBO would pay for the party.

Enter Sumner Redstone, who would initiate the fourth chapter of Biondi's business life. Redstone grew up in Dedham, Massachusetts, where his father managed a handful of drive-in movie theaters. Sumner Redstone attended the Boston Latin School and excelled academically, graduating first in his class. He breezed through Harvard in two and a half years, and his fluent knowledge of Japanese and near-photographic memory brought him to the attention of a clandestine group in the War Department's Military Intelligence Division, which was trying to break the Japanese code. After the Second World War, he earned a degree from Harvard Law School and served as a clerk on the United States Court of Appeals, and in 1948 he joined the Truman administration's Justice Department as a special assistant to the attorney general in the tax division. Then, as now, he called himself a liberal Democrat. Then, as now, he loved litigating; when he left government, he joined a law firm in Washington, and he spent the next three years there.

In 1954, Redstone left the law firm to join his father's company, National Amusements. His litigating skills came in handy, for theater owners are in a constant battle with Hollywood studios over the movies they get and what percentage of ticket sales they may keep. (Their share ranges between 35 percent and 70 percent and averages about 50 percent.) Redstone turned to lawsuits on more than one occasion. He also became known as a manager who would yell at executives when he was dissatisfied. He was a loner. He had no business heroes, then or now, he says. He proved to be a superb entrepreneur, and became a pioneer in multiplex cinemas. By the beginning of 1994, the company owned more than nine hundred theaters and had become the country's preeminent privately owned chain.

Redstone's strong will served him well in 1979, when he suffered a horrific accident. To escape from a fire at a Boston hotel, he says, he climbed out a third-floor window and clung by his fingertips from a ledge. Although his body was engulfed in flames, he did not let go. When he was rescued, after ten minutes, the prognosis was that Redstone would not live. He later underwent sixty skin grafts. While Redstone still battles the pain, the sole visible reminder of the fire is his right hand, which is bright purple-red, with three gnarled, nail-less fingers.

With the advent of pay TV and the videocassette recorder, Redstone realized that growth in the theater business would level off, so he began to invest in movie studios. At one point, he owned 5 percent of Twentieth Century Fox, nearly 10 percent of Columbia Pictures, and 8 percent of MGM Home Entertainment. He subsequently sold these shares for a handsome profit. In 1985, he began investing in Viacom, a cable- and media-programming company that had been formed in 1971 after the government forced CBS to divest itself of its syndicated programming holdings. Redstone found Viacom alluring, and watched intently as it grew—forming pay-TV services to rival HBO; branching into broadcast and radio stations; buying MTV Networks and Nickelodeon from Warner. In September of 1986, an investment group led by Viacom's management, which was worried about a takeover by, among other possibilities, Sumner Redstone, proposed to take the company private, in a $2.7 billion leveraged buyout. Redstone, enraged at what he considered an insider's price for the stock, launched a bidding war.

The bidding spiraled upward, and Redstone suspected a deal between Viacom's board and its management to deny him victory, even though he was offering shareholders a better price. He wrote to Allan R. Johnson, the chairman of the Special Committee of Viacom's board of directors, threatening litigation. He says he told Johnson, "I want you to know there is not enough money in the world to indemnify you from the suit I will bring." Johnson called Redstone the next day and assured him that the bidding process would be fair and open. On March 4, 1987, the board announced that Redstone had bought Viacom for $3.4 billion.

Redstone, searching for an experienced CEO, called his friends in the entertainment business, including Barry Diller, of Fox, and Lew Wasserman and Sidney Sheinberg, of MCA. Many of those he consulted told him that the ideal candidate had just accepted a promotion and was in the process of relocating his family to the West Coast. Redstone called Fay Vincent and asked for permission to talk to Biondi.

In truth, Biondi was unenthusiastic about moving. He sensed what Fay Vincent knew: If Vincent left, Victor Kaufman, not he, had the inside track for the top job. He also sensed that Coca-Cola was pulling back from the entertainment business. In September of 1986, when Fay Vincent was given an opportunity to buy CBS, Coke's lethargic response signaled they weren't interested. "I wasn't feeling terrific about the company anymore," says Biondi. So after talking with Redstone for several hours on the phone, Biondi told him, he remembers, "If the deal is right, I'll do it."

The next morning at breakfast, Redstone lathered him with compliments. He told Biondi he was new to the cable and TV business and needed help, that Viacom could be a great company. "The missing link, Frank, is you," Redstone remembers saying. "I'd like you to join." The meticulous Biondi had come to the breakfast with a list of what he would require to join. "He said yes to everything, except ownership," recalls Biondi.

Before the weekend ended, a memo outlining the terms of a proposed agreement was drafted by Philippe Dauman, who at the time was at Shearman & Sterling, and was Redstone's attorney. It was on Biondi's desk Monday morning. Redstone offered Biondi a five-year contract, starting at six hundred thousand dollars a year, and stock that Redstone assured him would be worth at least fifteen million dollars in five years. A major impediment stood in the way of Biondi's accepting the offer: Could he be sure that Redstone would not micromanage him? "Everybody I knew who knew him said he was impossible to work with," Biondi says. But finally Biondi decided to accept. "I'm fairly flexible in being able to work with people," he says.

Now Biondi had to tell Henry Schleiff to cancel his going-away party. No, Schleiff said. "We'll have the party to celebrate your staying in New York."

Michael Fuchs phoned from California and told Schleiff he had heard that Biondi was staying in New York to run the company that owned Showtime, HBO's chief competitor. There was no way that he would fly across the country on a weekend to attend a party for a competitor, Fuchs said. Eventually, however, he agreed to do so.

The party was, by most accounts, a success, but it added to the friction between Biondi and Fuchs. As soon as Fuchs arrived, Biondi recalls, he pulled Schleiff aside for what appeared to be an animated discussion. Although Biondi didn't know at the time that HBO was covering the costs of the party, he now says, "Michael was upset that he was paying for the party and I was not leaving. Two, he's giving a party for a guy who's becoming a competitor. And, third, I would probably convince Henry to leave HBO."

Fuchs confirms that he was incensed. "I made a gesture. I told

Henry HBO would pay for the party," he says. "And that night Frank offered Henry a job. Henry was one of my closest associates. I thought the timing was rather inappropriate." Although Biondi denies that he had yet spoken to Schleiff, Fuchs had reason to feel aggrieved. Viacom had recently started conversations with Tony Cox, who was leaving HBO, about running Showtime, and soon Schleiff did leave HBO, to become chairman and CEO of Viacom's entertainment and broadcast groups.

In his first months at Viacom, Biondi fired a top layer of managers. He labored to open executive lines of communication, because he realized, he says, "that no one knew anyone else." Mark M. Weinstein, who was then a senior vice president and general counsel and today is senior vice president for government affairs, says, "For the first time, Frank brought together the top twenty-five or so managers. He got everyone together to understand one another's business. He ran the company in a very collegial manner, with weekly executive-committee meetings of those who reported to him." He wanted people to operate freely, as long as they met their budget goals and did not hit him with surprises. Geraldine Laybourne, Nickelodeon's president, says, "Even though I don't own Nickelodeon, I felt as if I did."

The technique was deliberate, and Biondi says he learned it mostly from the team culture at Coca-Cola. At Time Inc. he was accustomed to fifty-page strategic and budget presentations. At Coke, they did three-page presentations containing nothing but numbers. Sometimes a three-year plan was squeezed on a single sheet of paper. He'd go to Atlanta to review his television budget with Coke executives and they would spend up to five hours talking about the numbers and the implicit strategic issues. "That was it," he recalls. "And it worked. They did the same in their beverage and food business. It struck me that this was a totally different way of running a company. It was a lot easier on everybody. In part, it was necessitated by the global reach of the company. Even if they wanted to micromanage Australia, they couldn't. It was a powerful contrast to the so-called well-managed companies. I said to myself, 'If I ever get the chance to do it again, I'd do it that way.' It was a way of managing people by picking good people and by the numbers."

Viacom had some financial problems at first. In 1988, the company reported a loss of $189 million. Its debt burden—$2.3 billion—was steep. Redstone, having bet the store to win Viacom, now acted conservatively. "In some respects, business is not a game in which you hit nine hundred," Biondi says. "If seven out of ten decisions are good, you're probably not taking enough chances. At Viacom, because of the debt, we did not have the luxury of taking risks."

The company did make a few good, inexpensive deals, but luck was also a factor. In 1988, the year after Redstone and Biondi arrived, Viacom sold the rerun-syndication rights to *The Cosby Show* for $515 million, the highest gross in the history of TV syndication. Redstone's litigiousness also helped, particularly when it was aimed at Michael Fuchs and Time Inc., which would merge with Warner Communications in July of 1989. In May of that year, Viacom filed a $2.4 billion antitrust action against HBO and its parent, Time Warner, claiming that by making exclusive movie deals with several major Hollywood studios HBO was denying movies to Showtime and Movie Channel subscribers, and that Time would not allow its cable systems to carry Showtime or the Movie Channel.

At first, Biondi and other Viacom executives resisted the idea of bringing an embarrassing lawsuit. Redstone would not be deterred. "He is an extremely tough negotiator," Tom Dooley says. "Sumner is a guy who goes for the last marble on the table. Frank doesn't need the last marble. He has a reputation in the community as a very fair person to deal with." Another Viacom executive describes the difference this way: "Business for Sumner Redstone is personal. Winning and losing is personal. That's what being an entrepreneur is all about. For us professional managers, it's more an intellectual challenge."

But Redstone knew that lawsuits—as had been true in National Amusements' battles with the Hollywood studios and in his fight with the Viacom board—could bring adversaries to the bargaining table. By late 1989, the two sides were already talking privately about a settlement, and had begun negotiations.

Meanwhile, in April of 1990, Viacom inaugurated a channel called HA!, to compete with HBO's Comedy Channel. Eight months later, the two companies—which, by splitting the market, had both been losing money—agreed to merge their comedy channels into a joint venture, now known as Comedy Central. Fuchs had been Time Warner's chief negotiator. The day after the agreement was made, Biondi wrote to Time Warner's chairman, Steve Ross:

> Ordinarily, I would simply call to say it's nice to have part of our differences behind us and that we look forward to great success in Comedy TV with HBO and Time Warner.

I would be remiss in not mentioning that this simply could not have happened if it were not for Michael's effort, which was bold and thoughtful on a corporate and personal level.

Two years later, Biondi would make a similar peace gesture toward Nick Nicholas when he was fired from the merged Time Warner. Biondi wrote a "Dear Nick" letter:

Maybe, you will get to the point where you are more philosophical than angry, but I am sure you are a long way from there at the moment. All I can tell you is, if my history is any guide, these events have silver linings under the pain and disappointment.

If you feel like discussing the world in terms that do not include global visions, give me a call. I'll buy!

There are two competing explanations for Biondi's letters. The first is that, as friends insist, he is a generous man, who sees it as a mark of maturity to be ecumenical. This interpretation seems to be supported by his statement "The two most difficult things I had to learn in life were to deal with losing and not to hold grudges." The second explanation is that Biondi was just being practical, and that cunning should not be confused with virtue. This interpretation seems to be supported by his statement "I learned that what goes around comes around. You treat people well, and though you may not see them for nine, ten years, the next time you do they'll feel good about doing business with you."

By 1993, Viacom's revenues topped $2 billion; its consolidated earnings from operations were $385 million and had grown for the sixth straight year. In 1993 MTV Networks alone had worldwide earnings from operations of $273 million; since 1988, MTV's revenues had climbed at an average annual rate of 25 percent. And in October of 1993, *Forbes* magazine calculated the net worth of Redstone, who owned 83 percent of Viacom, at $5.6 billion, making him one of the world's richest men.

Still, Redstone felt that Viacom needed the film-and-TV-production capacity and the library of a Hollywood studio. When it had those, he believed, his software factory would be so potent that no distributor could deny it shelf space. To that end, Redstone was secretly courting Martin Davis, the chairman of Paramount Communications. This heightened interest in a studio coincided with a desire to become more active in running Viacom. In this, Redstone was no different from Rupert Murdoch, who announced in 1992 that he was moving to Los Angeles to be more active in Twentieth Century Fox. But Biondi's reaction was different from that of Barry Diller, Fox's chairman and CEO, who left the company as a consequence of Murdoch's decision. Biondi appeared to accept Redstone's desire. "As Paramount became more of a reality, three, four years ago, Sumner wanted to be more involved, and he said so," Biondi recalls. "I said, 'Fine. It's your company.' There are a few CEOs who have autonomy, but they are very few. The flip side of it is that it's very nice to have your shareholder sit next to you. I never came into this thinking, It's all mine. I've always reported to someone." Here, too, there are two interpretations of Biondi's reaction. Some believe Biondi is simply expressing the limited ambitions of a competent clerk; others think that he was demonstrating cool logic by not succumbing to false pride.

In September of 1993, Redstone and Martin Davis announced a friendly merger. Redstone would be chairman and majority owner of the enterprise, and Davis, not Biondi, would be Paramount's CEO. Redstone and Biondi both suggest that this arrangement was a precondition of the friendly deal. Before they agreed, Redstone says, "I went to Frank and said, 'If we want to do this thing, it's in the interest of Viacom to make Martin Davis CEO.' " Since Davis was sixty-five, Biondi says, he knew that he would not to have to wait long to become CEO of the merged company. "Sumner said, 'Trust me on this,' " Biondi recalls.

Soon Redstone was calling the merger "the deal from Hell." Eight days after the announcement, Barry Diller, now chairman of the QVC home-shopping network—echoing the charge that Redstone had made against Viacom's management in the mid-1980s—claimed that Redstone was paying too little for Paramount, and made a higher bid. His backers included the Comcast Corporation and Liberty Media Corporation, a subsidary of TCI. If Diller felt that Redstone was shortchanging Paramount shareholders, Redstone felt that Diller, the man he had once called "my best friend on the West Coast," was betraying him personally. Meanwhile, through Viacom, Redstone had been readying a familiar weapon—an antitrust lawsuit—to attack a key Diller ally, John Malone, who had been looking at Paramount himself for a year. The ninety-one-page lawsuit contained claims that Malone's company had used its power as the number one cable distributor to force competitive programming networks like Viacom to pay hefty fees—or lose their channel space.

QVC filed its own lawsuit, claiming that Viacom and Paramount had made a deal that deprived shareholders of the best possible price.

The battle turned exceedingly ugly. Diller supporters charged that Redstone had inflated Viacom's market price by purchasing his own stock through Viacom's parent company, in July and August, and that Redstone's motivation was to drive the stock up in order to keep the purchase price down. In early December, *The New York Times* reported that WMS Industries, a video-game company that Redstone personally owned a quarter of, might have propped up Viacom's stock price by aggressively buying shares in late September and early October. Redstone issued a statement saying that he had traded in Viacom stock, as he had in the past, in July and August, when he thought the Paramount deal was dead; he insisted that he was ignorant of WMS's trading.

Diller was incensed. He knew that Viacom was using Kroll Associates, a leading investigative company, to look at him and his business partners. The charges against Malone were personal, portraying him as a business gangster. Malone's friend and business associate Peter Barton, the president of Liberty Media, whose chairman is Malone, says, "They did tremendous damage to us in Washington, and to John personally. It was mean and ruthless." As for the personal attacks on him in Viacom's lawsuit, Malone says, "The attack was more personal than it should have been, in my opinion. Sumner later apologized to me for making the suit so personal."

"I never apologized to him," Redstone says. "I might have said, 'John, I'm sorry you took this as a personal attack.' " But, since "TCI is John Malone," he adds, there was no way for Viacom to protect its interests and at the same time "insulate" Malone.

Now, nearly a year and a half after this bitter fight commenced— after Malone's wife urged him to get out of the business rather than endure such attacks, after aides predicted that he would never again do business with Redstone—TCI is about to participate in the purchase of Viacom's cable systems. Was it difficult for Malone to swallow his anger, or even pride, and negotiate with Redstone? "No," Malone replies, as if surprised by the naïveté of the question. "I'm a businessman."

Redstone's differences with Malone and Diller were not the only ones that surfaced during the battle for Paramount. A disagreement that arose between Redstone and Biondi was reported in *The Wall Street Journal* on September 23, 1993. Viacom's stock price had been falling, and the company feared that it was because of investors' belief that Diller had made the better bid and their concern that Redstone was so emotionally engaged in the contest that he would pay any price to win. At a lunch of the company's top four executives, Biondi and Dooley recall, it was agreed that Biondi would do a series of press interviews in which he would make it clear that Viacom's approach was dispassionate. "I thought I'd take the pressure off by showing that we might not get Paramount," Biondi says. One paragraph in the *Journal* quoted Biondi as saying that he "did not rule out a sweetened bid, but said 'that is not the only option.' " The next paragraph quoted Redstone's response: "Maybe Frank is not as precise or as articulate as I am but there has absolutely not been any discussion or contemplation of increasing our bid." Biondi felt humiliated. He remembers thinking, "Who needs this shit?" He says that he thought of quitting.

Looking back, Redstone says it was no big deal—"much ado about nothing." In fact, it was a big deal to Biondi. He thought that maybe he had made a mistake when he rejected a feeler asking whether he might join Capital Cities/ABC as a potential successor to its president, Daniel Burke, and chairman, Thomas Murphy. But Redstone came in to see Biondi the next morning and apologized, say both Biondi and Dooley (but not Redstone). In any case, harmony was restored.

While this was perhaps the most serious discord between Redstone and Biondi, it was not their only difference. Philosophically, the two often diverge. To Redstone, business is a contest where winning is the only way to keep score, while Biondi believes both sides must walk away feeling they have won. "He's much more inclined to go for that last nickel in negotiations," says Biondi. He remembers one day hearing Redstone scream loudly from the small conference room next to his office. Later he asked what the commotion was about. "He said, 'They were trying to screw me out of eighty thousand dollars!' This is not about money. This is about sport." On a few occasions, Redstone leaned on Biondi to replace executives. He admits that in 1993 "I pushed Frank" to relieve Henry Schleiff as chairman of Entertainment and Broadcasting. And Redstone pressed to replace Mark Weinstein as counsel with his own outside counsel, Philippe Dauman.

Philippe Dauman, who became senior vice president and general counsel in early 1993, became a source of tension between Biondi and Redstone. In the spring and summer of that year, he played a pivotal role in the Paramount negotiations, sometimes circumventing Biondi. Biondi says that he was angry because Dauman was acting more like Redstone's personal attorney than like Viacom's counsel, talking directly to the investment bankers, and working around him and Tom Dooley. "Tom and I said, 'Time out. It's not the way we play the game here,' " Biondi recalls. "Philippe came in and said, 'Can we talk about this?' From that point on, he did a one hundred and eighty." Today, their relationship is outwardly harmonious, and Biondi says of Dauman, "He's a very smart guy and a terrific lawyer." (Dauman recalls "a brief conversation in which I made it clear that it had been my intention to work as a team." Regarding his relationship with Biondi, Dauman says, "We get along spectacularly well.")

Dauman, who is forty, had reason to see his role in a special way. "I have a close relationship with Sumner that goes beyond the corporate relationship," he says. Having represented Redstone in the acquisition of Viacom, Dauman was invited to join the boards of Viacom and, later, its parent company, National Amusements. He is one of the executors of Redstone's estate. Although it is generally not known, Dauman is listed in the will as Redstone's successor as chairman. According to Redstone, the will makes Dauman chairman for a transition period of between two and three years, and stipulates that the company will not be sold during that time. Redstone also makes it clear that he doesn't envision his son, Brent, or his daughter, Shari—both are attorneys, who joined the Viacom board in 1994—or even Dauman as a replacement for Biondi. He adds, "I don't want to see them in Frank's role—for now, at least." Biondi's role was very much on the mind of NYNEX, which joined the Paramount bidding as Viacom's ally, investing $1.2 billion in Viacom. Before it did, NYNEX executives wanted to make sure that the company would not be disrupted or sold upon Redstone's death; they were told, a NYNEX negotiator says, that Dauman, as a family representative with knowledge of the Redstone estate and Viacom's operations, would ensure continuity and could effectively block an unlikely family putsch.

Redstone won the battle for Paramount by bidding nearly $10 billion—an offer that was made possible, in part, by Viacom's plan, announced on January 7, 1994, to acquire Blockbuster for $8.4 billion. Blockbuster's chairman, H. Wayne Huizenga, had surprised people by agreeing to sell his company, which he had expanded from nineteen video stores in 1987 to nearly four thousand stores by the time it officially merged with Viacom, in late September. By then, Blockbuster was expanding at a rate of one new outlet every seventeen hours and hiring five hundred new employees a week. Huizenga says that despite Blockbuster's brisk growth—and Wall Street's projected net cash flow for the company (minus the cost of purchasing inventory, paying taxes, and building outlets) of $3 billion from 1994 through 1999—its stock multiples "were half what they should be." The market, he feared, would never reward Blockbuster with the stock price it deserved, because investors had come to feel that Blockbuster was already outmoded—a middleman that would be supplanted by video-on-demand services.

While Redstone at first saw Blockbuster as a source of cash to pay for Paramount, Biondi saw Blockbuster as a strategic "layup." He was certain that, along with MTV Networks, it would become one of Viacom's two most profitable divisions over the next five years. "I was much more certain of Blockbuster than Sumner was," he says. "Sumner was much more certain of Paramount, maybe because he has lived in the motion-picture business, and I haven't." Despite all the talk about how video on demand and an information superhighway would weaken Blockbuster, Biondi thought that this new world would dawn more slowly than, say, Time Warner and most telephone companies imagined. And while overall movie-theater revenues had plateaued at $5 billion a year, the video-store business in North America was generating nearly three times as much, and its revenues were being projected at $21 billion a year by the end of this century. Overseas, growth opportunities were even more abundant, since 70 percent of all households with VCRs were outside the United States, and as yet there were relatively few video stores. Blockbuster planned to plug this gap, doubling its overseas stores to more than two thousand by 1998. The opportunities for leverage were also enormous. Blockbuster was Hollywood's largest single revenue source, spending about $1 billion annually to buy videos. Blockbuster combined with MTV would be the foremost customer, as well, of the $10 billion-a-year domestic record industry. And, through a joint venture with IBM, Blockbuster hoped to become tomorrow's electronic jukebox, sending video on demand to customers' homes.

The problem for Biondi and Redstone was not tomorrow, however; it was today. The merger with Blockbuster was to be financed by Viacom stock, but investors, feeling that Viacom had paid too much for Paramount, were skittish, and Viacom's stock price was dropping. Blockbuster shareholders would be receiving paper that was worth less than they had anticipated; debt-laden Viacom would drag down Blockbuster. The Blockbuster merger was announced in the fall of 1993, but the deal remained uncompleted through the summer of 1994. If Viacom's stock price didn't rise, the deal was off. Huizenga urged Redstone to offer shareholders a better price; Redstone refused. Redstone, Biondi, and other Viacom executives went on the road, peddling the message to Blockbuster shareholders and Wall Street investors and analysts that Viacom was a quiet giant, whose strengths would become apparent after it owned both Paramount and Blockbuster. It didn't hurt that the investor Kirk Kerkorian, who happened to be an old business acquaintance of Redstone's, was buying Viacom stock, and inviting speculation that this was not by chance. (Redstone insists he knew nothing of Kerkorian's activity.) Viacom's stock price steadily rose, from nearly twenty-eight dollars in June to about forty dollars by the end of September, when Blockbuster's and Viacom's shareholders voted to approve the merger.

The successful road show that Redstone and Biondi put on reflected, in part, the varied talents of each man. "I do the so-called passion bit at the beginning of a presentation," Redstone says. Whereas Biondi makes a persuasive case by marshaling facts and figures, Redstone evangelically waves his arms and gets red in the face. He never tires of reciting the same script. "I think I brought a vision," he says. "It's hard to believe that we started with a theater chain—we had maybe two hundred and fifty screens at the time—when we went for Viacom. My vision has always been the same. I didn't know Viacom's business. But my basic instinct was right, which was to ride the escalator of home entertainment." Redstone's friend George Abrams, who is also his attorney, observes that he is "more a catalyst than a manager." Redstone himself says that he contributes something else: he thinks like an entrepreneur. "We needed that," he says. "But without the management team Frank assembled, we could not have done it."

The chairman has been known to explode at executives, but no one recalls his yelling at Biondi. Perhaps the sense of security Biondi conveys explains the trust between the two men. "In some people bluntness is unattractive," says Senator Bill Bradley, who remains a friend of his college teammate. "In Frank's case bluntness is attractive because it conveys a condition absent in a lot of corporate relations."

The day after Redstone described to me his seven-year business relationship with Biondi, a three-page, single-spaced memo marked "dictated but not read" arrived from Redstone. It said, in part:

Among Frank's great skills that I may not have mentioned is his ability to secure the support and loyalty of the entire management team. A part of this flows from his general capacity as a leader, a part flows from his management style (which I totally endorse and, indeed, participate in), namely, the style of open access and communication with a special emphasis on giving great operating leeway to the division heads. . . . By supporting them, Frank secures their loyalty and trust.

Redstone and Biondi share some common attitudes. Both live relatively frugally. Perhaps the trust between the two men can also be explained by Biondi's candor. In the two months I spent around Viacom, Biondi never ducked a question about his six-year contract, which will pay him a million dollars annually (and far more in bonuses), about whom he voted for (Republican George Pataki against New York governor Mario Cuomo), or about Michael Fuchs.

To put it another way, Biondi provides ballast. While Viacom executives are often frustrated by Biondi's cryptic, unemotional manner—unlike Redstone, he is stingy with praise and is not physically demonstrative—he does convey stability. Redstone is "the king of good news," as a Showtime manager says. "He doesn't know how to deal with bad news. If we came in and said, 'Subscribers are down for August,' Sumner would get very excited and say, 'Maybe we ought to sell Showtime?' " Frederik V. Salerno, a vice chairman of NYNEX, who joined Viacom's board in 1994, describes Biondi as an "unbelievably even-tempered person, who also has a remarkable sense of humor." He adds, "Yet he always has time to talk to you. He never seems to be overburdened with work."

Not that Biondi has no temper—particularly when his sense of Team Viacom is involved. One blowup came around the time of President Clinton's inauguration, when Biondi learned that MTV was giving an inaugural ball that would be attended by Hillary and Bill Clinton. Biondi says that he found out when a printed invitation arrived. Referring to MTV Networks' chairman and CEO, Thomas E. Freston, and an outside public-relations adviser, he explains, "Without asking, they did up an invitation saying, 'Tom Freston and Ken Lerer invite you . . .' I was really pissed. Lerer is a consultant. Where were Sumner and myself? Where was Viacom?" Although he holds both men in high regard, each received the verbal equivalent of a whipping. These outbursts probably carry more weight because they are relatively rare.

Geraldine Laybourne, of Nickelodeon, who is a former school-teacher and founder of the nonprofit Media Center for Children, is lavish in her praise of Redstone. She is more reserved about Biondi, but says, "Frank's more stable, analytical approach is reassuring. You know you won't go too far off. The real tribute to him is that he allows us to run our businesses without conformity. He won't impose his style on us." The only point Biondi insists on, she says, "is that Viacom's divisions make the numbers."

Making the numbers was partly what Biondi wanted to emphasize during the visit to Paramount in early November. He and Redstone would meet with Jonathan Dolgen, of Viacom's Entertainment Group; Sherry Lansing, the chairman of Paramount Pictures; Kerry McCluggage, the chairman of the Paramount Television Group; and their key executives. Biondi and ten other Viacom executives from New York sat on one side of an immense horseshoe-shaped table in Paramount's boardroom, directly across from Dolgen and his team. While Biondi rarely spoke, the meeting demonstrated both his laid-back style and a technique that he had used to help make Viacom the world's fastest-growing communications company. He had come to California with several goals. He wanted to get to know his team better, he said, to exchange information, to talk about strategy, and, above all, to send the same message he is trying to send throughout Viacom. The message is that everyone has to concentrate on growth and the bottom line.

The problem for Viacom, he explained to me before flying to California, was that it had guaranteed a certain stock price to Blockbuster and Paramount shareholders. If Viacom's stock price, then approximately forty dollars a share, rose about 25 percent in the next six months, under the terms of the merger agreements with Blockbuster and Paramount Viacom would owe nothing. But if the stock price did not rise, Viacom would be compelled to pay shareholders a penalty, in cash or stock, totaling anywhere between a few million dollars and more than a billion. Because of this requirement, Viacom saw 1995 as the most important year of the next several years. One way to raise the stock price, Biondi felt, was to demonstrate that Viacom would continue to be a growth company, and would increase its operating cash flow, from Wall Street's estimate of $1.9 billion in 1994 to at least $2.3 billion this year, when Blockbuster is part of the company. The idea was to impress Wall Street analysts and the financial press. One means of doing this would be for Viacom to exceed its 1994 profit performance this year, and in particular to exceed the predictions of Wall Street analysts.

Some of the numbers that Paramount executives presented were not displeasing to Biondi. Since the merger of Paramount and Viacom, the number of employees of the combined entertainment division had been reduced by 10 percent, without a single layoff, Dolgen reported, and he had pared administrative overhead at the studio by $49 million. These were not particularly big numbers, of course. Paramount spent more money to market the fifteen films it produced in 1994—an average of between $12 million and $15 million a picture—than it spent on its employees. Since the film studio's revenues were estimated by Wall Street to be $1.4 billion in fiscal 1994, with profits of $120 million, the big targets lay elsewhere.

The average cost of a Paramount movie is $32 million, a Paramount executive explained to the group, and if the cost of the five big, or "tent pole," movies scheduled for 1995 was subtracted, the average cost would drop to $23 million. The rub, the executive went on, was this: Big movies are less risky. That was another way of saying that just as best-selling books and hit albums carry most publishing and record companies, big movies do the same in the movie business.

Studios, said Sherry Lansing, who sat beside Dolgen at the meeting, usually lose the most money on the so-called medium-priced movies, which don't have big stars. These big stars—Tom Cruise, Harrison Ford, Tom Hanks, Julia Roberts, Michael Douglas, Arnold Schwarzenegger—tend to have a guaranteed box office overseas, where about 50 percent of a studio's profits are made (a little less in Paramount's case, since they have tended to do fewer of the action-adventure movies that more easily cross language barriers). But when a studio pays a star $15 million just to appear in a picture, as some studios have done, it siphons profits. And when, after Paramount declines to pay a noted director $4 million for his next film, another studio jumps in and pays him $7 million, the costs escalate for all studios. And when some stars are given from 5 percent to about 15 percent of every gross studio box-office dollar, in some cases the star winds up making more than the studio. Paramount, for example, is expected to receive net profits of just over $100 million from worldwide sales of *Forrest Gump*, including product tie-ins, video-store sales and rentals, pay TV, and a sale to network television. But because actor Tom Hanks and director Robert Zemeckis received a significant slice of each gross dollar collected, Paramount's profits on the fourth-largest-grossing movie in history will not be appreciably more than theirs.

Paramount would love to continue making movies with Hanks and Zemeckis, but everyone on Dolgen's side of the table knows that the movie business is not a logical business. And while they would spend part of the next four days trying to arouse the enthusiasm of Redstone and Biondi with extravagant claims about how this or that movie would be the next *Forrest Gump*, they also hoped that New York understood they were not scientists. The entertainment business is "unlike any other business," Sherry Lansing would later say. "It's not like you can apply Harvard Business School rules. It requires instinct and talent." And, she didn't say, luck. Four of the five top-grossing movies in history—*Home Alone, E.T., Star Wars,* and *Forrest Gump*—were each first rejected by at least one studio, Paramount Pictures' vice chairman Barry L. London told Redstone one day at lunch in the commissary.

Paramount's strategy, Lansing explained to the group, was to increase the number of movies the studio made in 1995 from fifteen to a minimum of eighteen, and to reduce Paramount's costs by spreading the risks to other investors (who will also share the profits).

Paramount Television presented a different set of problems. Wall Street analysts project that in 1995 Paramount TV—its library, its TV stations, the programs it produces for both independent and network stations—will take in revenues of just over $1 billion, with profits of $100 million. Like all studios, Kerry McCluggage said, Paramount financed the shows that it sold to the networks at a deficit, hoping to recoup the loss if the shows were hits and the studio could sell them to local stations for reruns.

"The Street is worried about deficits," Biondi told McCluggage. "There are two prisms we are looking at—reality and perception. The perception of the Street is key." The price of Viacom stock might be kept down, according to Biondi, if investors felt Paramount was wasting money by running deficits to finance network shows.

McCluggage argued that the money was hardly wasted, noting that since 1967 the various activities of Paramount's Television Group had generated profits of $1.3 billion. *Star Trek* alone had produced more than a quarter of a billion dollars in profits.

Biondi was not unsympathetic to the facts, but the point was perception, not reality. A top Viacom executive confided to me near the end of the four days of budget reviews that the Viacom executives from New York hoped to induce Paramount, as well as the other divisions, to improve its profit goals. This executive said that division heads tended to be conservative in their forecasts, since their compensation rises as they exceed their goals. Biondi's task, this executive suggested, was to coax more out of Paramount without ordering it.

The week in California underlined Biondi's management style. Jonathan Newcomb, the president and CEO of Simon & Schuster, who was accustomed to the more hierarchical corporate culture of Paramount, says of the Viacom executives, "They put a premium on give-and-take, on dialogue. It's an informal culture. There's not a lot of paper or memos." Once division heads have agreed on their revenue and profit goals for the year with Biondi and Dooley, they are left alone. Except for a secretary, Biondi has no staff aides. His appointment calendar is surprisingly light. He makes time to see old business associates, like Robert Bookman, of the Creative Artists Agency, and Lou Weiss, the chairman emeritus of William Morris, when they call for breakfast or lunch. Jonathan Newcomb says that he speaks to Biondi on the phone maybe three times a week, and Tom Freston, of MTV Networks, says that he speaks to him perhaps twice a week; Jonathan Dolgen speaks to him more often.

Before the Paramount and Blockbuster deals, Biondi conducted weekly executive-committee meetings, but he now plans to have just one a month. Hoping to promote collegiality and to generate ideas for new synergies among the various companies, Biondi invites the head of each division to the budget reviews of all colleagues. At the Paramount budget review, for example, Geraldine Laybourne sat in on all four days. At one session, Sherry Lansing asked Laybourne—who Lansing says "knows more about family films than anyone in the world"—if she would read a film script and share her thoughts on it. "There's a very family style to this company," Lansing says. "There's a really sharing, informal style. People don't feel nervous around Frank or Sumner, because they make you feel comfortable."

This relaxed, sharing culture is, of course, the same one that summarily dismissed Richard Snyder, the chairman and CEO of Simon & Schuster, in June of 1994. No one accused Snyder of lacking vision or of producing inadequate profits. "He did a great job—and I mean a great job—of building a publishing company," Redstone says. Then what happened? Perhaps the first clue surfaced when Snyder and Biondi had lunch at the Four Seasons earlier that spring. After lunch, Biondi said goodbye outside the restaurant and began

walking west, toward Viacom's offices. Snyder headed toward a waiting chauffeured Lincoln Town Car, but as he neared it he paused,
turning toward Biondi. Just then, Biondi turned and saw Snyder
standing beside an open door.

Snyder asked, "Can I give you a ride?"

"Just drop me off at your office and I'll walk," Biondi said. Biondi
recalls thinking that the situation was like a microcosm of the different styles of the two organizations. At Viacom, no one, including
Redstone, had a car and driver (though Redstone now uses Martin
Davis's car, driver, and airplane). Biondi drives a 1991 Acura to work.
Viacom had had no corporate jet, in contrast to Paramount, which
had two. So unconcerned with money was Redstone that he didn't
even draw a salary from Viacom or National Amusements. Viacom
counted its pennies—one reason for its obsession with relatively
petty expenses like a chauffeured car.

The stylistic contrasts became even more pronounced that spring
as Viacom executives came to feel that Snyder was resistant to their
collegial ways. Snyder had toiled in Martin Davis's fear-inspiring hierarchy, and had exhausted a lot of energy trying to protect his people from Davis's whims; perhaps he somehow learned the wrong
lessons. Psychologically, Snyder felt liberated when Viacom bought
Paramount—so much so, a former colleague says, that "Dick saw it
as an opportunity to run his own shop." If Snyder had read the signals correctly after the merger, he might have been free to run his
own shop.

Snyder says that he was not trying to circumvent Viacom: "I
wanted to be part of a team, because I was coming out of ten years
of purgatory." Viacom executives saw Snyder as guarded, and protective of information, insisting that information for Viacom from
his people be filtered through his office. Last winter, Biondi says,
when he asked whether a rumor that the high-profile editor Judith
Regan was leaving was true, Snyder told him that he expected her to
stay. When she did leave, Redstone and Biondi were annoyed that
their no-surprises policy had been flouted, although Snyder says he
had warned them that she might leave. Dauman and Dooley reinforced Biondi's concerns. "We're informal," Dauman says. "We call
up someone and ask for information. We found that after we called
we'd get a call back and we were asked, 'Why do you want that information?' " Snyder, however, says he felt that Viacom wasn't letting him in. "This was a club that no one was welcome in," he says.
"We were bought, and no one called us. We were flabbergasted.
Since they didn't know the publishing business, I suggested weekly

or biweekly meetings to Biondi. He said it wasn't necessary, because they would have regular meetings with our lawyer, our chief financial officer, and others. These meetings were held, and I was not present at them." This didn't rankle Snyder, he says, but the feeling of being excluded did: "As a Paramount executive said to me, 'We weren't bought. We were conquered.' "

Complicating these differences was the matter of Snyder's contract. Last spring, Snyder told Biondi that although his contract would not expire for nearly five years, he and many of his managers had been treated inequitably by Davis, and he wanted to renegotiate it. "Biondi said to me, 'Why don't you give me your wish list.' " So Snyder came up with a list and hired the noted attorney Joseph Flom to renegotiate his contract. Viacom, which had just recently announced that it was going to reduce its costs and sell certain assets to lower its $10 billion debt, was now confronted by contract problems with an executive whom Redstone and Biondi had not warmed to. They thought that Snyder's requests were excessive: a longer contract, the potential for more generous bonuses, stock options. Snyder thought he was producing the "wish list" that Biondi had requested.

By chance, Snyder's contract was mentioned at the weekly luncheon of the company's top four executives. "There were some issues that Dick had raised about his contract that caused us to start talking about Dick Snyder and publishing," Dauman recalls. "As we talked about it, we realized that we had all come to the same conclusion about Dick Snyder." Tom Dooley recalls, "It sort of hit us all at once." Although Redstone had recently praised Snyder publicly and was scheduled to have dinner at his home in a week—a small dinner, to which Biondi was not invited—the group concluded that Snyder would not change. Biondi suggested that they end the relationship. Redstone's secretary canceled the dinner. A few days later, on June 14, Biondi asked Snyder to join him in his office. When he got there, Snyder says, Dauman was with Biondi. In his matter-of-fact way, Biondi told Snyder that the relationship wasn't working and that he planned to announce that afternoon that Snyder was leaving to pursue other interests. Biondi handed Snyder an already drafted press release.

Snyder objected. He said that he was giving an engagement party for his son that weekend and wanted to delay the announcement. Biondi refused. He didn't tell Snyder that he had already told Jonathan Newcomb, who was Snyder's deputy and would be his successor. He did tell Snyder that people already knew and that the news would leak out anyway. Some of Snyder's friends suspected that Biondi wanted him out because he didn't want a publishing company run by someone who wished to spend more of Viacom's resources to expand. Despite Viacom's public denials, some at Simon & Schuster feared that Biondi might want to sell the company. That may be true. When Biondi is asked about the publishing company's future, he says, "Can publishing grow more than single digits? And, if not, what does that mean?" Does it mean that Viacom might sell? The answer is "not clear," Biondi says, candidly. What is clear, he says, is that "people won't pay twelve, thirteen times cash flow" for a business that grows only in single digits.

The book world was startled by the firing of Snyder. Whatever his flaws, he was one of America's preeminent publishers. Then, there was the way in which he was fired. Barry Diller, who was once a Paramount colleague of Snyder's, was one of many people angered by the brusqueness of Snyder's termination. He says, "They could have said to Dick, 'It's not going to work out. You're here thirty-three years and we're not going to hurt you. This is a secret. Take five, six months. Quit on us. It will be your idea, your dignity.' "

In response, Biondi asks, "How do you do it more gracefully?" He says that he derived no pleasure from the task. Yet he described this termination to me, as he had others, in such a matter-of-fact way that I was prompted to ask what he was thinking when he was sitting across from Snyder that day. "We were very comfortable with the decision and the rationale," he answered. "But what you don't know is how emotional this will get. Whether someone will pick up a phone and throw it at you or scream and beg. Those are the worries you have. Having been on the receiving end, I know."

While the words most often used to describe Biondi these days include "serene" and "nice," associates also speak of his detachment. He is said to be guarded with everyone but his wife, Carol. "You don't get close to Frank Biondi, and a lot of people struggle with that," says a senior Viacom executive. He can be opaque, as inscrutable as a stereotypical Oriental diplomat. He is unfailingly polite to everyone, but his compliments tend to come sideways. Norma C. Katcher, who has been Biondi's executive assistant since joining him at Columbia a decade ago, and who adores him, says, "I guess I really knew he liked me when he asked me to come with him to Viacom."

What is Biondi passionate about? Asked that question, Redstone untypically pauses before answering: "I don't think you can use the word 'passionate' with Frank. That's not his style. 'Committed' is a better word for Frank. Frank is committed to winning." Asked the same question, attorney Paul Grand, Biondi's close friend and frequent tennis partner, remembers how Biondi could re-create every pitch in a baseball game he once played in. "His athletic career was a passion," says Grand, "as is his family. He is deeply involved with his daughters in a wonderful way. If Jane was playing tennis, Frank would cancel his own tennis game to go watch her."

If sports was a Biondi passion, politics is not. By contrast to Redstone, who is a passionate liberal Democrat and hosted a fund-raising dinner for Bill Clinton last fall, Biondi has ambivalent views: "I tend to be a Democrat on most social programs. I'm probably, at best, a conservative Democrat or a liberal Republican on fiscal matters." He sees no ideal presidential candidate on the horizon—save former Princeton classmate Bill Bradley of New Jersey. "Even when I disagree with him, he's enormously thoughtful," says Biondi. Asked her husband's passion, Carol Biondi simply says, "Family. And tennis."

Biondi's cool detachment was apparent in late October, when the American Museum of the Moving Image honored Biondi and Howard Stringer, the president of the CBS Broadcast Group, at a dinner at the St. Regis—an event at which Michael Fuchs, now HBO's chairman, was serving as the master of ceremonies, and had been asked to introduce Biondi. The room grew still as Fuchs rose and went to stand behind a lectern framed by white silk drapes. Biondi sat facing the lectern, looking serious, a forefinger pressed over his upper lip. "When I was offered a job at HBO, in 1976, the first person I called was Frank Biondi," Fuchs said, adding that after he took the job the first person he tried to recruit for HBO was Biondi. Fuchs recounted a few of their exploits. "We were brothers," he said. But by the mid-eighties "the brotherhood turned into Cain and Abel," he added. "I'm proud to say we're past the Cain and Abel stage."

Fuchs paused to look at Carol Biondi, who was smiling. She admits that while her husband openly discusses most things with her, Fuchs remains "a sore subject." Yet, even when the rift between the two men developed, she regularly took her daughters to lunch with Uncle Mickey. Fuchs now said of Carol Biondi, "You cannot have a more loyal friend." He rhapsodized about Janie Biondi, his goddaughter. "Even the dog, Mickey, was named after me. The Biondis, all of them, have been part of my life for twenty years."

When the program concluded, Carol Biondi rushed up to Fuchs

and gave him a prolonged hug. Her husband, a step behind, patted Fuchs on the arm and shrugged, saying, "It doesn't seem like twenty years." Fuchs leaned forward, as if to embrace his old friend, and though their cheeks brushed, both men stiffened. When Biondi was asked afterward to describe how he felt, he said, "I don't think that I was emotional." Fuchs was plainly hoping for a catharsis. "Frank asked me to introduce him," he had said before the event. "So maybe this night will be like the end of a chapter."

Listening to Biondi, friends thought that the night was more like the end of the book. "This may be a failing in me, but I can't figure out how to go back to where we were," Biondi reflected a few days later, in California. "Too many things happened. There is less time to hang out. And the fact that we're somewhat competitors makes it harder." We were in a car, on the way to a Paramount meeting, and Biondi looked out the window at leafy Beverly Hills. "Those were good times," he said. "I often think about them the way I think of college: You can't stay there."

Now that he had been the president and CEO of Viacom for nearly eight years, how would Biondi describe himself as an executive?

"I'm somewhere between a coach and a point guard," he responds. "I hand the ball off a lot. On a personal level, I'm much more of a counterpuncher. If you watch me play tennis, I play a very good defensive game. It plays to my strengths. I'm good at analyzing a situation. I'm less confident taking the initiative. When I'm aggressive, it's because I'm very confident of the reasons and the facts on our side of the table." Friends say that Biondi's tennis game tells much about the man. Tom Bernstein, a friend and business associate who on most winter Saturdays plays tennis with Biondi and a foursome that includes the defense attorney Paul Grand, the federal judge Pierre Leval, and the labor lawyer Jerry Kauf, says of Biondi, "He has total self-confidence and poise. He's a gentleman. He never loses his cool. It's a little unnerving." With his glasses and short stride, Biondi does not suggest an athlete. But: "He is deceptively fast and plays an extremely consistent game; he rarely double-faults or gets excited. He cares about winning more than he lets on."

It remains to be seen whether Biondi wants to do more than win. "Frank is now in a position where he has enormous influence," a good friend says. "Will he make his presence felt outside the business world? Will he broaden his reach and his concerns?" Biondi and his wife have been involved in civic and school activities in Riverdale, and have been generous financial supporters and active members of Leake & Watts, a child-welfare agency that works with inner-city children. But so far there is no evidence that Biondi wishes to use his prominence to advance a cause. He approved, but did not champion, MTV's role in urging young people to vote in the 1992 presidential election. When Biondi is asked to describe his public purpose at Viacom, he becomes uncharacteristically vague and uncertain, as if business were divorced from his person.

The same is true of Sumner Redstone. One morning during the November visit to California, Redstone said that he was reading Doris Kearns Goodwin's best-selling biography of Eleanor Roosevelt, *No Ordinary Time*, and he marveled at how the media had allowed the Roosevelts a level of privacy no longer possible. Today, he lamented, the secrets that the Roosevelts kept would be trumpeted by a sensationalist tabloid press. Redstone spoke these words just hours before a meeting at which Paramount's television group presented plans for a new gossip show called *Scoop* and extolled the success of its tabloid newsmagazine *Hard Copy*, whose ratings and profits were up this season. During the lunch break that day, while we were waiting in a buffet line at Paramount, I asked Redstone if he had qualms about *Hard Copy*, especially in light of what he had said earlier about privacy.

"I've got to be careful," he answered slowly. "I have to go and watch these shows. I'm not a great television watcher. Occasionally, I watch them. In general terms, we have to be financially successful." He added that "there needs to be a balance" between making profits and being "socially responsible," but he conceded that he had not thought about programming in those terms as he listened to the presentation that morning. He was clearly uncomfortable with his response, for he spotted Kerry McCluggage and beckoned him over. How, he wanted to know, would McCluggage answer the question?

"I have a quick answer," the low-key McCluggage said. "We're in the entertainment business." Many viewers, of course, think they're watching news.

In evaluating Viacom's future, one may count this opaqueness in the debit column, along with Viacom's occasional cry of victimhood. Sumner Redstone is quick to complain that John Malone and TCI or the six major record companies have organized a "cabal," while excusing Viacom's quest to create risk-free capitalism by becoming vertically integrated. Internally, Blockbuster is hailed as a potent distribution system that can give preference to Viacom's products and force other companies to pay "tolls"—exactly the charges that Redstone lodged against Malone, Time Warner, and the six major record companies.

At an all-day session with Viacom's top four executives at Blockbuster's headquarters, in Fort Lauderdale, in October 1994, the president and CEO of Blockbuster, Steven Berrard, discussed the interactive-game business. He said that Blockbuster would soon have the ability to make instant copies of games or movies in each store, or to send movies out over a giant IBM server, right to television sets. "The next big horizon for us is to get a royalty out of every game made," Berrard said. "Therefore, we have to participate in everything made. We have to run that server and get fifteen to twenty percent of everything that's made. My position with Sega"—the video-game company—"is that we put up all this development money, but we want a piece." Microsoft could have said it no better.

A few moments later, Berrard added that because Blockbuster was a unique distribution system—owning nearly a quarter of all video stores, and generating more revenues than its next 550 competitors combined—"you can force people to bring you their projects, or else they can't get on your highway."

Berrard isn't alone in speaking this way. In addition to "growth" and "software," perhaps no word is uttered more frequently at the new Viacom than "leverage." Executives at Paramount television speak of the leverage they will gain when their new network is open for business and the other networks will have to either pay more or fret over whether they will yank *Frasier* off NBC, or *Melrose Place* off Fox. At Simon & Schuster, executives speak of the advantage of choice shelf space at Blockbuster. MTV Networks wants to team up with Blockbuster to promote its shows. The Paramount theme parks want to tap into Blockbuster's fifty-million-person database to direct-market their wares. Redstone gives fiery speeches about the monopolistic practices of predatory cable or record-company distributors, and he may be right. But Viacom is not just a software company. It seems to aspire to be something more—a software company that uses powerful brand names like "MTV" and "Paramount" to assure distribution for its lesser brands; a distribution company that uses Blockbuster to leverage competitors. According to Alan Schwartz, an investment banker for Viacom at Bear, Stearns, what is going on today in the communications business is a "new paradigm shift," in which brand names like "MTV" will be used to spawn new games, new books, new interactive forms. "Blockbuster does give you cash flow," Schwartz says. "But more important, it gives you distribution leverage."

There are obvious questions about Viacom's future. It still has a large debt burden, which currently totals $10.2 billion but will be reduced to about $7 billion after Viacom completes its sale of Madison Square Garden and sells its cable systems, as it hopes to do this month. Annual interest payments on the debt this year will amount to about $750 million, Dooley says—a manageable sum in a company that Wall Street estimates will have an operating cash flow of up to $2.5 billion in 1995.

There is also the question of size. A major Hollywood figure says, "Size is not a friend to the creative process." Nor is a company that is managed by the numbers necessarily a friend to creativity. There are those in Hollywood and at Viacom's New York headquarters who think that to avoid duplication of effort and to create synergies, a more interventionist, more creatively driven, hierarchical management is called for. Unlike a Diller or an Eisner, however, neither Biondi nor Redstone has demonstrated a creative bent. Michael Fuchs says of Biondi, "He sops up information. He's very smart. He is versatile. But I don't think creativity is his long suit." And if Biondi maintains his current arm's-length approach, a senior creative executive at Viacom says, the central question will be whether he and Redstone understand the messy, sometimes expensive creative process. Or do they understand only numbers?

There are obvious holes in Viacom's portfolio. In addition to launching a fifth network, which is intended to target male viewers with action-adventure shows, both Redstone and Biondi say they would jump to purchase either NBC or CBS—but not at the prices the owners of these networks seek. They would also like to own a large music company, and are currently debating whether to build such a business, as Tom Freston of MTV urges, or to buy it. The biggest hole, however, might be the eventual departure of Redstone and the loss of his passion. Will Frank Biondi, alone, be able to duplicate Redstone's zeal? Will he be too much the "counterpuncher," and not enough the visionary leader? Would Redstone's departure fuel the same type of speculation that has bedeviled Time Warner since Steve Ross died, with some investors complaining that his successor, Gerald Levin, lacks the charisma to lead?

Not long ago, Redstone received a hint of what things might be like for him if he were no longer running Viacom. Returning from their daylong visit to Blockbuster in Fort Lauderdale last October, Redstone, Biondi, Dauman, and Dooley had every reason to celebrate. Wall Street analysts were lavishing praise on them. "With the possible exception of Time Warner, they are the best-positioned communications company in the world," David Londoner, the entertainment analyst and managing director at Wertheim Schroder, told me. Redstone and his executives had just been at a dinner with Blockbuster's top managers, at which Redstone had announced, "Take my word for it, it won't be long before we're number one." Viacom would soon eclipse Time Warner as the world's biggest communications company, he predicted. Even his old nemesis John Malone tipped his hat. "I think Sumner ends up with the biggest pyramid in the desert," he told me. The balance of power in communications had shifted from distributors, like TCI, to providers of content, like Viacom. As the plane leveled off on the way back to New York, everyone relaxed. Redstone stretched out to finish Tom Clancy's latest novel, while the others flipped open a table and played poker.

About an hour outside New York, Redstone remembered that Wayne Huizenga had given him a tape of his farewell appearance before Blockbuster shareholders, on September 30. Although Huizenga was now vice chairman of Viacom and chairman of the Blockbuster Entertainment Group, as well as the sole owner of two professional sports teams and co-owner of the Miami Dolphins, he had become noticeably subdued since agreeing to sell Blockbuster. At the meeting at Blockbuster's headquarters, for instance, Huizenga, looking pale, attended only a few of the presentations and said barely a word. On the plane now, Redstone asked the stewardess to put the tape in the VCR.

When the tape came on, Redstone was still reading and the others were talking or reading. "Viacom is the greatest collection of entertainment assets the world has ever seen," Huizenga declared to the assembled shareholders. The four Viacom executives looked up at the bulkhead screen. They watched as Huizenga, who read from notes, extolled the strengths of the combined companies. He made a point of saying that Viacom had granted long-term employment contracts to more than fifty Blockbuster managers, and had asked Berrard to continue to run Blockbuster and to oversee Paramount's theme parks and Viacom's Showtime. These points were meant to be soothing, but Huizenga's manner was tense.

He admitted that Blockbuster was trapped: by technology, which "will threaten our video business one day"; by a stock market that wrongly "penalizes our stock"; and by economics, which made it prohibitively expensive for Blockbuster to diversify and buy a Hollywood studio. Huizenga started talking about the early days of the company. As he did, his eyes began to mist, and he started stammering. When he acquired Blockbuster, in 1987, he reminded the shareholders, many of whom had been his partners then, it was just a small company with a few stores. No one had to remind Redstone and Biondi that this was the same year they joined Viacom. In the seven years since, Huizenga said, Blockbuster's revenues had grown at a compound annual rate of 90 percent, and its market value had soared 22,000 percent.

The voice of Louis Armstrong singing "What a Wonderful World" flooded the auditorium. As a succession of pictures capturing important moments in Blockbuster's history flashed on a screen above Huizenga's head, tears streamed down his cheeks. He paused to regain his composure, then singled out individuals for praise, then began sobbing again. Redstone closed his book, and Biondi put down his magazine. All four men sat watching Huizenga on the small screen, uttering not a word, transfixed by the sight of a winner in their yearlong quest speaking as if he had lost. And, indeed, Huizenga had—he had lost his baby. This moment was about more than business victories. Huizenga's obvious pain made them feel vulnerable at a moment when they had been feeling invulnerable.

The celebratory journey home from Florida came to resemble a wake as each man retreated into his own thoughts. Biondi said later that he thought of how he had quarterbacked HBO's triumphs, only to be discarded. Redstone remembers that he thought of his own mortality. The next morning, he immediately phoned Huizenga. "When you were crying, I absolutely understood it," he recalls telling him. "If I were you, I'd think the same thing could happen to me. Your whole life is with a company, and then it's over."

POSTSCRIPT:

Success, at least as measured by stock price, proved ephemeral for Viacom. Its share price, which Redstone and Biondi hoped would soar above fifty dollars, did so briefly in 1995 and then sank. It remained in the mid-thirty-dollar range throughout much of 1996 and into February 1997. Blockbuster's growth has slowed, and the much-touted synergies have produced little. By 1996, Viacom had decided to close its Interactive division and scale back Simon & Schuster's new media ventures.

But there is another issue here. As companies like Viacom and Disney and Time Warner and News Corporation grow larger, it becomes more difficult for them to maintain a focus, to make quick decisions, to stay creative. General Electric under Jack Welch is a rare exception: a sprawling conglomerate that continues to act like a small, quick company.

A principal reason Redstone and Viacom agreed to allow me to roam their offices for several months, to scan documents and attend meetings, including a Viacom board meeting, was their desire to demonstrate that Viacom enjoyed a depth of management below Redstone, who was past seventy. Redstone knew that this business profile would focus on Biondi, and it was not easy for him to accept this, as his minions often reminded me. But he went along. However, the appearance of this piece, *The New York Times* would later report, marked the beginning of the end of the relationship between Redstone and Biondi. A year after this article appeared, Redstone summoned Biondi to his office, much as Biondi had summoned Richard Snyder, and fired him.

6

No Longer the Son Of

Seagram's Edgar Bronfman, Jr.

(The New Yorker, June 6, 1994)

As the day wore on, the rumor that Seagram was about to mount a hostile takeover of Time Warner echoed loudly. Perhaps the rumor began with stockbrokers, who tipped their clients, who placed their bets and then alerted reporters, who jumped at a scoop and phoned Seagram officials, who were not in, and then reached jittery Time Warner sources, who could not deny that their company was a target, because they did not know, and naturally made their own inquiries, thereby confirming hunches that something was afoot. By the end of the day—Tuesday, May 3, 1994—the whispers became a roar. The stock market stampeded: nearly four million shares of Time Warner stock were traded that day, and its price was racheted up by almost 11 percent. Time Warner, the world's largest communications company, was besieged. "The press calls started to come fast and furious," says Ed Adler, a Time Warner spokesman who was relaying the rumors to senior executives.

The rumors said that Edgar Bronfman, Jr.—the "star-struck president" of the Seagram Company, according to the next morning's *USA Today*—was closeted with his bankers readying a fifty-five-dollar-a share hostile takeover of Time Warner. One rumor claimed that Bronfman was in cahoots with QVC's chairman, Barry Diller. Another said that Bronfman's co-conspirator was the entertainment entrepreneur David Geffen. Rival bidders—Bell Atlantic, Tele-Communications, Inc., U S West, Bell South—were said to be armed for battle. Bronfman, who had not been in his office for two days, was rumored to have scheduled a press conference for seven-thirty that night, to announce that Seagram, the six-billion-dollar spirits, wine, beverage, and investment company, would raise its Time Warner stock holdings from almost 15 percent to outright ownership.

The reality of the situation was considerably different. Until four that afternoon, Bronfman, the thirty-nine-year-old president of Seagram, had no hint of the rumors. Dressed in faded blue jeans, brown suede half boots, and a knitted navy-blue shirt, Bronfman had spent the day at the Doral Arrowwood Resort and Conference Center, in Westchester, New York. He had been sequestered for two days not with bankers but with more than two hundred Seagram managers, who were attending the company's first worldwide marketing meeting. The morning of May 3 was shared with the managers of such brands as Chivas Regal and Glenlivet Scotch, Martell cognac, Crown Royal whiskey, Meyers's rum, Perrier-Jouët and Mumm champagnes, Absolut vodka (Seagram markets and distributes it worldwide), Sterling Vineyards, and Tropicana beverages.

When the flood of rumors crested, that afternoon, Bronfman was seated in a windowless basement meeting room, where he was being interviewed by me. He had no evening press conference planned; instead, he was scheduled to return to Manhattan to attend a dinner that his grandparents Francis and John Loeb were giving for him and his wife, Clarissa, whom he had married just ten weeks before. When the interview ended, at four o'clock, Bronfman wandered upstairs to his suite and phoned his office. Only then did he learn, from his secretary, of the rumors and the avalanche of calls from reporters and business associates. He checked with his people, and they offered the guess that the rumors had been provoked when it was noticed that E. I. du Pont de Nemours & Company, which is one-quarter owned by Seagram, had filed with the Securities and Exchange Commission the previous Friday to raise three billion dollars through a bond issue. The market may have believed that the three billion dollars was meant to finance a Seagram war chest.

Bronfman had known that the market was jumpy about his intentions. Ever since it was announced, exactly a year earlier, that Seagram had secretly bought 4.9 percent of Time Warner's stock as a "friendly" investor, and planned to increase that stake to nearly 15 percent—a total that had just been attained, on May 2—the market had been convinced that there was a gap between Bronfman's soothing words and a naughty intent. At a time when Seagram's core business had become mature in North America and Europe, while communications companies' stocks were sizzling everywhere, investors and reporters were certain Bronfman was scheming to control Time Warner. They knowingly recited from his biography: that he chose to become a movie and theatrical producer rather than attend college; that he wrote the lyrics for a number of recorded songs; that he was tall and lean and had the dashing good looks and the graceful manner of a movie star; that, like many stars, he spoke almost in a whisper, so one strained to hear him; and that he was a pal of Barry Diller, of the Hollywood superagent Michael Ovitz, and of the actor Michael Douglas. They were certain that he wanted a bigger playpen: one offering the world's most profitable studio, pay-TV company (HBO), record group, and publishing company; such magazines as *Time, People, Sports Illustrated,* and *Fortune;* a position as the second-largest owner of cable systems; and keys to a vast entertainment and information library.

Like the market, Time Warner itself was suspicious. Although it had issued a statement that "welcomed" Seagram's initial investment, an executive who is close to Time Warner's chairman and chief executive officer, Gerald M. Levin, admits, "This was just corporate talk. We were just responding to the 'sincerity' of their statement. Ever tell a woman you love her?" A truer measure of Time Warner's corporate attitude toward Seagram came last January, when its board—mindful that in 1986 Laurence Tisch and the Loews Corporation had gained effective control of CBS when, after buying just 25 percent of the stock, he was welcomed on the board and used his position as leverage—unanimously adopted a poison pill stipulating that any entity that acquired more than 15 percent of the company's stock would have to pay a hefty bonus, unless it made an all-cash offer for the company. While Levin and Bronfman have had lunch about a dozen times in the past year, insiders at both Time Warner and Seagram concede that the courtship has been strained. Bronfman, whose original purchase made him Time Warner's biggest shareholder, wants to be treated as a partner, a co-owner. Levin, however, treats him as an investor—one of many in the company.

The Bronfman family, which has owned nearly 40 percent of Seagram stock since Edgar Jr.'s grandfather Samuel bought out Joseph

Seagram & Sons, in Canada, in 1928, interprets Levin's attitude as one of condescension. "If I were Jerry Levin, I would have at some point asked somebody over here to get on the board and start talking about a standstill agreement," Seagram's chairman and CEO, Edgar M. Bronfman, Sr., says. (He meant a pledge not to buy more than a certain percentage of Time Warner stock.) This is the first time a Seagram official has acknowledged publicly that that was the company's wish. Edgar Jr. says of a standstill that, if asked, "I'd be happy to discuss it with them."

Time Warner is unlikely to grant Bronfman's wish. A ranking Time Warner executive says, unequivocally, "This board has no intention of inviting the Bronfmans on the board." Nevertheless, a year after Seagram went public with its investment, both parties are still in the handshaking phase of their relationship. As a courtesy, Bronfman Jr. phoned Levin from Arrowwood soon after he heard the rumors on May 3. "Listen, Jerry, I'm here at a marketing meeting," he said. "I just called my office, and I heard about all these rumors. I don't even know what the rumors are, but I'm denying all of them." Levin thanked Bronfman for his call. What Bronfman didn't say to Levin, but said to me shortly after making the call, was this: "The closer Time Warner and Seagram get, the fewer these rumors."

"How do you get closer?" I asked.

"That's up to Jerry," he said, clearly exasperated.

The man whom Levin and Time Warner prefer to keep at a safe distance is the second son of Edgar Bronfman, Sr., and his first wife, Ann, who have three other sons and a daughter. Until Edgar Jr. was eight, the family lived in the leafy Westchester suburb of Purchase. They were the grandchildren of Samuel Bronfman and John Loeb— the former the gruff Russian Jew who created what became America's foremost spirits company, the latter an elegant descendant of the nineteenth-century German Jews memorialized as Our Crowd, who founded Wall Street's Kuhn, Loeb & Company. Edgar Jr. and his brother Samuel Bronfman II, who is nineteen months older, have been close since childhood. (The two other brothers are Mathew and Adam; their sister is named Holly.) "Edgar was always older than his years," says Sam Bronfman, the president of the Seagram Classics Wine Company, which is based in northern California. "He wanted to be grown up. He liked doing things adults did. I always enjoyed being the age I was." Their father invested money in various theatrical and movie projects and had a screening room installed in their home. After viewing a movie, he often invited guests to discuss it. No matter that Edgar Jr. was still a teenager, the father wanted to hear his opinions. "He was never bratty," Edgar Sr. recalls, adding, "I remember having more mature conversations with Edgar than with almost any other member of the family."

Edgar Jr. was forced to grow up a bit more quickly than he had expected when his parents went through a bitter, protracted divorce during his teenage years. In 1963, the family had moved to a co-op on Park Avenue, and before long Sam was attending Deerfield Academy, while Edgar went to the Collegiate School, where he received reasonably good grades. In addition to his eagerness to attain maturity, a certain rebellious spirit took root. "From the time I was ten, I always knew, but always, that I would never go to college," he says. "School bored me to death." He was sent to summer camp, but when he was fifteen he refused to go back. He had meanwhile picked up from a family coffee table a movie script, *Melody*, and sensing that his father would continue to indulge him, he now persuaded his dad to put up $450,000 to help finance it. In early 1971, Edgar Sr. hired a first-time producer, David Puttnam, and a first-time screenwriter, Alan Parker, and they planned to go into production the next summer in England. Edgar Jr. talked his father into getting him hired as a gofer. He was paid ten pounds a week, and got his "room and board from some friends of Dad's"—the actress Joan Collins and her husband, Ron Kass. "I loved living in London," Edgar Jr. says. "I loved being on my own. And I liked dealing with the creative process."

He went back the next summer as an assistant to Puttnam, reading scripts, among other chores. "I fell in love with a script called *The Blockhouse*," he recalls. It was about Poland during the Second World War. Puttnam thought it dark and depressing, but Edgar Jr. saw it as a drama of dignity. He raised the money for the project as co-producer, and took a couple of months off from high school to complete the film. "It was everything Puttnam said," he recalls. "It was dark and it was depressing. And absolutely nobody wanted to see a movie that took place underground with seven men slowly starving to death."

Edgar Jr. came back to finish Collegiate, and during his senior year he forged a partnership with the Broadway producer Bruce Stark, to whom he had been introduced by his father, one of Stark's backers. "Edgar Jr. waltzed in one day after Collegiate," Stark recalls. "Remember, I'm then thirty years old and he's sixteen. He brought along a script he wrote. For a script by a kid, it was fantastically well written." They liked each other instantly and decided to team up to develop movies and Broadway plays.

After graduation, Edgar Jr. told his parents that he would not be going on to college but instead would continue his partnership with Stark. "I can't make you go to college," he remembers his father saying. "But I'm not going to support you in order for you not to go to college." Nevertheless, Edgar Jr. admits, "in fairness, he never let me starve." Stark, however, had two children to feed, and he was soon devoting himself full-time to producing a musical.

Young Edgar headed for Hollywood, and there the Bronfman connection opened doors closed to other young men. He met Barry Diller, who was then running the Paramount Studio. He also met with Robert Redford and Sydney Pollack. He impressed people. "I saw he was smart," says Diller, who became a mentor. "When I met Edgar, I was thirty-two. As a teenager, I always acted older than I was. You see that in someone else and gravitate to it."

Diller gave Edgar Jr. intellectual succor but no work. And Edgar Jr. hated the interminable waits to get anything done. Though he was cushioned by his father's wealth, he did not want to become his ward. "It became a matter of pride not to go to him for help," he says. Still, he sometimes did receive help. John Bernbach, whose father was the legendary advertising man William Bernbach, and who is a close friend of Edgar Jr., says that what is hard about being the child of a famous parent "is that you never have a piece written about you that doesn't say 'son of . . .' "

Edgar Jr. and his mother, who has not remarried and lives in Washington, D.C., have always been close. "She was unjudgmental," Sam Bronfman says. "She loved you unconditionally." Their father was more remote. Edgar Sr. spent much of the seventies and eighties getting into and out of four marriages, and his children had sometimes felt neglected. "I was always much more standoffish," Edgar Sr. concedes. Through the first twenty-five years of young Edgar's life, his relations with his father were often stormy.

They pulled together as a family in 1975, when Sam, who was then selling ads for *Sports Illustrated,* was kidnapped; a few days later he was retrieved unharmed. Father and son became estranged after Edgar Jr. decided to marry Sherry Brewer, a beautiful black actress. They had met after he wrote a pop song, "Whisper in the Dark," for her friend Dionne Warwick. His father was furious. "He didn't want me to marry Sherry in the worst way," Edgar Jr. says. "I was too young to get married. He genuinely worried how difficult an interracial marriage was. We never had a problem. I never saw Sherry as a black woman. I don't see my children that way at all." In one of the many quarrels between father and son, the son remembers erupting at him: "Are you telling me that I couldn't be president of Seagram if I married Sherry?" The father said no, but the son guessed that this was his father's true concern. "I don't care about that!" the son exclaimed.

Ann Bronfman intervened and persuaded her son to wait a year before marrying. He did so, but his mind was unchanged, and in November of 1979 the couple eloped to New Orleans. "Mom closed ranks immediately," her son recalls. "It took Dad a little longer. He threw a cocktail party for us, but I could see he was not happy. We remained estranged."

The first of Sherry and Edgar Bronfman's three children was born in 1980, on location for *The Border*, a film starring Jack Nicholson, which Bronfman was co-producing for Universal Pictures. Edgar Jr. and his father rarely saw each other in the two years between the marriage and an invitation he sent his father and his wife to attend a screening of *The Border*. It was a pro forma invitation, a courtesy, extended after Edgar Sr.'s longtime assistant, Maxine Hornung, suggested to Edgar Jr. that his father would be upset if he was not invited. The son reserved a table for four at La Côte Basque, for dinner after the screening. The father praised the movie profusely. They talked about the entertainment business, and Edgar Jr. said he was pondering a move back to New York, thinking that he preferred producing plays to producing movies. They had a pleasant evening. The following day, Edgar Sr. asked his son to come to his office.

Edgar Sr., who was then fifty-one and had run the company since his father's death, in 1971, had a purpose in asking to meet with his son. "It was always important to me, and I'm not quite sure why, that one of my children would take over this company after me," he says. Although Edgar Sr.'s younger brother, Charles, was deputy chairman, he had diverted his energies elsewhere: he had become the principal owner of the Montreal Expos. Their two sisters, Phyllis and Minda, had stock in Seagram but were otherwise uninvolved in the company. Sam Bronfman was now an executive in the wine division of Seagram, but his father did not believe that he was the natural heir to the company. When Edgar Sr. looked at Edgar Jr., he saw more of himself than he saw in Sam, he says. He thought his son could make the kinds of business decisions he had made. He took pride in having made more than two billion dollars by having had the good sense to sell Seagram's oil and gas assets in 1980, just before their value crashed. With a combination of stealth and luck, he then steered the profits toward the one-quarter ownership of Du Pont, and Du Pont had since tripled in value. The expansion of Seagram's basic spirits business was in no small measure financed by two billion dollars in cash dividends harvested by Du Pont.

Days before the dinner at La Côte Basque, Bronfman Sr. had consulted his friend John L. Weinberg, a director of Seagram and a senior chairman of Goldman, Sachs & Company. "Edgar Sr. respected the fact that his son did his own thing," Weinberg recalls. "They both had a creative bent." The day after the dinner, father and son met in the fifth-floor waiting room adjoining Edgar Sr.'s office in the Seagram Building. The father told his son that by now, at the age of twenty-six, he had proved that he could work with difficult people and had honed his skills as an executive. There was no reason his creative thirst couldn't be sated in the business world. "Will you join the company with a view to eventually running it?" he asked. Those are the words that Edgar Jr. remembers hearing from his father. And they came as a shock, for his father is not an expressive man. "Dad doesn't really like to talk about stuff," Sam Bronfman says. "He writes letters."

Edgar Jr. was moved. "In a sense, it was a great personal victory to be asked," he recalls. "In a sense, it was like having approval bestowed on you." He knew that his father was offering to entrust the family's welfare to him. The son said he wanted to go back to California, talk it over with Sherry, and think.

To Edgar Jr.'s surprise, Sherry urged him to accept. He tried to separate the minuses and pluses. The minuses: he would be working for his father; he would be ceding some of his independence; and, he remembers thinking, "In joining the company, I would have to accept for all time never being viewed as making it on my own. Because I didn't. That was part of the leap of joining the company." The pluses: it was a fresh challenge at a time when his patience with the movie business was worn; it was a bigger stage from which to advance his activist views, as his father had done as an early fund-raiser for Hubert H. Humphrey and other Democratic presidential contenders, and as the president of the World Jewish Congress. "Besides," he says, "I think I wanted to prove to Edgar that I could do it. And I wanted to prove it to myself."

The overriding plus, however, was family. Bronfman Jr. says, "There are a lot of parts to the answer, but the first part is family. There is an incredible pride in this company and its family, in who we are and what our grandfather created, and what our father and uncle enlarged." The writer Barbara Goldsmith, who was a neighbor of Ann and Edgar Bronfman in Westchester and has been a lifelong friend of Edgar Jr., says of him, "He is a born caretaker. He's always taking care of his siblings. A typical Edgar thing is to call me and say, 'Holly might move to New York. Where should she get an apartment?' He is very sensitive to emotional nuances. He has an artist's temperament. His nerve endings are always exposed."

To his father, Edgar Jr. put forth three conditions that had to be met: First, whatever job he started with at Seagram had to be temporary; if he didn't like corporate life, he was gone. "Second, I have an older brother who's already in the company whom I love very much, and if it's not OK with him I'm not joining," he recalls saying. "The third condition is if I'm joining the company, I'm running it. I can't, ultimately, work for someone else if I prove myself." His father remembers accepting two of the conditions but does not recall that his son said he wanted to check first with Sam. According to Sam, there was some initial tension. "When he first joined the company, I had been there awhile. A rivalry was created by others," he says. Finally, young Edgar said yes.

"I was shocked," Bruce Stark, his former Broadway partner, says of the decision. "It was the last thing I ever thought he'd do. He was a renegade. He was an artist. He didn't have a corporate attitude." David Freeman, who was a scriptwriter on *The Border* and had shared many meals with Bronfman, says, "I thought, This guy will be back. He loved movies and theater and music."

Those close to Seagram were shocked for different reasons. "There was absolutely no reason to think he had any ability," says Richard M. Clurman, a former chief of correspondents at *Time* and *Life*, who served as a counselor to Edgar Sr. during the eighties.

Edgar Bronfman, Jr., joined Seagram in early 1982, as an assistant to Phillip Beekman, who was then the president and chief operating officer. Edgar Sr., perhaps remembering his own experiences in the company with his authoritarian father, was determined to give his son room to grow. Soon after assuming his post, Edgar Jr. impressed skeptics. "He was astonishingly self-assured without seeming arrogant," Clurman says. "He was extremely capable. He handled himself beyond anyone's comprehension. The question was: How was he so good with so little preparation?"

Within months, Edgar Jr. asked to run Seagram's European operation, which, like the company's other international operations, was then losing money. "He wanted to work for me," recalls Edward Francis McDonnell, who at the time was the president of Seagram

International and ran it with an iron hand. "I was very reluctant to take him." McDonnell had left the presidency of Pillsbury International the year before to join Seagram, and he was determined to transform Seagram's inbred culture—to induce Seagram to think of itself as more than a North American company and of Europe as more than a playground. He worried that Edgar Jr. would strengthen forces that he was seeking to change, or would get in the way.

Young Bronfman moved to London and soon impressed his colleagues. "It was not that easy to change that culture," McDonnell recalls. "But once I had a Bronfman it was easy. He's a very good listener. After he'd been with me a few months, I came to the conclusion that he was the youngest fifty-year-old I'd ever met. He inspired tremendous loyalty. He had great humor. He was part of the team. What he didn't know, he asked about. Which is what bright young people born with a silver spoon in their mouths don't do." Today, McDonnell says—tracing the turnaround from the time he and Edgar Jr. joined the company—Europe contributes about a third of the company's revenues.

When Edgar Jr. started at Seagram, the company had three basic businesses—spirits, wine, and nonalcoholic beverages. In 1984, Edgar Sr. asked his son to return to the United States to run the largest segment, spirits. Having done so for a year, Edgar Jr. set out to expand the wine-cooler business. "The night we became the closest was when he wanted to hire Bruce Willis to do the wine-cooler ads," Edgar Sr. recalls. His son had come to his apartment to say that his instincts told him to go with Willis and an aggressive ad campaign. "Then go with your instincts," his father said. Edgar Sr. recalls, "It was one of the great successes of all time. We went from number five to number one." The father believed that marketing was crucial, and his son was proving adept at it. "The only difference between Chivas Regal and something else is the creativity of the advertiser and the aura about it. That's what marketing is all about. And he knew how to do it," says Edgar Sr., inadvertently subverting Seagram's claim that Chivas is different because it's better. There are, of course, limits to good marketing: today, Seagram's wine-cooler business, which was once projected to have annual sales of a hundred and twenty-five million cases in the United States, sells just twenty-five million.

Edgar Jr. next wanted to streamline and reorganize the company, terminating people, but he encountered internal opposition. Sam Bronfman lobbied his father to support his brother. "That was the transition point for Edgar and me, where we went from being rivals—at least, in some people's minds—to being mutually support-ive," Sam recalls.

The brothers had one more test of their relationship, and this came in 1986, when Edgar Sr. decided to choose Edgar Jr. to even-tually succeed Beekman as president and chief operating officer. "When I did make the choice, it was very hard," Edgar Sr. recalls. "It was hard to do what I had to do to Sam, who expected to get it because of primogeniture. I knew he was going to be angry and dis-appointed. And I didn't want to hurt him." The father nevertheless says almost casually, "I'm not as close to Sam as I am to Edgar."

"It was hard for me at the time, because I was the oldest son," Sam admits. The decision was just as hard to swallow for the company's deputy chairman, Charles Bronfman, who was upset when he saw his darkly handsome brother and his then chubby-cheeked nephew peering from the cover of the March 17, 1986, issue of *Fortune*. To reporters, Charles Bronfman objected to the disclosure of the choice. This was the first public rupture within the family, but it was quickly repaired. Meanwhile, Sam Bronfman was anguished. He and his wife, Melanie, who had met when they were freshmen at Williams College, had long talks. "I had to examine, deep down, what was most important in life," he says. He came up with the same answer his brother had come up with in 1982: family. And he came up with something else: "After a while, it became clear in a variety of ways that Edgar was better suited for it. That's something I agree with from both a lifestyle and an ability standpoint."

The crisis drew the brothers closer. "I admired him for many, many things, including his ability to be supportive of me," Edgar Jr. says of Sam. "I'm not sure that I could do that if the situation were reversed." He successfully pushed for Sam to become a Seagram di-rector, and when Melanie Bronfman was dying of breast cancer, in late 1991, he moved out to California to live with the couple for the last month of her life. Today, when the brothers encounter each other, they exchange a five-step handshake that is a cross between a secret fraternal greeting and a high five.

One of the first major corporate decisions that Edgar Bronfman, Jr., advanced came in 1988, when he acquired Tropicana Products from Beatrice Foods for $1.2 billion. The price was too high, critics said. Seagram officials concede that Tropicana has not done as well as they had hoped. Stephen E. Banner, the senior executive vice president of Seagram, says, "The cash-on-cash return is in the ten percent range. And it would be twelve percent if we were not in-vesting in internal expansion. Now, is ten percent great? No. But we're making money." Most of Tropicana's sales are in North America, and Bronfman Jr. says, "We believe that we are well placed to make Tropicana a world brand." Last year, he notes, Tropicana netted $100 million, or about 13 percent of Seagram's profits. And, according to Ellen R. Marram, the president of the Beverage Group, which includes Tropicana, the juice business throughout the world is expanding by 5 percent annually. "So every year the entire juice business grows more than Tropicana, which claims three percent of the world market," she says. She believes it is a real growth business.

Bronfman's next major acquisition owes something to a 1985 trip he made to Asia as one of about thirty leading business executives— a trip sponsored by Time Inc. In south China and elsewhere in Asia, Bronfman was stunned at the popularity of cognac—a legacy of the French colonial era—which is usually drunk with ice and water. So in 1987 he pursued the Martell family, which had been in the cognac business since 1715, and persuaded it to sell to another family-owned concern. "We'd be nothing in Asia if it weren't for Martell," McDonnell says. In ten years, he predicts, Asia will represent 40 percent of Seagram's business, with North America, South America, Europe, and Africa evenly splitting the remaining 60 percent.

Soon after they purchased Martell, Bronfman and McDonnell traveled to China, Singapore, Tokyo, Hong Kong, and Malaysia to bond with the Martell organization. Their hosts introduced them to a local custom, which was to fill a glass with scotch or cognac, sometimes cut with soda or water; raise the glass in a toast; and then chugalug. "It's an absolute insult not to finish," says McDonnell. During the flight, however, McDonnell had informed Bronfman that he was in his annual two-month dry-out period, to rest his liver. "You'll have to do all the drinking," McDonnell told him. Every night, Bronfman recalls, he downed an entire bottle of Passport scotch, and every morning he woke with a pounding head.

It was worse in Korea the next year. No street-corner culture can rival the chest-thumping macho toasts of a Korean nightclub. A glass is filled, says Eric Nielsen, who was general manager of Seagram in Korea, and the honoree holds aloft his right hand as a show of friendship. "Then you down the shot glass. Then you give your shot glass back to another person and he does the same. If there are ten other people at the table, each does the same to you." Which means the honoree consumes alcohol at a rate of ten to one. When Edgar Bronfman, Jr., arrived at the upscale Dae Won Gak restaurant in 1989, he sat on the floor, braced with pillows, and was soon challenged to the local sport. "Now, nobody can put it down," says

Nielsen. "If you don't drink the shot glass offered, it's called 'illegal parking.' It means you have to drink an extra one." A bottle a night is nothing, he says. His predecessor in Korea, Ian Streiker, a tall, beefy man, was once dared to a toast contest and triumphed when he consumed two and a half bottles of Chivas Regal. "The culture is different," explains Nielsen. "Here you have moderation. It's seen as not being good for you if you're too drunk. In Korea, it can be a positive. If a person passes out, he gave his all. It's very masculine to do this." How did Bronfman do? "Edgar carried it with dignity, though he was drunk," says Nielsen. How did Nielsen survive this ritual almost nightly? "I eat well," he says, "and drink a lot of water in between!"

By the end of the eighties, Bronfman was pushing to identify Seagram as the home of premium brands. He realized that the profit margin for inexpensive spirits and wine is low, while better quality— or at least the aura of it—translates into a higher retail price and a steeper profit margin. He shed high-volume, low-end, low-profit brands. In all, by 1992 Seagram had sold twenty-three American brands, including Wolfschmidt vodka, Ronrico rum, Calvert whiskey, and the Leroux line of cordials. Seagram's North American profits broke company records last year, even though the volume of sales was down by nearly 2 percent, and one reason, according to the president of Seagram North America, Steven Kalagher, was that profit margins jumped. Now Seagram sells fewer cases but makes more money.

Before the 1980s were over, one other momentous event occurred in Edgar Jr.'s life: he and Sherry separated. "Sherry made a perfect Hollywood wife but not a perfect corporate wife," one of her close friends says. She was said to be an attentive, caring mother of their son and two daughters, but was often uncomfortable in business and social gatherings. The 1991 divorce was amicable; most mornings, Edgar Jr. picks up and chauffeurs the children to their private schools, and on alternate weekends he drives them to his weekend home in Pawling, New York.

By 1991, Edgar Jr. had managed to establish himself in a new career, to solidify a relationship with his father, to become the leader of the Bronfman clan. But he was lonely. Beneath a veneer of reserve, he is a hopeless romantic; he has written song lyrics, all of them nakedly emotional. A psychiatrist, he says, helped him discover that while he could be full of ardor, he was emotionally buttoned up. "You have to open yourself up, and in doing so you create a vulnerability," he says. "I wasn't terribly good at that." He remembers that on a trip to

Caracas, in late 1990, he said to the wife of a Seagram local partner, "If there's anyone as pretty as you, I'd love to meet her." That night at dinner, the local couple brought along Clarissa Alcock, a tall, lithe beauty with long dark-blond hair. Born in Venezuela and educated in New York, Paris, and Caracas, Alcock has a college degree in industrial relations; her father was the chairman of Petróleos de Venezuela, and her family had sufficient resources to maintain a New York apartment. At the time they met, Alcock was an executive in the strategic-planning division of Sivensa, the second-largest private company in Venezuela. "We didn't like each other," she says of their initial meeting. "It was so prepared, and it wasn't natural." They talked about Seagram's business, which bored her. Afterward, while the other couple danced, Edgar and Clarissa talked about other things. "There was a chemistry there then," she recalls.

Two weeks later, she made a previously scheduled visit to New York, and Bronfman took her to see *The Phantom of the Opera*. Bronfman recalls, "We held hands. It was electric." For the next year, they carried on an affair long distance. Within a month of their theater date, Clarissa says, he asked her to marry him. Every day for an entire year he sent flowers to her office—initially, two dozen roses, and later two dozen orchids. He didn't dare to send them to her home, for her parents, he recalls, were definitely cool to him. "First of all, an American," he explains, with a sigh. "Which means that their baby, if she falls in love, leaves home. A Jewish man. That's very tough on a devoutly Catholic family. And, even more difficult than that, he's divorced—which makes it impossible for her to be married in the church, or so they would have thought. Who has three children! I think for any parents, watching a woman getting involved in a marriage and having to take *that* on as well is troubling. This is not exactly what every South American Catholic mother dreams of for her children!" To test the durability of the relationship, Alcock enrolled in a master's degree program at New York University, starting in the winter of 1991.

Bronfman called her Chica, and she called him Chico. He eventually asked her to marry him a total of three times, and each time she declined. The last time he asked, she recalls, he vowed not to ask her again until she brought the subject up. Their relationship seesawed, for she was seeking the same kind of independence that he had once sought. She did not want to be known as "the wife of . . ." She recalls, "It was scary. Always, in a foreign country, it's difficult. No matter how much I grew up in the States and how Americanized my parents are, I was very Latin in my being. I was here alone. My family, my friends, my job were in Venezuela. Here in New York, I was an unknown person going out with someone who was very well known. Here it was always 'somebody with Edgar.' Not 'Edgar and Clarissa.' "

They broke off for three months in the summer of 1992, and then resumed their relationship. He wrote her a song, "If I Didn't Love You." Bruce Roberts, who is a friend of Bronfman's and his longtime musical partner, composed music for it. It contained these over-wrought lyrics:

> *I wish I wasn't breathless at the way you move*
> *I wish I wasn't blinded by the lightning when you do*
> *I wish you couldn't reach me in the place in my heart*
> *Belonging only to you*
>
> *If I didn't love you, I wouldn't lose control*
> *The danger loving brings would never touch my soul . . .*
> *If I didn't love you, perhaps I wouldn't die*
> *If you should ever say to me I'm sorry, goodbye*
> *If I didn't love you, if I didn't love you.*

Finally, at the bar of the Carlyle Hotel in 1993, she stammered and stuttered and asked him to marry her. He flew to Caracas to ask her father's permission. They were married in Caracas in February 1993, by a priest and a rabbi; Sam Bronfman was the best man. Bruce Roberts sang "If I Didn't Love You" to the newlyweds as they danced before twelve hundred guests in the garden of Clarissa's grandmother's estate.

Bronfman felt that his professional and personal lives had fused. "It is her personal stature that allows Edgar to sit at the table with her and for there to be two Bronfmans at the table," observes Paul B. Ford, Jr., who is a partner at Seagram's law firm, Simpson Thacher & Bartlett, and represented Bronfman Jr. when he was in the entertainment business, and who remains one of his closest friends. "He has a partner. While she is very much a part of Edgar's life, she has her own life. In a very Confucian way, you must have your house in order to be great. Edgar has his house in order."

By late 1991, Edgar Bronfman, Jr., had begun to search for another kind of partner—one for Seagram. The business was profitable: Seagram had an operating income last year of $754 million. With years to go before the spirits and beverage businesses in Asia and elsewhere overseas appreciably boost corporate profits, and with the overall wine-and-spirits business already stalled, and with Du Pont earnings flat, Bronfman undertook a search for the kind of investment his father had made in Du Pont. "We looked at a lot of areas," says Stephen Banner, who had shepherded Seagram's business as a Simpson Thacher partner before joining the company, in 1991, and becoming its chief financial, legal, and strategic-planning executive. "First, we looked at our basic business. We concluded that it was unlikely that there would be an opportunity of a megabillion size." They also looked at the fragrance and luxury-goods businesses, among others, he continued, and then "we became convinced that the communications-media-entertainment area was one in which a lot of money would get made." It didn't hurt that Edgar Jr., like his father, had an emotional commitment to this business.

Michael Ovitz, the chairman of the Creative Artists Agency, whose father had worked for a Seagram distributor for more than forty-five years, and who had met the Bronfmans years ago, was asked to help analyze this sector for Seagram. In the summer of 1992, over dinner at Bronfman Sr.'s home in Sun Valley, Idaho, Ovitz made an informal presentation on the future of communications. The dinner was attended by Sam and Edgar Jr. Ovitz later made a more formal presentation to the board. The following winter, Herbert Allen, Jr., of Allen & Company, who has specialized in entertainment-industry mergers, was invited to Seagram's Park Avenue office to share his thoughts. Bronfman Jr. also explored the matter with several of Seagram's longtime investment advisers— Felix Rohatyn, of Lazard Frères, and bankers at Goldman, Sachs.

The four appraisals pointed in the same direction: the communications business would grow very fast, and the future would belong to those information-and-entertainment companies which controlled content. "There was a fair degree of unanimity that this industry would have dynamic growth for the foreseeable future," Bronfman Jr. recalls. "Probably more than any other single industry." Ideally, Seagram should invest in a company that manufactured for a variety of entertainment/information outlets—from movies to television to music to print, Seagram concluded, and that specification narrowed the choice to a Hollywood studio. Neither of the two Japanese-owned entities—Sony's Columbia and Matsushita's MCA—was for sale. Rupert Murdoch was not about to sell Twentieth Century Fox. Disney had strong management, a formidable stock price, and the protection of a nearly one-fifth ownership by the investor Sid Bass. Each was impregnable to a takeover. Besides, according to Ovitz, "there was never any discussion of a hostile takeover." MGM was too frail. This left two candidates: Paramount and Time Warner. While Paramount (later acquired by Viacom) was attractive, it was thought to lack the range of assets enjoyed by Time Warner. Paramount didn't own a music business; it didn't own comparable cable-programming assets; and it didn't own magazines. And its management was generally thought to be inferior to Time Warner's.

Steve Ross, the Time Warner chairman, had decided not to sell assets to help reduce his company's $15 billion debt load—a strategy that his former co-CEO, Nicholas Nicholas, had advocated. Ross wanted to sell minority ownership in the core businesses. In 1992, Ovitz offered to visit with Ross, whom he knew, to explore an equity investment. Without ever mentioning the name of the company he was representing, Ovitz went to see Ross at his house in East Hampton. (The date of the meeting is disputed, with people at Seagram believing that it was in September 1992, and Time Warner sources insisting that it was closer to the spring of that year, because by September Ross had been severely weakened by cancer and was not receiving visitors.) Nothing came of the meeting.

By December of 1992, Bronfman Jr. had come to a conclusion so obvious that more experienced investors overlooked it. Investors agreed that Time Warner owned a brilliant array of diverse assets. They agreed that it was a global colossus, and that acquiring the entire entity was out of the question—it was too expensive. What Bronfman Jr. and his advisers saw was that Time Warner need not be swallowed whole for an investor to benefit from its potential growth. Time Warner didn't have an owner; it was run by managers. None of its various shareholders held a dominant minority position; none was represented on the board. Seagram saw an open parking space just waiting to be filled.

In February of 1993, the Bronfmans persuaded the Seagram board to approve the secret purchase of as much as 4.9 percent of Time Warner's stock, or just under the 5 percent threshold that would require public disclosure. One reason to buy quietly, Herbert Allen, Jr., says, is simple: "Why tell the market you're buying stock and drive the price up against yourself?" Another is that the predator did not want the prey to know. From the moment Seagram began buying stock, that month, until its purchases were disclosed, in May, Time Warner knew nothing of this activity. Nor did it know of another decision made by the Seagram board, later that winter— to expand the initial investment to as much as 15 percent of Time Warner shares.

"The moment we crossed"—acquired more than 5 percent—"I called Jerry," Bronfman Jr. says. "I told him we thought he was running a terrific company. We thought we'd like to buy up to fifteen percent of the shares. This is a benign investment." It was the same kind of "passive investment" that Seagram had made in Du Pont, he told Levin. Although Levin was dubious, he thanked Bronfman for the courtesy, and released a statement welcoming the investment. In the coming months, Seagram continued to accrue Time Warner stock.

The presence of an uninvited guest provoked an intense debate within Time Warner: doves said that the company should take the Bronfmans at their word and seduce them into the tent, and hawks said that it should not act on their hopes but should instead prepare for battle. At first, Levin did not choose sides. He had just finished reducing the membership of the Time Warner board from twenty-four to eleven. (It is now fifteen.) He had discarded many insider directors. He had always opposed poison pills. It wasn't that Levin was soft, for he had not hesitated to snatch power and supplant first Nick Nicholas, and then Steve Ross's retinue. But he hesitated.

In the end, the hawks prevailed, led by Warner studio chairmen and co-CEOs Robert Daly and Terry Semel, and HBO's head, Michael Fuchs. They were joined by representatives of investment funds like the Capital Group, which owned 9 percent, and by U S West, which had invested $2.5 billion in Time Warner Entertainment. Each made the same argument: If we avert our eyes, the Bronfmans can pull a Tisch, gaining control of the company without paying for all of the stock. The Bronfmans had a plan to dominate the company, one hawk insisted. "No one spends one billion nine hundred million dollars to sit by the side of the road," another said. "And this is show business."

The more Levin and Time Warner studied Seagram's past, the more nervous they became. They noted that Seagram had not always made "benign" investments. In 1981, Seagram launched two takeover bids that turned hostile. The first was for the St. Joe Minerals Corporation, and resulted in a bidding war won by the California-based Fluor Corporation. The second was for Conoco, Inc., which was bought by Du Pont, with Seagram's Conoco shares converted to 25 percent ownership in Du Pont. The Bronfmans claim that the Du Pont investment is passive, because they do not intrude in management of the company, but many at Time Warner have maintained that it is not so passive. What unnerves Time Warner about the Du Pont investment is contained in the fine print of Seagram's latest proxy statement: "The corporation has the right to designate 25% of the members of the Du Pont Board of Directors." The fine print also requires the approval of Seagram before Du Pont may sell more than 20 percent of its voting stock, or dilute Seagram's ownership by issuing more stock without Seagram's consent.

Time Warner's suspicions were further fueled by Michael Ovitz. Executives at the highest levels of Time Warner believe that Ovitz, who has not been paid by Seagram (although a Creative Artists Agency official does not rule out a future financial arrangement), has set out to "destabilize" their company, because he wants Levin's job. According to well-placed Time Warner sources, Daly, Semel, and Fuchs have complained to Ovitz in private meetings.

"Let's separate rumor from reality," Ovitz says. "The rumor mill has had me running first Sony, then MCA. Next came MGM. And now Time Warner. None of this has come to pass, because the reality is I'm in a great situation here at CAA. Our work with these other companies helps to stabilize markets for our clients and, by the way, for everyone else's clients as well. The reality is: We are building our own company, CAA, for the future."

Few in the Time Warner hierarchy believed him. Their suspicions—what one friend of Bronfman's refers to as their Iago-like paranoia—prompted Time Warner's board on January 20 to adopt the poison pill that might make it more difficult for Seagram to swallow more than 15 percent of the company's stock. When Bronfman Jr. heard rumors the night before the board was scheduled to vote, he phoned Levin. "What's going on?" he remembers asking.

"I can't discuss it with you," Levin replied.

"Better if you tell me, so that I don't read it in the paper."

"I'm in a difficult position," Levin responded. "I can't discuss what my board may or may not do."

The next day, Levin phoned to tell Bronfman that the board had instituted such a takeover defense, but noted that it would not affect Seagram because Seagram had said it didn't plan to purchase more than 15 percent. Bronfman responded that he did not believe in poison pills. Bridling because Seagram had not been consulted—as Levin had bridled because he was not consulted when Seagram decided to buy his stock—Bronfman issued a public statement. He says that in it he tried "to be as gentle and noninflammatory as is possible under the circumstances." He did not succeed. While the statement noted that Seagram had not yet read the details, it declared, "As a general matter, however, we believe that Rights Plans, or so-called 'poison pills,' can interfere with shareholder choice and adversely affect shareholder values" by depressing the stock price.

Now Levin was angry, and he phoned Bronfman. "He was incensed that we would make a statement that was anything but wholly supportive of Time Warner management," Bronfman recalls.

"Jerry, I did and do disagree about the Rights Plan. But we have disagreed about nothing else," Bronfman said.

That is plainly wrong. They disagree, fundamentally, about Seagram's intent. And about whether Seagram is just another shareholder or a partner of Time Warner's. While Seagram hadn't previously divulged this, it might be willing to sign a standstill pledge, as it did with Du Pont. But in return it would want to be accorded the respect of a principal, and to receive at least a couple of seats on the Time Warner board. That's what Edgar Bronfman, Sr., says he would demand, and he adds, "If they want to have an honest discussion about shareholder values, I'm perfectly willing to have an honest discussion, with him or anybody else. If I were in his shoes, that's probably what I would do. But then I'm not as scared of me—or of Edgar—as Jerry Levin is."

Time Warner officials are at present adamantly opposed to inviting a Seagram representative to join the board. "He has to demonstrate to us he is worthy," a Time Warner board member said of Bronfman Jr. in February. "He has to prove it. He can't buy his way on." In the months since, that view has hardened. A Time Warner executive who is not a member of the board but reflects the prevailing view at the company says, "We have bigger partners than Seagram. U S West has more money invested than Seagram, yet they're not on the Time Warner board." And a standstill? Time Warner's attitude is that a standstill is unnecessary, because it implies a negotiation, and there is nothing to negotiate.

Inevitably, the absence of true dialogue encourages sniping. Some Time Warner officials sneer at "the alleged" kidnapping of Sam Bronfman in 1975. (The two men accused of abducting him were convicted of grand larceny, not kidnapping.) It is said by people at Time Warner that Edgar Bronfman, Sr., is "a loose cannon." Edgar Bronfman, Jr., is said to be "a rich kid" enamored of Hollywood. Seagram executives, in turn, assert that Nick Nicholas was right to want to sell off some of Time Warner's lesser assets to pare its debt, as Viacom is now doing to help digest its acquisition of Paramount. If Levin had asked Edgar Jr. to support him before a proxy fight over whether directors' terms should be staggered, a top Seagram official says, Edgar Jr. might have acquiesced rather than align his shares with dissidents at Time Warner's annual shareholder meeting on May 19. And Seagram executives ask incredulously, "Did you see Time Warner's proxy statement?" They note that although the company reported a loss, before "unusual and extraordinary items" of $94 million last year, nevertheless Levin received a $4 million bonus, up from $2.5 million the previous year. Bronfman Jr.'s bonus last year was $962,600.

Despite this climate of potentially corrosive mistrust, both Seagram and Time Warner say they are trying to improve what Steve Banner calls "the comfort level." Neither Bronfman nor Levin wants a public squabble. Each is a gentleman. But there remains a gulf between them, because each wants a distinctly different relationship: one wants to marry, the other would like to end the affair but can't. So they date, chatting occasionally on the telephone; they have lunch, the last two times in the very public arena of the Grill Room at the Four Seasons. Their discussions, both men have told others, are somewhat stilted.

Both camps report that this is how the discussion proceeds: Levin tells Bronfman he has no intention of inviting him to join Time Warner's board; Bronfman says he hasn't asked. Levin says one of his partners, U S West, is concerned about Seagram's intentions; Bronfman offers to meet with U S West but gets no response. Levin mentions that the federal government's rollback of cable prices has retarded Time Warner's ability to invest in undervalued cable properties; Bronfman says Seagram might be willing to put up money to assist these investments, but Levin lets it slide. Levin extolls Warren Buffett as a model investor, mentioning his passive investment in Capital Cities/ABC, where he ceded his proxy to ABC's chairman, Thomas Murphy, and its president, Daniel Burke. But "that's because he had a dialogue with Burke and Murphy," Bronfman told me. "I have not had that kind of dialogue with Jerry."

Also at variance are the two sides' perceptions of the current stalemate. Time Warner feels that it has blocked Seagram, spurning its effort to be treated as a partner. "They're frozen at fifteen percent," an important Time Warner executive asserts. "If Bronfman wants it now, he won't get it." The prevailing view among Seagram executives is that the company has already established its leverage. Levin may not like it, but he is stuck with Seagram as a partner, company executives believe.

Perhaps in the short run Levin has stymied Seagram. Even if Sea-

gram executives wanted to—and they insist they don't—they concede Seagram cannot afford the more than thirty billion dollars it
would cost to ingest Time Warner and its poison pill. But if this
places Levin temporarily in the driver's seat, it is also a hot seat. The
core question lingers: Can Levin and Bronfman establish a relationship of trust? The risk for Levin is that he may have made himself
hostage to events he can't always control. If Time Warner stock
drops, if anticipated profits don't materialize, if a controversy or
scandal in one of the company's divisions erupts, Levin's seat gets
hotter. And if Seagram is on the outside and can say "I told you so,"
then shareholders might turn to it as a savior, or might insist that
Levin bring it inside in an effort to boost the stock price.

The simple truth is that Levin cannot yet advance the argument
that Steve Ross's supporters once did—that he's indispensable.
Levin is a man of keen intelligence and has an attractively modest
manner. He can articulate a vision better than Steve Ross did. But,
unlike Ross, he does not have a personality that fills a room. His
management style is to delegate to assertive, capable executives. One
result is that Levin is perceived as being less indispensable than, say,
Robert Daly or Michael Fuchs. If Time Warner's stock price drops,
a prominent entertainment executive who respects both Levin and
Bronfman says, "At that point, poison is something that management will drink." The relationship is combustible. As Levin keeps
Bronfman at a distance, Seagram officials sense that he is condescending. Inevitably, the feeling grows that he is treating Edgar
Bronfman, Jr., as "the son of . . ."

In his comfortable office, five stories above Park Avenue, Edgar
Bronfman, Jr., displays two framed letters signed, simply, "Tree."
One, a short, handwritten letter, stands on a table beside a brown
leather couch, and the other, which is typed, stands on a credenza
under a Joan Miró canvas and faces an antique writing table that
Bronfman Jr. uses as a desk. The letters are from his father, whom
Edgar Jr. sometimes calls Tree. (At their Westchester home, when
young Edgar sat in his father's chair, Edgar Sr. would bark, "Get out
of my tree!") The letters warmly thank Edgar Jr.—for justifying a father's business faith in his son and for an affectionate and glowing
speech that Edgar Jr. made in May of 1991, on the occasion of Edgar
Sr.'s fortieth year at Seagram. "Those are the kinds of letters *he*
never got from *his* father," Bronfman Jr. says. In Edgar Sr.'s office,
just down the corridor, a bronzed copy of Edgar Jr.'s speech hangs

on an oak-paneled wall. When Bronfman Sr. is asked how he felt when his son stood before the company and declared his love and admiration, tears flood his eyes, and he does not—cannot—speak.

Father and son are united, and Edgar Bronfman, Sr., says that at the board meeting and then at the annual assemblage of Seagram shareholders in Montreal on June 1, 1994, "I will tell the directors that I intend not to stand for reelection. I will say this at the annual meeting. And then I will recommend—and they will do as they please—that my son be my successor."

There will, undoubtedly, be sneers that Bronfman Jr. inherited his position. "People say he is a playboy. I remind you of those who used to call Ted Turner the idiot boy when he took over his father's billboard business," says his friend Henry Scholder, who met Bronfman Jr. when he was an investment banker in London, and who now is writing a novel. Elizabeth Rohatyn, who serves with Bronfman on two boards he has recently joined—those of the New York Public Library and WNET-TV—is struck by his poise and self-assurance. "I was knocked out when he had President Clinton at his apartment for a fund-raiser," she says. "He was so relaxed and cool and elegant. Edgar chatted with each person. He wasn't nervous that the president was coming to his apartment." Asked about his son's self-possessed manner, Bronfman Sr. says he is a good actor: "Some of it is insecurity. He covers up a certain amount. Which is fine. He has a great presence. I am sure he asks himself a lot of questions."

One question Bronfman Jr., who turned thirty-nine on May 16, need not ask is whether he will be noticed. Can the kid who craved the creativity of the entertainment business be happy as an adult who spouts business lingo and runs Seagram? Bronfman Jr. attacks the assumption behind the question.

"I reject the notion that business and creativity are mutually exclusive," he says. "I think I've been a reasonably creative president of this company. I think we have fundamentally altered the direction of this company in the last five years. . . . I don't lack for creative outlets."

Nor, he goes on to say, does he lack for a higher purpose, and that is how the adult Edgar squares with the young Edgar. "It really goes back to what my grandfather built. And what my father and uncle enlarged," he says. "I knew that when I grew up I would be financially secure. Having that sense of financial security made it possible for me to be 'brave,' to go out and try new things, to be 'different.' It's easier to be different if you know that there's a net underneath the wire. For reasons that I suspect my psychiatrist can explain better than I, I find it enormously important that my brothers and sisters and cousins, my children, my nieces and nephews, feel that same sense of security. Because I want them to have the freedom that I had. And I get enormous personal satisfaction in being a part of that for them. And so, when I try and square where I am and where I was, I think I'd be having a lot of fun there—but it would ultimately feel frivolous. It would ultimately feel that I had shirked a larger and more important responsibility, and that would sour whatever pleasure would exist in that other life."

POSTSCRIPT:

Edgar Jr. was not content to remain only in the spirits business, despite his claims about the "creativity" of the business world. He would sate his entertainment appetite by acquiring MCA in 1995. Although the two companies' relationship remained frosty, Seagram retained its 15 percent stake in Time Warner, waiting to sell it or to exchange it for some Time Warner assets. Levin had succeeded in stymieing the Bronfmans, just as he boldly outflanked Rupert Murdoch, John Malone, and Jack Welch of GE by acquiring Turner Broadcasting in 1996. But by early 1997, Levin had not succeeded in boosting Time Warner's stock price, and thus he remained a hostage to events—and perhaps to hostile shareholders like Seagram.

7

THE CONSIGLIERE
Herbert Allen, Matchmaker
(*The New Yorker*, May 22, 1995)

Herbert A. Allen's official biography contains just five sentences, and they reveal almost nothing about the man. The biography says that Allen is the chief executive officer of Allen & Company, a privately held investment-banking firm; is a member of the board of directors of the Coca-Cola Company; was once chairman of the board of Columbia Pictures; and graduated from Williams College. There is no hint that Allen has become the unofficial investment-banking consigliere to many of the world's communications companies. Allen calls himself "an owner" and "a buyer," and notes that on behalf of Allen & Company, the Allen family, and the firm's partners, he has several hundred million dollars invested in a hundred and fifty companies. "I'm not an investment banker," he says.

Nevertheless, clients seek his investment advice. In November of 1994, he was summoned to a meeting in Osaka with Yoichi Morishita, the president of the Matsushita Electric Industrial Company. In 1990, Allen represented Matsushita when it purchased MCA, which is the parent of Universal Pictures, Putnam Publishing, a television company, and a record company, and in addition has an interest in theme parks. Now, having tired of the extravagant costs and the narcissism of Hollywood, Matsushita wanted Allen to help it sell MCA. What unfolded over the next five months will no doubt contribute to the Allen legend.

Allen is fifty-five but appears to be older. He has pouches under his eyes, which are deep-set and dark; his forehead is high and is topped by thinning, short black hair. His face looks gaunt. He is wide-shouldered and lithe, however, and laughs easily, and listens intently, which is part of his charm. Sitting across from Allen in Osaka, Morishita asked him how much MCA might bring in an outright sale, and Allen said seven to eight billion dollars. Morishita asked him to prepare what Allen calls "a menu of choices" regarding MCA.

Back in the States, Allen visited MCA's executives and listened to their analysis of the business and to their complaints about dealing with Matsushita. In December, at a second meeting in Osaka, with eleven Matsushita executives present, four members of Allen's team offered a financial analysis of MCA's strengths and weaknesses—the kind of computer-generated presentation that Allen prefers not to do himself. He is not a spreadsheet specialist, not a banker who is constantly taking notes. Although his clients all agree that he is smart and financially adept, when they are asked to describe his talents they usually mention not professional qualities but personal ones—candor, independence, loyalty, and a shrewd sense of people and their motives.

Only after the focus of the second Osaka meeting shifted from hard numbers to matters of judgment did Allen speak up. The Matsushita executives said that they were puzzled by the behavior of MCA's chairman, Lew R. Wasserman, and its president, Sidney J. Sheinberg, who for months had been talking openly to the American media about their grievances against their Japanese parent, their principal complaint being that Matsushita would not allow them to expand into cable, to buy a television network to assure the distribution of the studio's TV product. Expressing one's grievances publicly was alien to the team-oriented Japanese management culture, and in the most indirect, polite ways the Matsushita leaders said so.

Allen's response was clinically blunt. The Matsushita executives had just three options, he said. One, they could retain Wasserman and Sheinberg and other MCA executives and support their expansion plans; two, they could sell MCA; or, three, they could keep MCA and invest more resources but change the management. Allen said that if they choose the first or the third option they should do as Wasserman and Sheinberg wanted and invest more money to buy a broadcast network, perhaps, or expand into cable and other new media.

For weeks after the December meeting, Allen heard nothing from Osaka, but the silence was no surprise. Nor did it particularly bother him. Allen is a stoic, and he knew that Japanese companies tended to be secretive about their decision-making and tried to reach an internal consensus before sharing their thoughts with outsiders. Dealing with the Japanese, he liked to say, was akin to pouring ketchup from a new bottle. You have to be patient, but once the Japanese have reached a consensus, everything pours quickly. By March, Matsushita had reached a consensus to sell MCA. By April, Allen had helped engineer a sale.

What was unusual about the sale was its inside nature: without a public auction, Allen deftly delivered MCA from one client to another. In some ways, it was the kind of inside deal that Allen criticized when he represented Barry Diller in 1993, and Viacom's chairman, Sumner Redstone, and Paramount's chairman, Martin Davis, agreed to a friendly merger without an auction. Nevertheless, with the help of Michael Ovitz, the chairman of the Creative Artists Agency, who had also represented Matsushita four years ago and was an adviser in this negotiation, Allen persuaded the Japanese to sell the company to Edgar Bronfman, Jr., the president of the Seagram Company. Bronfman advanced his own cause by twice flying to Osaka to have dinner with Morishita. The first of these visits was arranged by attorneys at Simpson Thacher, which had long represented Seagram and now, conveniently, was counseling Matsushita. After the second dinner, Morishita told Bronfman that he would make an exclusive arrangement to sell Seagram 80 percent of MCA, subject to price and timing. Word of the deal leaked out in late March, when Seagram raised the cash by selling its stock in E. I. Du Pont de Nemours & Company for nine billion dollars. And when word did get out, a friend of Bronfman's says, "a long, juicy list of those interested in MCA" materialized, including Ronald Perelman's New World Communications and the German-based media conglomerate Bertelsmann.

By then, however, Allen and Ovitz—along with Seagram's traditional investment-banking firm, Goldman, Sachs, which, like Seagram's usual outside legal counsel, Simpson Thacher, was this time representing Matsushita—had locked up the deal for Seagram. There would be no bidding war. "All our regular advisers were on the other side," Edgar Bronfman, Jr., says. "I hope that was helpful." According to Allen, the major credit for putting the deal together for Bronfman belongs to the attorneys at Simpson Thacher. In corporate boardrooms, however, Allen is thought of as the architect. When it comes to Hollywood and investment banking, "all roads lead through Herbert," says Brian Roberts, who is the president of Comcast, one of the nation's cable giants.

Now that Bronfman owns MCA, he has at least three major decisions to make—all of which could involve Allen. First, he must decide on an executive to run MCA—a decision in which Allen may play a pivotal role. Sources close to Bronfman say that his first choice is Ovitz, who is a friend of both Bronfman's and Allen's. "Bronfman has talked to him about the job," says a mutual friend. Ovitz, a friend says, has grown tired of the constant hand-holding and selling required of an agent, even one with his power. But before Ovitz would leave Creative Artists, a company he built and from which he is reported to receive thirty-five million dollars annually, he would want to net about two hundred million dollars from the sale of his stake in it. He would also want a promise of "reasonable autonomy" and other assurances, an Ovitz intimate says. The one man Ovitz relies on for counsel is Herbert Allen.

Second, Bronfman must decide whether to sell Seagram's 15 percent stake in Time Warner, which is worth two billion dollars, and which Allen helped Bronfman buy last year. Within the investment-banking community, one hears mild derision that Bronfman and Allen, as soon the MCA sale was set, did not exercise Seagram's leverage over Time Warner to extract a steep exit price. This is, undoubtedly, what the investor Kirk Kerkorian may be trying to do by threatening a takeover of Chrysler. The practice is called "greenmail," and Allen considers it a form of extortion. Still, there are Allen clients—NBC, which is owned by General Electric, and perhaps two or three others—who might like to be partners with Time Warner.

Finally, the road for Seagram to acquire or become partners with a TV network could pass through Allen. Seagram needs a network for the same reason Twentieth Century Fox, Warner Brothers, and Paramount started their own networks: studios want a guaranteed distribution system for their television-production factories. At the moment, however, Bronfman says he is not thinking about a network. "I have nothing on my plate but understanding MCA," he says. "This is a big enough bite for us. Once we understand the business, there will be a whole lot of opportunities for growing." When Bronfman is ready to make another move, Allen has a client—NBC, again—searching for a global partner.

. . .

Allen owns nearly 40 percent of Allen & Company, his family owns 35 percent more, and fifteen directors own the rest. Associates estimate Allen's private fortune to be a billion dollars—a sum that helps account for his independence. His firm employs a hundred and sixty people and occupies two floors of a building on Fifth Avenue in the Fifties. Allen has a modest office whose windows overlook Fifth Avenue, and which contains only a desk with a brown leather desk chair, a beige couch, two blue suede armchairs, and a third blue chair opposite his desk; the walls feature oil paintings of the American West. When he is not in the middle of a big deal, Allen says, his investment-banking duties consume about 20 percent of his time. The remainder, he says, is taken up with managing the firm and investing its capital.

Allen also spends time attending to a large and disparate collection of real friends, not acquaintances. He has homes in Sun Valley, Idaho, and Cody, Wyoming, and each February he flies about ten or so fellow Williams alums out West, where for a week they ski and enjoy themselves. Each August, he invites some of these pals and other old friends, including several from the office, for a week of fishing in Alaska. In the fall, he takes another group on an African safari. Allen's friends uniformly believe that friendship and loyalty—not deals or commerce—are at the core of his character. "If I were in trouble anywhere in the world and I had one phone call to make," says NBC anchor Tom Brokaw, a friend for two decades, "I'm convinced Herbert would stop anything he was doing to help me."

"Herbert's life wouldn't make a commercial movie," says the movie producer Ray Stark, who is seventy-nine, and has been one of Allen's best friends for nearly three decades. "He doesn't blow up. He is not an exciting person for the screen, because he's not volatile." Allen is intensely private. He has rarely been written about, and once explained this reluctance by telling me, "I don't need the kindness of strangers." He is set in his ways, like an old man. He does not smoke or drink or tell dirty jokes. He detests modern art, but loves American Impressionists, and he collects their work. He does not listen to jazz, or to music that has no recognizable melody. He hates all rock and roll, and is partial to Frank Sinatra and Broadway musicals of another era. Twice divorced, he does have a woman in his life—Gail Holmes, a Denver real-estate executive—but they are together only about ten days a month. He enjoys solitude and lives alone, except for four dogs, at the Carlyle Hotel. He rises early and walks to his office, about a mile away, arriving before seven. Most mornings, soon after he arrives at his desk, he speaks on the phone with Bronx Administrative Judge Burton Roberts, who has been a friend for three decades. Twenty years ago, Allen introduced Roberts to writer Tom Wolfe, who used Roberts as the model for the judge in *The Bonfire of the Vanities.* The call to Roberts is usually the first of more than a hundred calls. Daily, Allen speaks with his sister, Susan; with Ray Stark; and with Michael Ovitz; he is regularly in touch with other friends, including Brokaw; a San Francisco bar owner named Ed Moose; a Washington lobby-ist, Mike Berman; and ten or so of his friends from Williams College.

Around six, Allen walks from his office to one of two restaurants: Bravo Gianni or Gabriel's. When he dines alone, as he does one or two nights a week, he likes to read a book of nonfiction. At Bravo Gianni, he is often the restaurant's first customer. He always sits at the same table, next to a wall, orders a bottle of the same expensive Barolo Italian red wine, and drinks one glass, two at the most, with *bruschetta,* followed usually by a plate of pasta and tomatoes. Then he walks home and tries to get to bed by ten. Friends who want to spend time with him are expected to adjust to his habits. "He's the most eccentric person I know," Michael Ovitz says. Ovitz recalls a weekend that his family spent at Allen's ranch in Cody. Dinner was planned for six o'clock. "I made the mistake of being fifteen minutes late, since I was on the phone—unlike him, I have to earn a living," Ovitz says. "When I arrived, he and my entire family were into their second course. He had my courses stacked up like planes in a holding pattern at Kennedy. Herbert said, 'You're late.' "

In addition to the ranch in Cody and the home in Sun Valley, Allen owns a mansion in Southampton that his four grown children use but he hasn't visited in two years, a hilltop farm in Williams-town, Massachusetts, and several condominiums in Sun Valley, Idaho, where for the past dozen years or so he has held an annual weeklong retreat for leaders in the business community and their families. Allen has represented many of those who attend, including John Malone; Rupert Murdoch; Barry Diller; Edgar Bronfman, Jr.; Robert Wright, the president of NBC; two DreamWorks SKG co-founders, David Geffen and Jeffrey Katzenberg; Sumner Redstone, whom he helped acquire Viacom in 1987 and for whom he (and Stanley Shuman, a partner in his firm) sold Madison Square Garden last year; and Coca-Cola's CEO, Roberto C. Goizueta, to whom he sold Columbia Pictures in 1982. (With Allen's assistance, Coke then sold Columbia to Sony in 1989.) Allen always arranges for a photographer to take pictures of the guests at the Sun Valley retreat, and each day a new set of pictures is framed and hung on the walls—of the investor Warren Buffett fishing, for instance, or of Microsoft's Bill Gates and Disney's Michael Eisner hiking. Allen is always urging friends and clients to take time off and go on safari with their children, as he has. With a hint of envy, Michael Ovitz observes that Allen has achieved an ideal balance between business and leisure. "He has it all figured out," Ovitz says.

In terms of fees earned or number of deals closed, Allen is not the most prolific investment banker in the communications world—that distinction probably belongs to Steven Rattner, the head of Lazard Frères & Company's communications group. But Allen has certainly been involved in some of the largest media deals. Rupert Murdoch, who has relied on Allen & Company's Stan Shuman for twenty years but also talks regularly with Allen, observes, "They are the Hollywood investment bank of choice." Viacom's CEO, Frank Biondi, says, "He's the functional equivalent of a rainmaker."

Allen's influence stems not only from his knowledge of the entertainment business but also from long relationships with people in the business and years of giving advice. Associates describe doing business with Allen in metaphysical terms. Jeffrey Katzenberg says, "Why do we choose one doctor over another? We don't know what they know. What we do know is how they make us feel. Herbert makes you feel comfortable and confident. He's never pushing." Warren Buffett, who, like Allen, is a director of Coca-Cola, says of him, "He has none of the bombast and fad-of-the-month that you run into in the investment-banking world. It's a quiet self-confidence—but not a self-important style—that inspires confidence."

One reason clients feel comfortable with Allen is that in a business of salesmen he always behaves like a buyer. In part, this is because Allen & Company, unlike most investment banks, puts up its own money to support deals. "We own all or part of a hundred and fifty companies," Allen says. Most of the company's income—60 percent, he says—derives not from fees but from investments: "We are a buyer. Most of our business is buying. The essence of our business is ownership. We don't think like agents. We think like principals."

Allen has more money than many of his clients, and his wealth allows him to behave like a benefactor. Invariably, he picks up the tab at a restaurant. When friends or clients want to get away, Allen insists that they use one of his houses, and provides a full staff. When Katzenberg was thinking about leaving Walt Disney Studios, and again after he left and was thinking about what to do next, he sought Allen's counsel. When Michael Ovitz, several years ago, wanted to better understand investment banking, he asked Allen to serve as his tutor. When Senator Bill Bradley, who has counted Allen a friend for more than twenty years, wants disinterested advice, he seeks out Allen. Bradley says he trusts Allen, because "he'll always keep a confidence, and he'll never talk about other people to me."

Friends are often struck by two contrasting Allen traits: generosity and cynicism. "He has a way of building everyone up," says Jack Schneider, a director at Allen & Company. "He gives you confidence. Herbert doesn't take the credit. He dishes it out to everyone else." The cynicism manifests itself in the way he seems to view business as sport. For example, when asked whether he worried about the increased concentration of power in the hands of interlocking communications giants, Allen blithely says, "It's an Arab banquet. They may all eat together once in a while. But they'll kill each other in the morning."

With clients, Allen can be blunt, as he was with Matsushita. Edgar Bronfman, Jr., says that candor is the quality in Allen that most impresses him: "He tells you what he believes, whether you hear it or not." One witness recalls that when Bronfman was acquiring his stake in Time Warner, and Time Warner's CEO, Gerald Levin, refused to invite him to join the board and refused to solicit his advice, Allen was hawkish, urging Bronfman to get tough with Levin. Other Bronfman advisers were pushing conciliation, but Allen told him to insist on a board seat and to urge the sale of some Time Warner assets to reduce the corporation's debt. Bronfman recalls Allen's advice: "You need to decide what you want out of Time Warner. You've been patient enough. Now your patience is looking like weakness." Bronfman sided with the doves, and Levin succeeded in blocking Seagram—one reason that Bronfman decided to pursue MCA.

Contributing to Allen's allure is the way he treats fees. "It's your call," he often tells clients when they ask about the fee. "He gives you the feeling he's almost doing it for fun—and he is," says a client. In 1993, when Robert Greenhill, of Smith Barney, helped bring together Sumner Redstone, of Viacom, and Martin Davis, of Paramount, Redstone sent Allen and Alan (Ace) Greenberg, of Bear, Stearns, each a check for a million dollars. Redstone said it was for past services, since both Allen and Greenberg had tried over the past several years to put the two companies together. Greenberg kept his check, but Allen thought that Redstone was attempting to keep him on the sidelines in what might become a contest for the control of

Paramount, so he sent his check back with a note saying he had done nothing to earn it. (Redstone was not amused, particularly when, just days later, it became known that Allen was representing QVC president Barry Diller, who also was interested in buying Paramount. A year later, however, Redstone hired Allen & Company to sell Madison Square Garden.)

Allen refused another, even bigger check when the battle for Paramount ended. Diller says that in early 1994, after QVC lost to Viacom, he wanted to pay Allen a fee "well in excess of a million dollars" for Allen & Company's effort. On at least three separate occasions, Allen declined. Diller finally sent along a check for more than a million dollars to cover expenses. He recalls that Allen sent it back with a letter saying, "I committed to do this. I will not cost the shareholders of QVC a nickel." (Of course, Allen & Company was compensated handsomely when Diller later paid it a fee of ten million dollars to sell his stake in QVC.)

Some people feel that such acts border on the cavalier. "You can't pay the rent giving checks back," says Hans W. Kertess, a director of Salomon Brothers, who is a boyhood friend of Allen's and was the best man at his first wedding. Still, Allen's apparent largesse has become both a source of loyalty and a powerful marketing tool. One cannot talk to an Allen & Company director without hearing of how Allen spurned checks from Diller and Redstone. "There is no transaction that I'm involved in that he will not be a part of," Diller declares.

The disdain for fees adds to Allen's allure. Among clients, he is often perceived as a peer, a fellow buyer, not a grubby salesman. One night at Bravo Gianni, Allen had ordered a $130 bottle of Barolo wine and was sipping it when Rupert Murdoch entered the restaurant with a business associate. Allen went over to say a brief hello, and then sent a bottle of the same wine to Murdoch's table. Two days later, I was interviewing Murdoch about Allen and asked if he could think of a quintessential Allen anecdote. "The bottle of wine he sent over to my table," Murdoch answered. "That's Herbert. You don't get the feeling that you got that wine because you're a client."

Acting like a buyer is no doubt easy for one who is born into wealth. The firm that serves as Allen's power base was started by his uncle, Charles, and his father seven decades ago. Charles Allen was a high-school dropout who became a Wall Street runner and, in 1919, with his younger brother, Herbert, who was also a high-school dropout, began investing in stocks. Their initial capital was supplied by

Charles's first wife. From about 1922 on, the brothers deliberately stood apart from much of Wall Street, and by 1940 they were merchant bankers, relying not on fees to make money but on their own investment convictions. Allen says that his father, who is now eighty-seven and ailing, and his uncle, who died last year, "made a great partnership, in that they complemented each other's weaknesses." It was, however, a partnership of only two partners, and the company was a far less collegial place than it is today.

Herbert spent a good part of his childhood in the leafy Westchester suburb of Irvington. His father's mother "was Jewish—we think," Allen says. His father became a nonpracticing Unitarian—an agnostic, that is. Herbert's mother was an Irish Catholic, and she had her daughter and her son baptized in the church, but, like his father, Herbert abhorred the structure and discipline of formal religion. "It's like digging potatoes," he says. "You can do it one day, and then you get the idea."

Herbert attended the Hackley School, in nearby Tarrytown, boarding there in his junior and senior years. He had good grades, excelled in football, basketball, and baseball, and served as class president and president of the student body. He had a lot of friends, but rarely talked about his family or his feelings. "Herbert was always tremendously private," says Hans Kertess, who lived two doors away in Irvington. Phil Havens, who taught Allen history and English and coached him on football and baseball teams, remembers Allen as having been mature beyond his years. Havens had never coached baseball, and he came to rely on Allen, who played shortstop, as an assistant coach. "He was calm and cool and self-possessed," Havens recalls. "Often, fifteen-year-olds are scattered all over the lot, and struggling with who they are. He didn't struggle. He knew who he was."

In 1958, he entered Williams College but chose not to participate in team sports and, often, not to study. Herbert joked to friends that his four-year goal was to be first alphabetically and last academically. He dashed off term papers on the day they were due. He read the first ten pages and the last ten pages of a book before an English exam. He was restlessly searching for new experiences and applied himself to learning golf and squash, exploring the Berkshire mountains, and walking alone along the Mohawk Trail. He was something of a dandy, recalls a fraternity brother at Chi Psi, who saw in Allen traces of a spoiled rich kid. Once, Allen talked a professor into allowing him to take an exam early because his father owned a horse that was racing in the Epsom Derby and he wanted to fly to England. The fraternity brother remembers that Allen learned he was going to fail the course and "flew back on the next plane and persuaded his professor to give him a passing grade. Then he flew back." Allen's memory differs only in the details. "I flew back after the races and helped convince him to give me a D-minus," he says.

Although he was generally popular, Allen did not drink, smoke, tell dirty jokes, take drugs, or date much. He was self-contained. "I did not know that his family had a penny," recalls John A. Kroh, Jr., who was his roommate when they pledged Chi Psi together. "He never talked about business. He was very modest." As had been true in high school, he was thought of as an older man. "He was very much like he is now," recalls former baseball commissioner Fay Vincent, who is now an investment adviser and was then, as a junior, an adviser to the freshman Herbert Allen. "He seemed mature, and totally apart from the concerns most of us had."

Allen did have a social conscience, which surprised Gordon Davis, now a prominent New York attorney, who was a year behind Allen and one of only three Afro-Americans in the class of 1963. Davis had organized a committee at Williams to raise money to support the civil-rights work of the Student Non-Violent Coordinating Committee and the Northern Student Movement. At Chi Psi, Davis approached his brothers. "I got a mixed reception," he recalls. Nervously, he entered Allen's room. Contrary to John Kroh's recollection, it was no secret to Davis that Allen was a scion of Wall Street. Allen looked at him and said, "How much?"

"He wrote a check for two hundred and fifty dollars. That was a huge contribution," Davis says. "It was typical of Herb Allen. He was not giving me the bum's rush. He was totally committed to the idea."

Allen graduated from Williams in 1962, and in June of that year he married Laura Parrish, the daughter of a prominent Oklahoma City doctor, who had just completed her sophomore year at Smith. (Within nine years, they had four children—two girls and two boys; in 1971 they were divorced.) It was also in 1962 that Allen joined his father's firm, which was then on Broad Street. In 1966, at twenty-six, he became president of Allen & Company.

Like his father and his uncle, he had no interest in becoming a member of the New York Stock Exchange. He met Dan Lufkin, who was helping launch Donaldson, Lufkin & Jenrette, Inc., in the early sixties, hoping to challenge the established brokerage firms. Allen threw Lufkin some of his initial underwriting deals. They became friends, lunching regularly downtown at Eberlin's.

At the time, Allen thought of himself as a liberal Democrat. In 1966, when Walter Mondale, who had been appointed to fill Hubert Humphrey's Senate seat when Humphrey became vice president, was up for election, Harold Witt, a director at Allen & Company, asked Allen for a contribution. "If he meets me at Eberlin's at six-thirty A.M. and I like him, I'll make a contribution," Allen told Witt. Mondale appeared, watched Allen consume his then regular breakfast of bacon and coffee ice cream, and walked away with what was the first of many generous contributions from Allen over the years. In 1968, Mondale got Allen to help in Humphrey's run for the presidency. Allen met Humphrey's vice presidential candidate, Senator Fred Harris, and after Humphrey lost, Allen, who is a fiscal conservative, agreed to serve as finance chairman for Harris's presidential campaign, in 1972. Harris ran as a prairie populist who was going to avenge the wrongs of the rich investment-banking folks from the East, one of whom was his finance chairman. Allen came away with a sour taste for politics. (Though he remains a Democrat—"because the Republicans advanced no social agenda in my lifetime," he says—he nevertheless has contempt for President Clinton, who he feels is an economic illiterate. Allen concedes that Mondale and Humphrey were probably economic illiterates as well, but, as friends, they get a pass.)

One of the good things to come out of the 1972 Fred Harris campaign for Allen was a new friend: a social worker named Ed Moose, who had also signed on with Harris. Around this time, Allen began dating a succession of beautiful women, including the actress Jennifer O'Neill and the model Barbara Rucker. He invested in Moose's new bar, the Washington Bar and Grill, in San Francisco. Moose, who stays with Allen at the Carlyle when he visits New York, introduced him to his writer friend Tom Wolfe. Allen bought a large mansion on the beach on Gin Lane in Southampton, mostly as a place to go with his kids on weekends, but also as a place where he could entertain his friends. "It was like a fabulous country club," says Tom Wolfe, who, along with his wife, Sheila, was inspired to purchase a house in Southampton. "I think it was five acres. Everything you could want was there—except a golf course. And you didn't have to join. Tennis courts. A marvelous swimming pool. Right on the beach. There was a small house on the dunes for children. A sauna." The meals were sumptuous but always "served casually," Wolfe says.

Allen was accepted as a curmudgeonly host. One knew to go outside to smoke. One knew that come nine o'clock, Allen would disappear and go to bed. No one thought it odd or rude. It was just

Allen. "He's not the kind of person to whom you say, 'Oh, come on, Herbert! One more drink,' " Wolfe says.

Around this time, Allen met Ray Stark—an event that would alter his life. In 1957, after abandoning a successful career as a Hollywood agent, Stark became a Hollywood producer, bringing to the screen such hits as *Funny Girl* and *The Way We Were*, as well as his share of critical flops, including the musical *Annie*. Most of Stark's movies, good or bad, were made for Columbia Pictures, and in the early seventies he accepted stock in the studio, which was floundering, as payment for five million dollars he was owed. In 1973, Columbia's stock dropped to two dollars a share, and Stark bought more. He went to see Allen, to talk him into becoming an investor to reinvigorate Columbia (and his holdings), and he urged Allen to meet the people at Columbia. The more Stark talked about Columbia, the more interested Allen became. Against the advice of his father and his uncle, he decided to acquire the stock. Allen once explained his reasoning to Merrill Brown, of the magazine *Institutional Investor:* "It was show business; it was more interesting than heavy metals."

Allen got more deeply involved with Columbia than he had expected to. By 1981 he owned 6.7 percent of the stock, and became chairman of its board. Along the way, he had invited more than a few of his friends to join the board, including former vice president Walter Mondale; Dwayne Andreas, the chairman and CEO of Archer Daniels Midland; Robert Strauss, the former Democratic National Committee chairman; and Dan Lufkin, of Wall Street. To oversee Columbia, he recruited a friend, Alan Hirschfield. To run the studio, he selected David Begelman, Barbra Streisand's agent, whom Ray Stark had recommended.

The studio made a comeback, and Stark and Allen became close. Stark introduced Allen to the powers of Hollywood, and he introduced him to the pleasures of Sun Valley. Allen and Stark were incongruous friends: one brassy, the other private; one a shameless self-promoter, whose official six-page biography boasts of his "unique personality," the other so modest that in 1979, when he was dating the actress Ann Reinking, he did not tell her that he had tackled a robber on Madison Avenue, wrestled the gun away, and held the man down until the police arrived. Stark would have issued a press release. Reinking, who later became Allen's second wife, read about it in the *New York Post*. "If Herbert were a girl, I'd marry him," Stark says. "He's the perfect kind of companion."

Life at Columbia was not all glamour. In September of 1977, Allen learned that David Begelman had forged signatures on three studio checks worth a total of forty thousand dollars. Hirschfield wanted to fire Begelman immediately. Stark wanted to give Begelman, a friend, a chance to explain. Allen sided with Stark.

At first, Columbia executives quietly turned the matter over to the Securities and Exchange Commission and the Los Angeles District Attorney's Office, but when the publicity persisted, the board stripped Begelman of his titles and his options and removed him from the board. Allen and Stark were annoyed; they continued to consult Begelman, and in December Hirschfield tried to placate them by reinstating him as the head of the studio, only to ease him out again in February of 1978. In David McClintick's best-selling book, *Indecent Exposure*, Allen was portrayed as a callous cynic. Friends felt that he was just being loyal. To this day, Allen expresses no regret. He refuses to think of Begelman's act as thievery, saying that he had had a mental breakdown. He attributes the furor to a "post-Watergate morality," and says of the matter, "Given the morality of the day, and the uproar in the press, it was unrealistic to think he could be reinstated." There was one other casualty of this sorry affair: Alan Hirschfield. Before 1978 ended, Allen had replaced him with Fay Vincent, who was then a lawyer with the SEC.

In 1981, Donald Keough, the president of Coca-Cola, visited Allen. Domestic sales of Coke had peaked, Keough told him; the company was looking for ways to shore up domestic earnings and would be interested in acquiring Columbia if it was available. Keough recalls being stunned when Allen, instead of trying to sell him on the value of Columbia, warned him that the entertainment business contained unwelcome surprises, insufferable egos, and uninvited publicity, and that by entering it, Coca-Cola's wholesome name might be tarnished. "He raised a lot of objections that he thought we might like to consider," says Keough, who soon became a close friend of Allen's. "He was like Aquinas. It was sort of symbolic of Herbert." Despite Allen's warnings, Coca-Cola wanted to buy, and in March of 1982 he sold Columbia to Coca-Cola for seventy-two dollars a share, or nearly thirty dollars a share more than its stock price. Since Allen had paid roughly three dollars a share for the stock a decade before, for him the deal was a dream. Astutely, he sold only forty million dollars' worth of his Columbia holdings, converting the rest into Coca-Cola stock.

Keough and Coca-Cola's CEO, Roberto Goizueta, were so taken with Allen that they asked him to join their board. At first, he declined, but then, after Fay Vincent pleaded with him, he relented, agreeing to become a Coca-Cola director for one year to smooth the transition and to try to protect his people. Allen ended up controlling close to five million shares of Coca-Cola stock. Today, Allen calculates, each three dollars he paid for Columbia stock is worth about sixteen hundred dollars in Coca-Cola stock and dividends. But if he was prescient about the long-term value of Coke, he was dead wrong about the long-term prospects of Columbia under Coca-Cola. When the sale was consummated, in 1982, Allen was talking enthusiastically about synergies—much as Sony and Matsushita would do nearly a decade later, when they acquired studios. To Laura Landro of *The Wall Street Journal*, Allen predicted, "Columbia will be the General Motors of the movie business."

Allen describes Columbia as the most thrilling experience of his business life, and says of Coca-Cola, "I think Roberto Goizueta is the best chief executive officer of the best company selling the best product in the world." Allen, who drinks Sprite, now serves on the Coca-Cola finance committee and its executive committee. He is also chairman of the compensation committee, which makes recommendations about salaries and bonuses for Coca-Cola's executives.

It was during his tenure at Columbia that Allen met Broadway actress Ann Reinking, in 1979. They dated for three years and then married. "I didn't smoke, and I ate early because I was going off to the theater," Reinking says with a smile. A few family members, including his two sons and two daughters, attended the wedding, in Tarrytown; otherwise, it was kept as secret as one of Allen's deals. The marriage lasted six years. Reinking wanted a family, and Allen felt that he already had one. And, while she loved going to Sun Valley or Cody or Southampton for the weekend and enjoyed the comfort of living at the Carlyle Hotel, she felt there was a restless quality to life with Allen. "You'd be crazy if you didn't love the life," she says. "But I wanted to stay home, too."

Their friendship survived the divorce, as did Allen's friendship with his first wife, a decorator who, at his suggestion, was invited to fix up Fay Vincent's office at Columbia. What also survived was an idea that Allen launched soon after he and Reinking married: the annual summer "family" retreat at Sun Valley for entertainment-industry leaders. With this, and Columbia, Allen had achieved cachet. "Suddenly Herbert is a chic guy people respect," observes the actress Candice Bergen, a longtime friend. "People who used to sneer at him for having dinner at six-fifteen now line up. I feel like Louis Jourdan in *Gigi*, where suddenly I look at Herbert and I think, My God! He's grown up and become a great man. The Gary Cooper of high finance. Yet he has never changed."

. . .

Allen's firm has changed. It used to earn most of its income from investments, but now the fee side is growing faster, Allen says. Most of the traditional fee business comes from the communications sector. The firm remains small, however—just the hundred and sixty employees—"and we're not going to grow," says Paul Gould, who is the head of the arbitrage department. The firm employs no analysts and has no research department. The fifteen managing directors make less in salary than their secretaries, Allen says. Stan Shuman, who has been at Allen & Company for thirty-four years, says that last year his salary was twenty-six thousand dollars. This year, it was raised to thirty-five thousand, he says, so he could "qualify for maximum pension benefits." The directors are paid approximately 30 percent of the investment-banking fees they bring in. They are expected to invest in companies that become clients, so that the interests of banker and client are aligned. Profits become part of the firm's capital, and as shareholders the fifteen directors get a part of this reward. If one partner advises another on a deal, he may share in the ensuing fee. Each director has both his own franchise and the muscle of Allen & Company behind him.

One danger in such a fiercely entrepreneurial company is a lack of teamwork. Allen & Company has its internal feuds, its whispered asides about what a "name-dropper" this director is or how "lazy" that one is. To improve communications and lessen such friction, Allen has lunch with the directors each Wednesday. "It doesn't mean anything—it just keeps people from killing each other," Allen says, in his flip way. "The firm works because of the ecumenism of Herbert," Shuman says. Allen is the arbiter of differences, the benefactor. Partners praise him for delegating and for not second-guessing them.

They are also aware that Allen has weaknesses as an investment banker. A Wall Street figure who admires him says, "If you said to me, 'Who are the great corporate investment bankers?' I wouldn't say Herb Allen. I'd say Robert Greenhill or Bruce Wasserstein, but not Herbert." Stan Shuman says of his boss, "He knows relatively little finance." But he adds, "His strength comes from the fact that he has complete and maximum independence, and his judgment is unfettered by the fact that he has to get a deal done to support the organization."

Allen & Company does have some respected frontline bankers, but the consensus among both clients and competitors is that the firm lacks the depth and full range of services provided by larger banks. As the fee side of its business rises, its weaknesses in selling itself, in making presentations, may become more pronounced. "Allen & Company is really not a firm," a rival observes. "It's a group of entrepreneurs who work on their own deals." In the fee-business world of investment banking, Allen & Company is said to be better at relationships than at providing clients with armies of analytical bankers. One client asks, "Can Allen really put together a complicated situation, or is he only an adviser for deals that would get done anyway?" Allen's talents, this client says, are "much more qualitative than quantitative."

Both Barry Diller and Martin Lipton, of the law firm of Wachtell, Lipton, Rosen & Katz, who worked with Allen on behalf of Diller and QVC, make this point admiringly. "His talent is absolutely superb judgment," Lipton says. "He has an instinct for doing the right thing at the right moment. In the QVC bid for Paramount, his timing advice was impeccable: when we should move forward with a 'bear-hug' letter"—the original September 1993 QVC offer—"then when we should switch to a hostile tender offer. He had what I would call a perfect feel for the general market reaction of the holders of QVC." For Diller to succeed, it was important that the share price of his company not drop, and Lipton believes that Allen's feel for how the market would respond to their moves was perfect.

The "deal from Hell," as the long Paramount battle was dubbed, required diplomacy, which is not Allen's strong suit. Allen's client, Barry Diller, had to juggle the sundry interests of his partners in this takeover fight—Comcast, Cox Communications, Bell South, and Advance Communications, which owns *The New Yorker* and Random House, among other things. Diller respects Allen's financial prowess, but in this battle, he says, what impressed him most was Allen's personal qualities. At one point during the final days of contention, when a debate raged as to whether Diller and his partners should boost their bid price, there were perhaps thirty-five participants gathered in a conference room. Allen and most of the partners wanted to sweeten the offer, but Diller thought the price too steep and was resisting. Feeling besieged, Diller stepped out of the room, trailed by Allen. Diller expected his banker to explain why the price was feasible, and was surprised by what Allen said to him: " 'Don't listen to anybody. Listen to yourself. Go walk around the block and come back and tell us your decision.' " What Allen understood, Diller says, was that economic decisions were rooted not just in numbers but also in instinct. "I needed that," Diller says. Because Diller was unwilling to exceed an arbitrary ceiling he had set as the price he would pay, however, Sumner Redstone and Viacom ended up getting Paramount.

Friends say that to better understand Allen one must visit him at his house in Williamstown. "It's the most beautiful log cabin you've ever seen," says Stark, who is a frequent visitor. The house sits on the top of a hill, providing a commanding view of Allen's land—three hundred and fifty acres, with the Berkshires in the distance. A lawn and grazing land slope down from the main house past a small lake to a bright-red barn, where Allen keeps six horses. About a quarter of a mile away is a two-bedroom guest house, where he usually places visitors. In Sun Valley, Cody, and Southampton, Allen bought existing houses, but in Williamstown he bought the property, in 1982, and two years ago he built his own house.

He drew sketches of the home he wanted—it contains thirty-seven hundred square feet—and gave them to an architect. He specified that it should be constructed out of huge red-pine logs, which were found in Michigan. The house has a massive combined living-and-dining room with floor-to-ceiling stone fireplaces at either end. The walls are adorned with colorful photographs that Allen has taken of the changing seasons, and with the heads and the skins of animals that his sons shot in Africa—leopards, lions, sable antelope, cape buffalo—and a kudu he shot himself. There are only two large bedrooms—one upstairs, where he sleeps, and one downstairs—each with its own fifty-inch television. What if he remarried and his wife found this house too masculine? "If so, she better marry someone else," he says. "I don't believe people in their fifties should make major compromises as far as romance. That you should do in your twenties."

This harsh proclamation seems to clash with the sentimentality evident in the main room. Family photographs are everywhere. There are many pictures of Ray Stark—on horseback, in the woods, sitting on a stone wall with Allen's father and stepmother. "Ray is a type of father figure," Dan Lufkin explains. "He has the same sense of humor, the same devil-may-care attitude as Allen." Some people are puzzled by the relationship between the two men. Candice Bergen says, "Ray is what Herbert never had. He's openly affectionate. He's bawdy. He has given Herbert a Jewish center he never had." Allen speaks of Stark as a "total guide to me in the entertainment business," and adds, "Everything we've done in the communications business comes off Ray Stark."

Allen may have a jaundiced view of the motives of others, but he is rarely cynical about friends, as a visit to Williamstown suggests. Allen's home is just two miles from the college he attended, where he helped build an athletic center. Even after the primary fund-raiser for the center, John A. Kroh, Jr., was convicted of fraud, Allen insisted on putting his former classmate's name on a plaque by the front door. It reads: "The lead gift for this athletic complex was given by Herbert A. Allen in honor of John A. Kroh, whose good work helped make this facility possible."

Kroh, a real-estate developer, was convicted of defrauding banks to collect loans that would keep his failing firm afloat. The trial, in Kansas City, lasted nearly two weeks. Allen flew there to sit in the courtroom and support his friend. In 1989, when Kroh was sentenced, Allen was also in the courtroom. Allen & Company had invested in Kroh's real-estate ventures, losing an estimated $10 million. Although some of Allen's colleagues think Kroh was guilty— "Jack Kroh gypped us," Paul Gould says—Allen blames the banks, the court, anyone but his friend, who, he says, "never took a penny for himself."

Kroh served three years in prison, and soon after being released he visited Williams and the complex that he had helped build. He was surprised when he entered the lobby and saw the plaque. "I'm a pariah, but he put his allegiance in neon," Kroh says of Allen. "Not only did he do it, he didn't tell me about it. I cried."

It is no accident that the Williamstown property is separated from next-door neighbors by hundreds of acres. Allen likes being alone, and admits to having little patience for self-analysis, or the latest theories about discovering one's true inner self. "He has a lot of self-confidence," says his friend Senator Bob Kerrey of Nebraska. "He's not full of doubt and angst. He's mysterious in that he doesn't tell you about feelings and problems. I find that attractive."

Much as Allen loves his place in Williamstown, he has never spent more than three consecutive days there. He keeps moving. "I'm a hedonist," Allen says. "A person who does what he wants is a hedonist." Some friends wonder about the truth of such remarks. "There's a certain gypsylike quality in him that makes me wonder, Is he having a good time?" Hans Kertess says. "He does a lot of things. They're fascinating and exciting. I hope he's getting a lot of pleasure out of them. I'm not entirely sure he does. There's too much motion."

On Sunday, April 9, the sale of MCA to Seagram was announced from the law offices of Shearman & Sterling, in downtown Los Angeles. Seagram and Matsushita photographers darted about the conference room, on the twenty-first floor, snapping the principals as they signed documents and lifted champagne glasses. In the middle, at one side of a twenty-foot conference table, sat Matsushita's president, Yoichi Morishita, flanked by aides. Across from him sat Edgar Bronfman, Jr., and his father, Edgar Sr., who is Seagram's chairman. Although Ovitz served as Matsushita's counselor, he strode to the seat to the right of Bronfman Sr., facing his clients. Allen had flown in from his Cody ranch and was wearing cowboy boots, and had flung a blazer over a shirt and tie that he had hastily purchased for the occasion. Instead of taking a seat next to the younger Bronfman, as he was urged to, or smiling for a photographer, Allen stood off in a corner, acting like a buyer. "It seemed like the most comfortable place," he said later. "Get out of the way, and move on."

POSTSCRIPT:

Allen continues to be a formidable matchmaker. At his Sun Valley retreat in 1995, Michael Eisner of Disney and Tom Murphy of Capital Cities/ABC launched discussions that would lead to their merger several weeks later. Allen was the investment banker for Murphy in this deal. At his 1996 retreat, Murdoch and Ronald Perelman, who had been sparring, made up, and within days Murdoch agreed to purchase New World Communications, with Allen & Company doing the deal.

Allen's friend Michael Ovitz, despite boasts that he loved the agenting business and wouldn't want to leave the company he built, would overplay his negotiating hand in an effort to become Bronfman's CEO at MCA, before Michael Eisner lured him, in August 1995, to become his number two at Disney.

8

SYNERGY
The Mantra That's Bad
for Journalism

(*The New Yorker*, November 27, 1995)

Not since the eighties, when the buzzword was "strategic," has an empty word racked up as much mileage as "synergy." The word was invoked five times in four consecutive sentences by Michael Eisner when he announced the merger of Disney and ABC in the summer of 1995. "Synergy" was the mantra chanted by Westinghouse to explain its purchase of CBS, by Time Warner to defend its purchase of Turner Broadcasting, by Viacom to explain its expensive acquisition of Paramount. Increasingly, "synergy" is the word invoked by newspaper or magazine publishers to get the news department to talk to the sales department, to become more advertiser-friendly.

"Synergy" implies that one plus one will add up to four. This is the new math of the business world. Sometimes it adds up, as it has for the Walt Disney Company. A dozen years ago there were no Disney stores. Today these stores generate over two billion dollars by selling T-shirts and other products based on Disney movie characters. By contrast, Disney movies generate only one billion in sales. But Disney's experience is unusual.

There is scant evidence that synergy works as promised. Maybe the spate of business alliances and mergers in the communications business will one day strengthen companies. Maybe they will be able to reduce costs, improve their marketing, create new products, and boost both their profits and their stock price. Whether these wonders will come to pass remains to be seen. What is already apparent is that "synergy" is rarely journalism's friend. The business assumptions behind the word—cost savings, a "team culture," the "leverage" of size, the desire to boost profit margins—can be a menace to the business of reporting.

Witness the recent behavior of CBS as it dealt with a *60 Minutes* story about the tobacco industry. To avoid potential lawsuits, CBS's management ordered *60 Minutes* not to broadcast a Mike Wallace interview with a former tobacco executive who had signed an agreement pledging to keep his company's secrets. In the past, such fastidiousness never stopped CBS News from interviewing former intelligence agents supposedly bound by similar pledges.

To be fair, *The Wall Street Journal* produced evidence that the issues facing CBS's lawyers at the time were more complex than we at first assumed: The whistle-blower had been a paid consultant to CBS. And maybe, like many whistle-blowers, he was a bit paranoid. But we also know that the company that owns CBS owns the Lorillard tobacco company, and that lawyers generally feel safer saying no, particularly when they anticipate that this is what the boss wants. In truth, a determined executive can often find a way to support an editor or producer who wants to say yes to going ahead, as *The New York Times* did when it published the Pentagon Papers. The *Times* was advised that it would be taking on the president of the United States if it published the story, and that it might violate national security. But backed by in-house counsel James Goodale and editor A. M. Rosenthal, the publisher, Arthur O. Sulzberger, decided that the First Amendment meant what it said.

CBS, on the other hand, chose synergy and lowered its "risks" and "enhanced" its financial position by at first refusing to air an interview with whistle-blower Jeffrey Wigand. (Only after they were embarrassed by a wave of criticism did CBS air a *60 Minutes* piece.) According to Mike Wallace, the CBS lawyers were unnerved by Philip Morris's threatened ten-billion-dollar lawsuit against ABC. The network had spent seventeen months defending a hard-hitting account of tobacco company behavior on its *Day One* magazine program. Then, soon after the merger with Disney was announced, in August 1995, Thomas Murphy, chairman of Capital Cities/ABC, forced his news division to issue a humiliating public apology. *Day One* had made a factual error requiring correction—it wrongly accused tobacco companies of "spiking" cigarettes with additional nicotine. But its overall portrayal of a cynical tobacco industry that says it does not want to sell to minors and then aggressively markets to minors was powerfully true. The critical question is whether ABC's decision to settle was a matter of journalistic ethics ("We were wrong") or of corporate convenience ("We can't impede the merger"). It's hard to avoid the suspicion that the logic that prevailed was the logic of negative synergy.

Similarly, when Ted Turner sought to buy NBC or CBS he argued that by "consolidating" his CNN with a network news division he could realize as much as two hundred million dollars in cost savings. This "synergy" would be good for the bottom line, but not for news coverage—the consolidation would mean that instead of two newsrooms, there would be only one.

Of course, companies have always been preoccupied with cost synergies. Capital Cities had the cash to acquire ABC in 1985 largely because it had succeeded in generating 50 percent profit margins at its TV stations. This goal was met by skimping on costs, particularly the cost of local news, which is usually a station's profit leader. So while Capital Cities was earning more than fifty cents on each dollar of revenue, it usually failed to produce worthy local newscasts. So, too, distant group owners, who now dominate local TV and radio stations, improve their margins by gutting local news expenditures.

Synergistic thinking, now common among commercial-broadcast executives, is no stranger to the print world. Mark Willes, the new CEO of the Times Mirror Company, demands that the company's newspapers, which include the *Los Angeles Times* and *Newsday*, increase their profit margins from about 8 to 12 percent and then to 16 percent within two years. Surely there is fat at the newspapers. But even folks on the business side of the company privately acknowledge they are cutting muscle to achieve those profit margins. In doing so, they are measuring success by a single criterion: profits. (A year later, Willes announced his intention to sell Harry N. Abrams, its quality art-publishing house. Why? It wasn't because Abrams was unprofitable; rather it was because the publisher failed to achieve a 12 percent profit margin. Quality didn't matter. Nor prestige. Nor pride to be taken in good work.)

The reign of the accountants marches forward. Knight-Ridder has ordered its two Philadelphia papers to double their profit margins and looks enviously at such chains as Gannett, Scripps-Howard, and the Newhouse papers, which enjoy robust profits—partly because their newspapers too often skimp on such staples of news coverage as longer pieces and investigative reporting. Gannett's stock has risen largely because of double-digit profit margins. But Gannett sometimes succeeds as a business when it fails journalistically. Some editors at this newspaper chain now order reporters to skip press conferences and in-person interviews and instead retrieve information at their computers, thus improving "productivity."

The trends are discouraging. As communications companies get bigger, the role of the journalists within them is diminished and diluted. At Disney or Time Warner, for example, the news division rarely matches the profit margins of other divisions, such as entertainment or computer software or cable—and thus news often loses internal clout. To the top executives of these giants, the news operation looks more and more like a box somewhere out on the edge of the organization chart—a box whose occupants add less than their share to the bottom line and, if left to their own devices, have an annoying tendency to emit rude noises.

Although there may have been business reasons for the decision, there is reason to suspect that Kurt Anderson was fired as editor of *New York* magazine in September 1996 because he published rude pieces about Bob Dole and Felix Rohatyn—friends of owner Henry Kravis. Earlier in 1996, *Premiere* magazine killed a probe of Planet Hollywood after *Premiere*'s part owner, Ronald Perelman, complained.

As companies converge, occasions for journalistic conflicts of interest proliferate:

• Will the caution that drove CBS to lean on *60 Minutes* prompt ABC to soft-pedal a controversy about the site of a new Disney theme park?

• Will CBS News push stories on human-rights abuses in China when Westinghouse is bidding for an electronics contract there?

• Will NBC News go easy on its on-line partner, Microsoft?

• Will *Esquire* expose ABC, which is a partner of its corporate parent, Hearst, in the ownership of ESPN?

• Will *Time* disgrace itself as it did in 1989 by refusing to report on the merger of Time Inc. with Warner Communications? Or will it withstand corporate pressures and honor itself as it did by its full coverage of Time Warner's merger with Turner Broadcasting?

• Will corporate giants curry favor with foreign governments, as Rupert Murdoch did when he placated the Chinese government by dumping the BBC from his Star satellite system, or as NBC did when it apologized to China for Bob Costas's perfectly reasonable mention of human-rights abuses during the opening ceremonies of the 1996 Summer Olympics? ("We wanted to make it clear that we didn't intend to hurt their feelings," said an NBC spokesman. Nor, presumably, did NBC want to hurt the business interests of its corporate parent, General Electric, which has huge investments in China.)

• Will *Forbes* magazine drop another profile without disclosing why, as it did when it scrapped its story on Michael Ovitz without revealing that he had been retained as a consultant?

• Will America Online discard another editorial feature, as it did the Web site that debunked its sometime business partner Bill Gates?

An editor friend at *Time* argues that his magazine is relatively insulated from such pressures because its parent is too big to notice an individual magazine. As part of a large corporate entity, a magazine may be able to enjoy editorial independence, he says, and not be subject to the whims of a wealthy individual who wants his friends—and adversaries—to think he rules the magazine.

Surely, whimsical Henry Luce slanted *Time* magazine's coverage of the war in Vietnam. But today the greater journalistic peril in a large conglomerate comes from self-censorship, by timid and ambitious bureaucrats who want to be recognized. God, if we run this story will it harm our careers? Will we be seen as not team players? Won't our corporate bosses really be pleased if we run this cover on Bob Woodward's latest book, or Warner Brothers' newest film?

Self-censorship is usually a fairly subtle danger. The danger becomes more obvious when we ponder government's power over the communications giants. These companies want things from government, which is why they make massive campaign contributions. Newspapers don't want the telephone companies to be allowed into the classified-ad business. The networks want to have cross-ownership restrictions lifted. Cable wants to keep telephone companies out of their business. Local telephone companies want to keep long-distance deliverers out of the local market. Some companies want tariffs or content legislation; others don't.

When Speaker Gingrich dressed down Time Warner chairman Gerald Levin in front of other CEOs attending a January 1995 Capitol Hill summit convened to encourage relaxed government restrictions, the effect was potentially chilling. I don't assert that Levin would tell his editors what to print. The problem is anticipatory censorship. Will editors pull their punches so as not to embarrass the boss? We may have seen this kind of negative synergy this past year when the FCC and the Congress debated whether to make a gift of additional spectrum space to broadcasters or to compel them to enter a public auction and pay for it. Broadcasters vehemently opposed such an auction and advertised against it; the government claimed it could realize between forty and seventy billion dollars from an auction. This was an important story. Yet on their newscasts the networks and CNN and local stations almost uniformly ignored it. Too complicated, they said. Didn't lend itself to pictures. What they didn't say was that their corporate parents had mounted a massive lobbying campaign against the auctions.

Synergy produces other conflicts. *Business Week*, for example, scored a scoop in its November 6, 1995, issue by exposing financial columnist Dan Dorfman's alleged conflicts of interest because of "intimate" ties to a stock promoter. *Business Week* is a first-class publication. Yet it seemed to miss the irony that just a few pages after the Dorfman scoop there appeared a big advertisement for the magazine's annual golf outing and "Presidents Forum," at which CEOs are invited to meet editors. This, too, is "synergy," but doesn't it invite editors to be kinder to potentially newsworthy advertisers?

Inevitably, synergy leads to conflict between the corporate culture of the parent and the culture of journalism. The new megacorporations value "teamwork" and use "leverage" to boost sales. They dream of a "borderless" company that eliminates defensive interior barriers between divisions. Journalists, on the other hand, prize independence, not "teamwork," and should keep their distance from advertisers and sources, rather than seek "synergies" with them. Journalists need borders—to provide independence—so that they can do their jobs. Journalists want the advertising department to stay the hell out of the newsroom. The "leverage" they want is the kind that pries loose the story, not the kind that boosts the parent company's other "products." And the more corporate management retreats editors are sent to, the more they risk becoming good corporate citizens.

Even the new buzzword "civic journalism" can be an excuse to allow the marketing department to drive stories. If practicing civic journalism means getting closer to readers, if it produces news that citizens can use, if it invites reporters and editors to be less cynical and less reflexively negative, it can be a good thing. But if civic journalism becomes a tool of the business department, promoting stories aimed at appealing to key demographic groups, stories that are advertiser-friendly, stories that align the publication with civic boosterism, then civic journalism becomes chamber-of-commerceism.

Notions like civic journalism take root because publishers are worried about shrinking audiences. The competitive, fast-changing communications business is insecure. Network television became more preoccupied with ratings as it lost market share, and magazines and newspapers and, one day, on-line publications will probably do the same. To stand out from the many brands on the shelf, the pressure will be to do something distinctive, something hot and not necessarily principled. Why—the question will recur with depressing frequency—do journalists have to be such scolds? Why can't they be more advertiser-friendly?

This is not to suggest editors should be antagonistic toward publishers or that publishing shouldn't be run as a business. If we fail to make money—as *New York Newsday* did—investors will be chased, the stock price of the parent company will fall, and widows may lose their pensions. The Times Mirror Company can be faulted for the abrupt way it killed this fine newspaper, but if *New York Newsday* really lost more than one hundred million dollars and had no prospect of earning a profit, as its corporate parent claims, then Times Mirror eventually had no choice but to end it.

This brave new world offers some good news for journalists and citizens alike. Despite complaints that news choices are drying up, that conglomerates rule the world, citizens now enjoy more choices of news than ever. More cable channels have allowed competing networks to enter the twenty-four-hour cable-news business, and provided more reporters with jobs. There are more magazines to serve the special tastes of readers. About two hundred U.S. newspapers are now on-line. A citizen today has more, not fewer, sources of news: magazines, newspapers, on-line, CNN, MSNBC, Fox News, C-SPAN, all-news radio, and at least a couple of hours of local TV news.

In this respect, George Orwell was not prescient: technology will not imprison us, will not become a tool for totalitarian governments. Instead, as happened in Eastern Europe, where fax machines and cellular phones and satellites penetrated the Berlin Wall, technology is a friend of democracy. China can't police the Internet because individuals control it, not government.

Nevertheless, will technology and synergy improve journalism? Here I tend to be pessimistic. If the networks, for instance, decided to use their cable outlets as an excuse to pull serious and less profitable documentaries and newscasts from the network and shove these programs onto cable outlets, which have fewer viewers, this is negative synergy. Similarly, if the availability of more sources of news translates into more infotainment—as has happened with the prime-time magazine shows of the TV networks—the trade-off is not good either. And if the focus is more on the package than the content, this, too, is not good for journalism.

As an example, a recent visit to the newsroom of the Microsoft network, with its one hundred or so editors, revealed that it employs no journalists to go out and cover stories. Rather, editors sit in front of computer screens and collect information from others and then dress these stories up in fresh packages. They have no way to certify that the stories are true. And their criterion for what to include has more to do with packaging than journalism.

In a free society, unsubsidized journalism must generate a profit. But profits can never be the entire motive. There's another kind of synergy corporate chieftains might ponder before they spread the synergy gospel too far, namely: At what point do they harm their investment in the name of helping it? Considering the terrible publicity CBS reaped, it's hard to believe that it would today cuff *60 Minutes* the way it did.

It is easy to enumerate the sound business reasons to guard the credibility of journalism. A TV reporter who loses credibility will also lose both viewers and potential interviews. If the credibility of *60 Minutes* is diminished, its ratings will dip. If it becomes known that a serious publication pays its sources, it will lose readers. If *TV Guide* is perceived as a flak for its owner, Rupert Murdoch, and the Fox network, it will lose advertising dollars.

The public has a stake in the outcome too. If readers don't trust journalists, if they cynically believe we're beholden to our sources, or make things up, our politics will become even more shrill and uncivil. There will be no one to referee the facts. Journalism receives special protections under the First Amendment because in a democracy voters get much of their information from the press.

Despite the cynicism endemic to the trade, journalism—good journalism, anyhow—has a responsibility to tell the truth, "without fear or favor." Such talk may annoy some CEOs and seem pretentious even to some journalists. But if the task is to get the story, then permitting journalists to report "without fear or favor" is the only kind of synergy that works.

POSTSCRIPT:

I have added some examples of negative synergy since this piece was first published. After the piece appeared in *The New Yorker,* I was criticized by a journalist who works for the Newhouse chain, who argued that his employers had recently invested in improving journalism with more investigative stories. This is true. It's certainly true at the Newark *Star-Ledger,* where a former editor of mine at the New York *Daily News,* James Wilse, has stiffened the news and editorial spines. It may also be true at some papers in the Gannett and other newspaper chains. These are reminders to beware of generalizations.

But exceptions do not disprove a general rule: Synergy is usually bad for journalism. One notes, for example, that the stock price of Times Mirror has soared since Mark Willes proved that he would ruthlessly slash costs and would not be swayed by pleas on behalf of "the public trust." Disney's stock jumped in February 1997 after it annouced its intention to sell its newspaper and magazine division. The reason, said Wall Street, was that news distracted Disney from its entertainment focus and didn't produce the same profit margins.

9

THE REFEREE
The FCC

(*The New Yorker,* February 13, 1995)

To decision-makers in the communications business, government—not titans like Bill Gates and Rupert Murdoch—is the true powerhouse. Not since the Communications Act of 1934, which authorized the federal government to license radio and television stations to use the public airwaves, has the government become entangled in so many momentous communications issues. Reed Hundt, the chairman of the Federal Communications Commission, which was established by the 1934 act, says that 1995 "may be the most important year in the history" of American telecommunications.

There are at least three good reasons for believing that this is true. First, there is the FCC auction of wireless spectrum licenses, which began on December 5 last year, and includes everything from cellular phones, through pagers, to handheld wireless computers that will fax and phone and send e-mail. The many rounds of bidding—which initially were to conclude at the end of January—will perhaps stretch into the spring. "The auction is a metaphor for everything we're doing," Hundt said on the day the auction started. The auction will generate tax revenues, but the more wireless companies bid to win licenses, the more they must charge customers to recoup their investment.

The second momentous event will be telecommunications legislation that, if it is passed, could unshackle everyone in the business, freeing broadcasters of restrictions and inviting telephone companies into the video business, say, and cable companies into the telephone business; long-distance telephone companies into local service, and local companies into long-distance service; electric companies into telephone and video services. Such deregulation, however, is not a simple matter. A long-distance telephone company, for example, could not hook up local customers without access to the wiring system of a local or regional phone company, without agreement on the cost of such a connection, and without provision for a government appeals process to adjudicate disputes.

The third—and least noticed—event this year will be the FCC's issuing of guidelines for what is variably called advanced television (ATV), digital television (DTV), and high-definition television (HDTV). With the arrival of broadcasting equipment that transmits TV signals digitally, and of technology that allows over-the-air signals to be compressed into a smaller megahertz space than they now occupy, six or seven, and possibly as many as ten, new channels can materialize from every channel that currently exists. Technology offers new choices, but government must now decide among them. Do existing stations get the new channels or do the channels go to new broadcast bidders?

In Washington, rhetoric and reality often conflict. Publicly, every industry tells Congress that it favors rigorous competition—two or more arteries going into every home. Not for a moment does FCC chairman Hundt believe it. Despite promises from the telephone companies to jump into the cable business, and from the cable companies to jump into the telephone business, he says, "all the industry plans added together—and approved by us, if they are—really mean that two years from now the maximum number of households with more than one wire will be only eight percent."

Privately, every sector of the communications industry deluges the FCC with petitions to keep competitors at bay. Michael Katz, the FCC's irreverent chief economist, observes, "TV broadcasters say, 'We want free markets, but we don't believe that the government should use markets to sell us spectrum.' TV broadcasters have their own definition of a free market—'Give it to us free and we'll market it.' Radio broadcasters say they favor free markets, but they ask the commission to block competition from entrepreneurs who want to use satellites to deliver radio services. The local telephone companies say, 'Get out of our way and let us into the video-dial-tone business.' Then you say to them, 'How about letting the electrical utilities into telecommunications?' They say, 'Don't let them in!' The cable companies say, 'Don't let either of them in; they'll use their captive-monopoly rate bases to compete unfairly.' Yet they want no pricing regulations for themselves. The long-distance telephone companies say we should let them into the local telephone business, but they don't want the local companies in the long-distance business. So all industries say, 'Get out of our way—but regulate our rivals.' "

The only way to end the standoff and see the battle joined, Hundt says, is, first, for government to eliminate the legal impediments to competition, and, second, for it to create "a third force"—a force like an FCC auction of spectrum space. Yet Hundt knows that to create a third force Washington must be able to overcome a deep ideological division.

The ideological division was on display at the Postal Square Building, in downtown Washington, in December, when several hundred government officials and representatives of wireless companies appeared for the opening of the spectrum auction. Representatives of twenty-nine FCC-approved bidders—mostly alliances of cable and telecommunications companies—were also present. In person, or by computer or telephone, they would soon begin entering and amending their bids on ninety-nine spectrum licenses in fifty-one markets across the country. They listened to Vice President Al Gore praise the auction and propose that if the auction generated more than the ten to twelve billion dollars that the government was already counting on to help reduce the federal deficit, the additional money could be used to help transform American society by subsidizing a wire to link "every single classroom in the United States of America to the information superhighway."

Reed Hundt heartily applauded Gore's proposal. Edward Markey, Democrat of Massachusetts, then the outgoing chairman of the House Commerce Committee's Subcommittee on Telecommunications and Finance—who had suggested during the ceremony that funds from the auction could be used to subsidize public broadcasting as well—also rose to applaud. But, while Gore was speaking, Senator Larry Pressler, of South Dakota, who as a result of the Republican election victories last November is now the chairman of the Senate Commerce Committee, did not join the approving chorus; he kept his eyes fixed on the floor, yawned occasionally, and applauded tepidly. In his own remarks, Pressler stressed the idea that such spectrum auctions had been proposed by Republicans, and added that, in his view, "the Washington regulatory bureaucracy" remained the enemy of true reform. But communications policy, he implied, was no longer to be handled as a family affair between old school friends like Gore and Hundt (they had been classmates at St. Albans School, in Washington) and between a Democratic administration and Democratic congressional chairmen. No longer was there to be an unspoken agreement on shared assumptions, such as the assumption that government should actively referee market forces.

Listening to Pressler's remarks and spending a week as a visitor to Hundt's office suggested to me how dramatically the political landscape has shifted since November's Republican landslide. Within hours of the auction's start, Hundt and his staff were working the telephone to placate Republicans. On December 6, at a daily morning staff meeting at FCC headquarters, the commission's chief of staff, Blair Levin, drew up a list of calls that Hundt should make to business leaders urging support for Gore's proposal. Levin told Ruth Dancey, Hundt's executive assistant, that Hundt should also call Senator Pressler and Representative Jack Fields, of Texas, the Republican who was to replace Markey as chairman of the House Subcommittee on Telecommunications and Finance. Levin had learned that Pressler was angry that he had not been told beforehand of Gore's proposal to wire classrooms.

Hundt arrived at his office that morning at nine, looking as if he had got up late, dressed hastily, and forgotten to comb his hair and cinch up the knot of his tie. He conferred with Levin, took a seat at his desk, and dialed Representative Fields. Fields was not in, so Hundt asked for Michael Regan, the House Commerce Committee's counsel, whom he knew as a fellow St. Albans alumnus. Fields had been invited to sit on the stage during Gore's speech, but other business had kept him away, so Hundt started the conversation with Regan by recapping Gore's proposal. They needed to be able to speak frankly, Hundt said. "None of us should look to have this issue be the defining difference between Democrats and Republicans," he went on, and added that Gore wanted bipartisan cooperation for his proposal.

Hundt was preoccupied with Republicans. He realized that after the election sweep in November they had an appetite for battle with the Democrats, but he nonetheless hoped that telecommunications policy would be amenable to a truce. The 1994 Congress, in a bipartisan effort, had come close to passing legislation that would have lifted certain federal regulations in order to increase competition among the industries with communications access to the home—local and long-distance telephone companies, cable, computers, wireless, direct-broadcast satellites, broadcasters, and the electric utilities. Hundt's hopes for a bipartisan approach this year suffered a setback on January 9, however, when Republicans who spoke during hearings that Pressler was conducting to shape new telecommunications legislation suggested, as Pressler had in December, that government—and, by implication, Reed Hundt—was the enemy. The Senate majority leader, Robert Dole, disparaged the Democrats as too timid. "The marketplace, not government, should pick the winners and losers," he said. Senator Conrad Burns, of Montana, asserted that Republicans were for "equality of opportunity, not equality of results." These sentiments were expressions not of policy but of frustration.

In truth, both Republicans and Democrats exhibit what true conservatives might label "socialist tendencies." For example, a twenty-page draft blueprint of principles for new telecommunications legislation, which Senator Pressler has circulated among Republican members of his committee, requires that government guarantee "universal service" for telecommunications which is "affordable" and whose rates are "comparable" in rural and urban areas, and further insists that "rural and high-cost areas should have access to advanced health care, education, and economic development." This language could have been drafted by an activist Democrat.

Still, there are serious issues dividing the parties. "The whole burden of proof has shifted," Martin D. Franks, CBS's chief Washington lobbyist, says about the new political lineup in Washington. "Larry Pressler and Jack Fields are saying, 'Tell me why regulations should stay.'" While Democrats and Republicans agree on many things, there remains the basic philosophical divide between government as spectator and government as referee.

Reed Hundt himself is a believer in government. The GI Bill of Rights paid for his father's law-school education, at the University of Michigan. Hundt recalls that his parents were "solidly liberal," even voting for Socialist Norman Thomas in the 1948 presidential election. His father was a lawyer and later an administrative-law judge. The family lived in Falls Church, Virginia, and moved to Washington when Hundt was thirteen. They struggled to pay his tuition so that he could go to St. Albans, where one of his classmates, starting in the ninth grade, was Al Gore. Together, the two teenagers joined the 1963 March on Washington and heard Martin Luther King, Jr., give his famous "I Have a Dream" speech. "There was an intense emphasis on moral issues," Hundt recalls of his high-school years. "And the moral issue then was civil rights."

Hundt graduated from St. Albans in 1965, and in the fall of that year he entered Yale. He became the executive editor of the *Yale Daily News* and was the first to publish the cartoons of Garry Trudeau. After graduating, he taught school for two years. Then he enrolled at the Yale Law School, where he became friends with fellow student Bill Clinton. Of his selection as FCC chairman, in 1993, Hundt joked, "I owe this job to lots of hard work and to fortunate seat assignments in high school and law school."

In 1975, Hundt joined the Los Angeles office of Latham & Watkins, where he had an eclectic practice, devoted mostly to representing business in antitrust cases. In late 1979, he moved to the Washington office of the firm and there helped build a communications practice, becoming a partner in 1983. Hundt did pro bono work, taking on civil-rights cases for the NAACP Legal Defense and Educational Fund, and representing poor death-row inmates. "He's a child of the sixties," explains his wife, Dr. Elizabeth Katz Hundt, who is a psychologist. She always knew, she says, that the law did not satisfy "his wish to contribute and to give back in a larger sense to society"—a wish that was "imparted to all of us by John Kennedy."

At his core, Hundt was a believer in government as a force for good. In the 1992 presidential campaign, Hundt served as an economic adviser for his two former schoolmates, Clinton and Gore. When Clinton's original choice to head the FCC declined, Gore proposed Hundt. He was nominated on June 29, 1993, confirmed by the Senate on November 19, and sworn in by Gore as the twenty-fifth FCC chairman on November 29. The appointment was both praised and panned. Eli Noam, director of Columbia University's Institute for Tele-Information, and a former member of the New York State Public Service Commission, said of Hundt, "He's more likely to be open-minded and bring a fresh perspective, and [he's] not associated with any industry dogfight." On the other hand, Republican Jack Fields called him "a political puppet."

Hundt was initially more solicitous of the opinions of the four other FCC commissioners than many prior chairmen had been, but the FCC is organized in a way that discourages collegiality. Each commissioner has a separate staff of three professionals. There is one public meeting each month to address about one hundred large, or "precedential," issues, but voting on the remaining three hundred or so items the commission considers each year is done individually, by computer. The chairman's staff sends the commissioners what are called circulars, each commissioner's staff attaches a recommendation, and within a specified time period the commissioner logs on with a password and votes. The sunshine laws that were meant to prevent horse trading inevitably restrict dialogue, since they prohibit all private meetings of more than two commissioners. Hundt can schedule a weekly meeting with Commissioner Susan Ness, as he does, but if another commissioner pops in, one of them must leave the room. "It makes for an awkward situation," says fellow Democrat Ness. The system is inclined to produce Lone Rangers.

As chairman, Hundt is one part traditional Democrat and one part new Democrat. At first, the industries that came before the FCC heard only the voice of traditional liberalism—of government as activist. Soon after Hundt was sworn in, he angered broadcasters and cable programmers by criticizing TV violence, warning that if television did not police itself, government might do the policing. He also pledged that the FCC would aggressively penalize broadcast stations and cable systems that did not follow the commission's equal-opportunity rules for minorities and women. Unlike his Republican predecessors at the FCC, he emphasized industry's public-trust obligations—to teach children, to provide news. In February of 1994, after a congressional clamor over rising cable rates, Hundt and the commission voted to roll back the rates by 7 percent, on top of a cut of as much as 10 percent the year before—a move that prompted cheers from consumer groups and invective from industry.

Time Warner Chairman Gerald M. Levin, who spoke just moments before Hundt did at a Schroder Wertheim & Co./*Variety* conference in April 1994, unleashed both barrels at Hundt, declaring, "The glaring inconsistency of re-regulation reminds me of the bureaucratic bungling that Charles Dickens summed up in a government department he named 'the Circumlocution Office.' The Circumlocution Office, Dickens wrote, was the master of 'saying one thing while doing another.' So is the FCC."

Today, when Hundt looks back on his tumultuous first year, he says that while his goals were to encourage competition, the cable rate rollback in particular was "not a good vehicle to communicate my goals. It was like saying to an architect, 'I want you to build my building, but first clean up the garage!' The industry reaction was, 'This fellow must be anti-business.' I think what is true is that I am not pro any particular business or anti any business. But I am pro capitalism and pro markets."

Many in the communications business thought of him as pious, a smug traditional Democrat who believed government could micromanage the economy. Hundt acknowledged to Karen Dillon of *The American Lawyer* in September 1994, "I will admit this: For the first couple of months I came off more like a judge than a policy maker." Told that Hundt had provided me with a list of nearly fifty people to call for this article, one communications industry leader who was on the list observed, "He's completely concerned with his own PR. The fact that anyone would fax fifty names is astounding! He goes where the wind blows. Now he's pro deregulation. When he came in, ten percent rate cuts for cable were not enough."

Critics did not see what friends affectionately call Hundt's "goofy" side. Once, on the Diane Rehm radio show on WAMU in Washington, when a caller complained about a long-distance bill, Hundt responded, "If you have any questions about your bill, just call . . ." and gave out chief of staff Blair Levin's private phone number. More recently, Hundt was having a brown-bag lunch with mostly junior members of the FCC's Competition Division. He went around the room and asked each lawyer or economist to describe what he or she was working on. They had moved on to a general discussion when a middle-aged economist arrived late and stood, frozen, in the doorway. "The drill has been," Hundt told him with a serious face, "that we asked everyone to stand on a chair and tell a funny and dirty joke!"

What Hundt didn't see at first—until last November, at least, when the FCC voted to relax some cable rules—was that the goals of the commission and Congress were often in conflict. Both wanted to offer consumers relief from onerous cable fees, yet at the same time they expected the cable companies to invest in creating more channels and programming, although they would have less cash with which to do so. The FCC wanted to decrease the concentration of ownership in the cable industry and assure more cable channels for newcomers, but smaller cable companies, since their cash was reduced, couldn't borrow capital and would be bought out by the giants. J. Bruce Llewellyn, who headed a group of minority investors with a 20 percent interest in Garden State Cablevision, a small cable company, and who in January sold his stake in the company to Comcast and another partner, complains, "The small guys can't take a cut in cash flow. So the small guys are selling out and the big cable systems are getting bigger."

"Everybody talks competition and liberalization, but the reality is different," says Eli Noam, who has been disappointed in Hundt's FCC. "The question is not people's commitment to principle but what they actually do. When it comes to what they do, I don't see much deregulation that the FCC has accomplished." He notes that the original 1934 Communications Act was only forty-six pages long, whereas the proposed 1994 telecommunications bill, an amendment to the 1934 Act, was some two hundred pages long. Noam believes that the length of the 1994 bill had less to do with actual need than with congressional habit—"the approach of trying to forestall any conceivable negative effect, instead of waiting for bad things to happen and then dealing with them."

Of course, competition without rules can become anarchy, like a football game played without a referee. "Without technical rules to govern the radio spectrum"—for mobile phones and beepers and broadcast stations—"you'd have interference," says Robert Pepper, the chief of the FCC's Office of Plans and Policy. Could a wireless-telephone user from Chicago use her portable phone in Washington if wireless systems were not compelled to allow interoperability? If there had been no 1992 Cable Act, with its provision for "fair access to programming," cable would not have allowed its programs to be sold to a vibrant new delivery system—direct-broadcast satellites. Moreover, without government-required subsidies remote rural and suburban homes would not have relatively inexpensive telephone service, and the Internet would not have sprouted, either.

Both the inevitability and the clumsiness of the FCC's role in overseeing communications policy were apparent during a meeting that Time Warner executives had with Reed Hundt on December 8. The meeting concerned a proposal from Liberty Cable, in Manhattan. Liberty, which is owned by Millstein Properties and backed by local telephone companies, wanted the right to provide cable service to individual apartments by using cable that competitors—principally Time Warner—had already installed. FCC rules, which were written with single-family houses in mind, stipulate that any competitor wanting to tap into a household's wiring system can do so only "at or about twelve inches" outside the home, because the junction box that connects such a house to the main feeder line is usually located a foot from the building. Liberty had appeared before Hundt a week earlier to propose that the FCC amend this rule so that it could tap into an apartment building's cable wire, which is often connected to a junction box located in a stairwell, far more than a foot away from any given unit. If the rule was not amended,

Liberty argued, it would have to rewire apartment buildings, adding to the costs and making it more difficult for Liberty to attract new customers.

Led by Richard Aurelio, the president of Time Warner's New York City Cable Group, a delegation of five Time Warner officials showed up at Hundt's office. The Time Warner executives sat on two worn beige couches facing Hundt, who had taken off his jacket and slid into an armchair. After some polite banter, the subject switched to what Time Warner saw as unfair competition from Liberty Cable. "To me, the proposal to change the demarcation"—the twelve-inch line—"is the ultimate outrage," Aurelio, a bearlike man, with gold-framed eyeglasses and a dark mustache, said in a surprisingly hushed voice. "It bestows on our competitors, in Manhattan alone, a one-hundred-million-dollar gift." He ticked off some other outrages. "We feel it's the taking of property," he said, for Liberty would be piggybacking on Time Warner's wires. Liberty already had certain advantages, he said. Unlike cable franchises, Liberty beams its signals from transmitters on the roofs of tall buildings to satellite dishes on the tops of smaller individual apartment buildings, and therefore is not burdened by the cost of tearing up city streets to lay cable, or of paying a municipal franchise fee, or of agreeing to wire every school, library, and government building and to set aside public-access time and "provide universal service," so that every community in the five boroughs of New York City will have access to cable—all costs that Time Warner volunteered to pay under its franchise agreement. Meanwhile, Aurelio said, Liberty is seeking to wire only the best buildings in the richest borough—skimming the cream, as he put it.

"What do you expect them to do?" Hundt asked. "Isn't that what you're doing on the telephone side? You picked Rochester, not Elmira!" Hundt was referring to Time Warner's experiment to use their cable wire to transmit telephone service.

But what's pertinent, countered Aurelio, who seemed taken aback by Hundt's aggressive cross-examination, is that Liberty has all the other advantages already cited.

"I'm not sure who has the better list of advantages," said Hundt. "I urge you to focus on what I told them was an issue of this argument: Are they taking something you invested in and not compensating you?"

To prove that Liberty was taking from Time Warner, one of Time Warner's two Washington lawyer-lobbyists, Arthur Harding, of Fleischman & Walsh, said that a look at one of the junction boxes that house a building's cable wiring would strengthen Time Warner's claims. An engineering executive, Larry Pestana, thereupon stood up and displayed a wooden box that was about two feet high and three feet wide and deep. He showed that the thick wires inside would have to be pulled out and connected to Liberty's own box if Liberty's request was granted.

Hundt pointed out that Liberty had said it was willing to pay a fee to be connected to a building's main Time Warner cable line, just as Time Warner would have to pay a fee to a Baby Bell for access to telephone customers.

"You should know that they have taken over some of our wires and they've never offered to pay us," Time Warner's counsel, Robert Jacobs, said of Liberty.

"That's not what they told me," said Hundt.

If Liberty got access to a building, they would insist on throwing out Time Warner for ten years, because they are asking for ten-year contracts, Jacobs said.

"If Liberty takes away the wire, we need to get back to install another one," said Aurelio. The competition Hundt seeks—between, say, cable and telephone companies—would be retarded.

When the meeting had gone on for half an hour, Hundt said, "I'd like to make a suggestion. My staff needs to know how you plan to do the telephone business. We need to treat you as a new entrant into the telephone business, just as we treat Liberty as a new entrant into your business. Second, get back to us on how Liberty should compensate you." Hundt went on, "There's more here to learn than I can learn in half an hour. I can't take it all in. But I want you to understand that we're looking for some principle of fairness that allows them"—Liberty—"to compete. My sense now is that we don't have enough information to make a judgment."

"If competition is good, the only way to achieve competition is by having at least two wires," said Aurelio.

Not necessarily, Hundt replied. Some proposals to allow phone companies to enter the cable business would permit them to use the same wire. Separate wires would boost costs, and thus prices.

Two companies sharing a wire wouldn't work, Aurelio said. It would be "a nightmare for consumers," because there could be signal leakages, and if something went wrong, each company would blame the other.

After the meeting, Hundt told me that he thought the FCC would inevitably be drawn into such issues as Liberty's request. It was, in fact, much the same issue that the commission would face when long-distance telephone companies petitioned to connect to local wires, or local companies to long-distance wires.

Aurelio was uneasy about the meeting. "It boggles my mind that they get into such details," he said in the hallway after leaving Hundt's office. "Talk about micromanaging! But he's showing an openness. He wants to learn."

I asked how the FCC could keep from micromanaging, since both Time Warner and Liberty had requested a factual ruling.

He knew that the commission had to rule, Aurelio said, but what troubled him was that Hundt was getting personally involved in details that ought be adjudicated by the technical staffs. " 'Micromanagement' is the wrong word," he said. "What I'm really saying is that, given the global ramifications of this issue, it was not prudent for him to rush into a decision on this without his staff understanding all the technical details of what wire goes where."

Hundt obviously came to agree with Aurelio, because within days he decided to delay a decision, and had his legal office convene a meeting on January 18 of representatives from Time Warner, Liberty, and the various wired and wireless trade associations.

One of the most consequential matters that will soon cross Hundt's desk—what to do about high-definition television—was the subject of another meeting in Hundt's office, on December 8. The commission's director of technology policy, Saul T. Shapiro, entered with both good news and bad news. The good news, he said, was that there had been a historic breakthrough.

Early in 1987, the broadcasting community in the United States had resigned itself to the likelihood that HDTV sets under development in Japan would be marketed here, and that the quality of the picture would be so alluring that consumers would abandon the TV sets they had, compelling the TV stations and the networks to make huge new investments in high-definition cameras and other equipment. The issue then had been "pretty pictures," Hundt says, and the problem was that the image on an analog-television screen—the sort that we had then and still have—was composed of about five hundred horizontal scanning lines, whereas high-definition pictures would use double this number. To continue using the analog system to send high-definition pictures would have required both new equipment and more spectrum space. Broadcasters feared that some stations would lose their spectrum space to HDTV.

Alarmed, the Reagan-Bush administration, though it had denounced government-directed industrial policies, adopted an industrial policy to protect the broadcast industry. It wasn't called that, though. In September of 1987, the FCC established an Advisory Committee on Advanced Television Service to devise an American alternative. In May of 1993, companies backing four alternative systems formed an alliance and decided to develop a single system, called ATV.

By last December—this was the good news that Saul Shapiro announced to Hundt—there had been so many technological breakthroughs that "calling it high definition isn't even appropriate anymore," Shapiro said. The ATV system would consist of digitalized signals and, because of advances in digital compression, the system would be able to offer at least six or seven channels in the airwave space now occupied by one channel. Or a broadcaster could choose to offer remarkably crisp high-definition television over a single channel. "Technology is no longer a bottleneck," Shapiro announced. "The key for us is: What is our policy?"

That's where the "bad" news came in. The policy and its political implications would be far-reaching. The broadcast spectrum had traditionally been so limited that when the FCC was formed, its chief mission had been to ration it. The government had used this scarcity argument ever since to justify imposing certain public-trust obligations—news, children's programming, and community programming—on the holders of the rare licenses. But now Shapiro reported that instead of facing scarcity, instead of being forced to choose winners and losers, the nation was on the verge of abundance. "We must explore the policy implications of this," an excited Hundt told his staff at the end of the meeting with Shapiro. The decisions about allotting the extra channels would probably come to a head this summer, he said.

The competition could be fierce, because the stakes are enormous. By way of contrast, the aim of the wireless auction that began on December 5 was to sell a hundred and twenty megahertz of spectrum; ATV dwarfs that, with the equivalent of two hundred megahertz of spectrum. A day after the meeting with Shapiro, Hundt said, "The public-policy issues are who gets the extra spectrum and how do they get it. And, after you deal with that, the question is what do they get to do with it. In the course of asking these questions, someone has to stand up and say, 'Here's the public interest.' Yesterday's briefing told us we could take one channel and get six or seven channels. Should one of these be reserved for the local community? Should one be reserved for children's programming? Should one of these be for an all-news channel—a local CNN? These are all possibilities." So are channels reserved for culture, or transmitting data, or the Internet, or home shopping, or interactive distant learning, or games, or gambling. Just as newspapers print different sections, so TV stations could multiplex, using various channels to have one newscast from the western part of the state and another from the east, and still another in a foreign language.

The question remains: Is this extra spectrum space to be a gift, a sale, or a trade? Hundt refined his thinking and his list of questions in a speech he delivered at the Consumer Electronics Show on January 5, 1995. The conversion from analogue to digital signals, he predicted, will be "the Morse code of the twenty-first century." But first come the questions. Would the public interest be better served, he asked aloud, "if broadcasters paid spectrum fees or had to compete for spectrum in an auction?" Or, instead of paying for the extra channels with cash, perhaps broadcasters should pay "with commitments to devote time to children's programming, national and local news, and free time for political debate to occur in the electronic forum." How, he asked, can broadcasters "continue to reach their audiences' analog televisions" while converting to digital signals? How long should the FCC allow for this conversion? Then there was the question of how to resolve the collision between two values Hundt holds dear: using the extra spectrum space to promote market competition, and using it to strengthen weakened broadcasters. "Our two biases negate each other," admits FCC staff chief Blair Levin.

A more immediate clash preoccupied Hundt—a petition by NBC to force Rupert Murdoch's Fox television network, which was aggressively cutting into NBC's market, to reduce its level of foreign ownership or to allow American networks to seek foreign investment without limits. The NBC petition claimed that Murdoch's Australian corporation holds majority control of Fox, and thus violates the law that limits foreign ownership of a broadcast company with an FCC license to 25 percent. Murdoch, in a letter to Hundt, said that NBC's motive was to thwart "robust competition." Hundt's FCC, which had been reviewing the question of Fox's foreign ownership by the time of NBC's petition, issued a confidentiality order.

Already, there are signs that the challenge to Murdoch will embroil the FCC in a political fight. Senator Pressler and other Republicans have complained about the FCC's "gag order" (which has been softened to cover only certain documents). In the late fall, Murdoch made the rounds, meeting with Speaker-designate Newt

Gingrich and with other Republican leaders who share his conservative views. Two weeks before the November elections, Murdoch's *TV Guide*, the largest-circulation weekly publication in America, printed as an advertisement the Republican "Contract with America." According to Haley Barbour, the chairman of the Republican National Committee, Gingrich wanted to run the ad, and the committee paid three hundred thousand dollars for it. A move more widely noticed was that HarperCollins, which is owned by Murdoch, agreed to pay Gingrich a $4.5 million advance to write two books. (Murdoch insists that he was unaware of the book deal, and Gingrich claims he did not know that Murdoch owned HarperCollins.) With the stakes so high, Murdoch has not restricted his visits to Republicans. In November, he paid a call on Hundt and FCC staff members to speak about the prospects of ATV. In early December, Hundt displayed something he'd gotten from "my new best friend." It was a Christmas card showing Murdoch in a yellow sweater on his boat, flanked by his wife and their son, daughter, and son-in-law.

Beyond Murdoch, there are other questions that await the attention of Reed Hundt and of Congress—particularly subsidy questions. Should government permit the telephone companies to have current customers pay to cross-subsidize the cost of starting up video services? Should government impose universal-service obligations on video services, as it does on phone service? And what would universal service mean in that case—that customers can have access to all services, or that services must be affordable to all?

Should all taxpayers subsidize tax credits to increase opportunities for minority ownership? The FCC's seventeen-year-old program that offers tax breaks to companies that sell media properties to minority firms has a desirable social goal, for less than 3 percent of all radio, broadcast, and cable companies are minority-owned. But this tax break also enhances the wealth of giant companies and a mere handful of rich minority investors who line up wealthy white partners. This is what a black investor, Frank Washington, did when a partnership he headed bought Viacom's cable system, potentially saving Viacom $400 million in taxes.

Should government support public broadcasting—as Representative Markey and former FCC chairman Newton Minow, among others, have proposed—by diverting proceeds from the spectrum auctions? Or, with channel choices proliferating, should government get out of the public-broadcasting-subsidy business altogether? That is the view of Senator Pressler, who proposed that the Corporation for Public Broadcasting should be sold, to save the $286 million federal subsidy.

Reed Hundt says he hopes that after these and other questions are resolved, there will be so much competition that he will work himself out of a job. If the Republicans have their way, that day may arrive sooner than Hundt desires. One version of Senator Pressler's blueprint of telecommunications legislation contains a proposal to "downsize the FCC." And Gingrich's advisers have urged him to introduce legislation to abolish the FCC and rely instead on existing antitrust laws to prevent monopolistic practices.

The bipartisan consensus has unquestionably been weakened since the November election. Back on December 9, however, in a staff meeting that Blair Levin conducted in Hundt's office, the flame of optimism for a political coming together still glowed. Robert Peck, the deputy director of the FCC's Office of Legislative and Intergovernmental Affairs, reported that Vice President Gore had scheduled a meeting to discuss a common approach to telecommunications issues on January 9, and that Senator Pressler and Representative Fields would participate. The goal, he said, would be to enunciate a bipartisan statement of principles.

Neither Pressler nor Fields nor any other congressional Republicans appeared at Gore's meeting. That was the day Pressler had scheduled the hearing of the Senate Commerce Committee; the Republicans had decided to formulate their own legislation separately and then pressure the Democrats to go along with it. Senate Republicans who had protested the FCC's Murdoch gag order now imposed a gag order of their own, banning staff members from talking to lobbyists or the press about telecommunications legislation. On January 19 and 20, House Republicans held an off-the-record meeting with the senior executives of communications companies.

Although the Democrats were in danger of becoming spectators, Hundt remains undaunted. He told me in mid-January that he is optimistic. "The problems here are less likely to be between Democrats and Republicans than between entrenched businesses," he said. The real obstacle to reform, he believes, comes not from the government but from business—from captains of industry who complain about bureaucrats but then act like them. "Someone has to say, 'We want you all to take a chance,' " he says. "After the Brooklyn Bridge was built in 1886, a guy named Steve Brodie said he was going to jump off on a bet. There was huge press coverage. It became a big spectator sport, like executions in England in the sixteen-hundreds. He jumped off—and won. It became famous. The expression entered our lingo—'He took a chance.' What I'm saying is that these folks have to take a chance. . . . Ultimately, rhetoric has to be translated into reality."

POSTSCRIPT:

God, as they say, is in the details. I remember once appearing on ABC's *Nightline* with free-market enthusiast George Gilder after a Monday-night football game. As we waited in the studio we watched the referees whistle plays dead, call time-outs and offside penalties, rule pass interference and roughing the quarterback. The referees kept the game moving. When *Nightline* came on, Ted Koppel asked Gilder about government's proper role in telecommunications policy, and Gilder said it had none. Just let the combatants duke it out, he said, without rules or a referee. He seemed to have forgotten that business competition is often as brutal as football.

But Pressler and the fervent novice Republicans soon learned their lesson: After scaring the public with their extremist rhetoric, they retreated from more than a few of their pledges, including one to eliminate public broadcast funding. Sweeping telecommunications-reform legislation eventually achieved bipartisan support and was signed into law in early 1996. The assumption was that by liberating cable to compete with telephone, and long-distance companies to compete with the Baby Bells, and the networks to compete with cable, and vice versa, the legislation would bring about true competition. That may be so. So far it has not happened.

Communications companies invest millions to affect the decisions of government, and the bulk of the money is earmarked for the party in power. In the 1994 elections, 80 percent of the contributions from communications PACs went to incumbents; since the Democrats controlled both the House and Senate, they received about 60 percent of the money.

After the Republicans captured control of the Congress in 1994, contributions swung sharply to Republicans. AT&T, which gave 59 percent of its political contributions to Democrats in 1994, reported giving four times as much to Republicans as to Democrats in the first quarter of 1995. Ameritech, the Chicago-based Baby Bell, gave three and half times as much to Republicans as to Democrats.

By the 1996 elections, Republican tactics had frightened Americans, and the Democrats bounced back. As Democratic fortunes rose prior to the 1996 elections, communications-PAC monies followed, seesawing back to them and away from the Republicans.

10

LOCALISM CONFOUNDS TED TURNER'S GLOBAL VILLAGE

(*The New Yorker*, August 2, 1993)

"Unless American soldiers are there, American television is not there," the Cable News Network correspondent Christiane Amanpour observes about network coverage of overseas stories. In the world's various trouble spots, Amanpour says, she occasionally encounters a competitor from ABC, the American network that offers the most determined coverage of overseas news. Rarely, however, does she find herself competing with CBS or NBC, each of which saves money by having a correspondent in Washington or London do an authoritative-sounding voice-over with pictures from a video service. She more often notices reporters from the BBC or Reuters, or, sometimes, a team from Rupert Murdoch's British Sky Broadcasting, or the cameras of national broadcasters from France, Australia, Canada, or Spain. But Amanpour does not generally feel the heat of competition from other television-news outlets. "Everywhere I go, foreign news equals CNN," she says.

Since its birth at the hands of Ted Turner in 1980, CNN has come to dominate the world-news stage. Its ascendancy was confirmed in 1989 with its blanket coverage of the Tiananmen Square massacre. That was followed by coverage of the collapse of the Berlin Wall; of the first coup attempt on live television, in the Soviet Union; and of the first live war, in the Persian Gulf. Only CNN could dedicate its twenty-four-hour news channel to these events. Only CNN reaches two hundred countries and more than 16 percent of the world's eight hundred million TV homes. "By the time of the prodemocracy movement in China in 1989, world television had come of age," writes Lewis A. Friedland in *Covering the World: International Television News Services*, published by the Twentieth Century Fund. "Every major country in Europe, Asia, and Africa received CNN, including China and the former Soviet Union. And the world's heads of state and foreign offices had begun using CNN as a medium of diplomatic exchange."

When President Bill Clinton ordered the bombing of Iraq's intelligence center, in June 1993, and the armed services had yet to confirm a hit, David Gergen, who had just been appointed counselor to the president, phoned CNN's president, Tom Johnson, and got a confirmation. Only then did Clinton announce the mission.

Using the latest satellite technology, CNN has captivated decision-makers and the public alike. It proclaims itself to be "the creator of Marshall McLuhan's electronic village" and "the world's only twenty-four-hour global television-news network."

But Turner's CNN is now being challenged by competitors in the same way that he and cable TV challenged the dominance of the big three American television networks. The most prominent challenger is the BBC, which early last year announced the launching of BBC World Service Television, a twenty-four-hour global-news channel. And just this March the BBC forged a still vague alliance with ABC, saying that it would swap footage and pool foreign bureaus with ABC. This electronic alliance, the BBC proclaimed, would strengthen its hand against CNN—which the BBC disparages as an American news service, rather than an international one.

ABC itself is also challenging Turner. The network has shrunk its overseas news operation but still has more foreign bureaus or offices (a dozen) than either CBS or NBC, and its nightly newscast uses more stories from overseas than either of the two others. Capital Cities/ABC owns 80 percent of a third player, Worldwide Television News (WTN), a video service that often competes with the BBC. "The announcement that ABC is linking with the BBC is one we are tracking carefully," Tom Johnson says.

Another challenger is Reuters Television, which includes what used to be known as Visnews and now serves six hundred and fifty broadcasters, in eighty nations. Like WTN and the Associated

Press, Reuters has concentrated on being a wholesale supplier rather than a retail distributor of news over a dedicated channel. But this may be changing. In June, Telemundo, the Spanish-language television network, after deciding not to renew a five-year alliance with CNN, teamed up with Reuters and the BBC to create a twenty-four-hour Spanish-language news service to be disseminated throughout South America, Spain, and parts of North America. In contrast to CNN, whose international news is packaged out of Atlanta, London, and Washington, Reuters is decentralized, and is delivered in fifteen languages. Since Reuters provides customers like CNN with its video wire service, it does not want to antagonize them. Nevertheless, its executive director and editor in chief, Mark Wood, makes it clear that Reuters is open to other alliances. "If other opportunities come along in other languages, we'd look at them," he says.

The Associated Press is also looking at the possibility of using its bureaus—ninety of them outside the United States—to provide video as well as print and radio reports. "The board has authorized us to return with a plan for an international video service," AP's president, Louis D. Boccardi, says, adding that he hopes to reach a decision on this matter by the end of the year. Though Boccardi stresses the point that "we don't see ourselves in the retail business" (and thus as a direct competitor of AP clients), an AP video service could strengthen the hand of CNN rivals.

Then, there is Murdoch's Sky News. Sky has teamed with twelve state-owned broadcasters to form Euronews, which broadcasts by satellite in five languages and across the entire continent, as well as in Britain. And Murdoch is also trying to expand into Asia, the world's most populous continent.

Although the two less active American networks seem to have pulled back from the international arena—NBC sold its stake in Visnews to Reuters last year, and CBS can no longer claim to be the world's premier television-news outlet—there are signs that they, too, are stirring. NBC has kicked off a twenty-four-hour Spanish news service and is talking to Reuters and an Italian-owned superstation as it searches for partners. "Gartner did nothing overseas," the NBC News president, Andrew Lack, says of his predecessor, Michael Gartner. Lack goes on to say that the president of NBC, Robert C. Wright, who appointed him, is "so concerned that we are walking around with our pants around our ankles" that he formed a committee to explore new relationships overseas. CNBC, the twenty-four-hour cable channel owned by NBC, which now broadcasts mostly business news, has become a chip in the negotiations.

"It's a component, because for foreign partners it's another outlet for them," Lack says. "They're trying to crack the North American market." None of the big three networks will let them on their channels, but CNBC would. "CNN won't be there by itself anymore," Robert Wright predicts.

CBS, reflecting the aversion that its chairman, Laurence Tisch, has to diversification and costly risks, has been the most cautious of the big three American networks. Until recently, CBS has been content to maintain, like the other networks, a video-exchange program with various foreign broadcasters. But CBS is now talking to potential overseas partners, including CLT, the European media conglomerate, in Luxembourg, which has a potent satellite signal throughout Europe. "We're talking to people. We're looking for opportunities right now," says Jay L. Kriegel, a CBS senior vice president.

Each of these real or potential competitors comes at CNN frontally. Coming at CNN sideways and from below are local broadcast and cable outlets and national and regional news operations, all of which have blossomed as television has been privatized overseas and as satellite, cable, and fiber-optic technology has enhanced their reach. They are riding the wave of government deregulation and technology in the nineties, just as Ted Turner did in the eighties.

"The activity level is picking up dramatically in the global-news-service field," Tom Johnson says. He also says that he and Ted Turner welcome the competition. Nevertheless, Johnson worries that what happened to the big three American networks with the growth of cable TV could now be happening to CNN. "Sure, there could be some dilution," he says, but adds that he doesn't expect it for at least five years.

A nightmare possibility that CNN has so far refused to entertain is that Marshall McLuhan got it wrong. Maybe technology has leaped beyond McLuhan's conception and is no longer Ted Turner's ally. Maybe, instead of one wired global village, there will be hundreds of villages, each broadcasting in its own language, with its own anchor and news team, its own weather and sports and local slant. The idea of a global village was initially appealing because of its simplicity, observes W. Russell Neuman, the Edward R. Murrow Professor of International Communications at Tufts University's Fletcher School of Law and Diplomacy. "But it's a misleading simplicity," he says. "Look at Yugoslavia and you realize that all politics is basically local. In his original conceptualization, McLuhan envisioned Americans seeing what was going on live in an African village.

But Americans may not want to watch that. And perhaps vice versa. So something is missing from the formulation. The notion of instant bonding because a community is possible is the flawed naïveté of McLuhan's vision. What we have is a common space. It turns out to be dominated by Arnold Schwarzenegger and Bette Midler, not an African village. . . . I'd search for a new metaphor."

CNN has a nice head start in the race to be the dominant force in international news. Its global reach has multiplied seventyfold since Turner launched it. When it first appeared, in 1980, it had 1.7 million household subscribers, and in its first six months it had $7 million in revenue. In its first five years, it lost money. But by 1992, it reached 136 million viewers, brought in $536 million in revenues, and earned a profit of $155 million. In 1985, to expand overseas, Turner created a new channel—CNN International. He has now worked out cooperative agreements with 143 nations, rotating their news reports regularly on CNN, and he's established congenial relationships. CNN's credibility with the Iraqi government helped make it the only outside network broadcasting live from Iraq during the Gulf War.

Like soft-drink or cigarette manufacturers, who have looked to other markets as their domestic business leveled off, CNN is targeting its future growth outside the United States. Its broadcasts can now be seen in twenty-three Asian countries, and it plans to open a production center in Asia in 1994. In Europe, CNN is expanding its studio production facilities in London, and Turner has recently overcome a long-standing aversion to being involved in any business he could not put his brand name on. In the last year, Turner acquired a quarter of the shares of n-tv, a twenty-four-hour German-language news service, and he became an equal partner, with a group of former Soviet television executives, in Russia's first independent TV channel, Moscow 6 TV. In April, CNN announced an agreement to distribute its programs to households in five African countries— Nigeria, Ghana, Kenya, Zambia, and Uganda. Meanwhile, in Latin America, Turner has also amended his de facto policy that CNN's news will be transmitted only in English; CNN now produces four daily half-hour news programs that are broadcast in Spanish to nineteen countries in Central and South America.

"The BBC/ABC colossus could pose a big threat to us," Christiane Amanpour, who is based in Paris, says. "But we have now created such a niche for ourselves that we're a household word." As the television networks sought to save money by closing bureaus and relying on pictures supplied by news services, CNN built credibility by being on the scene. "It's better to provide pictures than nothing at all," Amanpour says. "But how can you know what's going on, particularly in a situation like Yugoslavia? When there's a shelling attack in Sarajevo, it's awful. But many times it's a response to an attack. You can't know that if you're sitting in London. You miss the details. How do you know what you're voicing over?"

CNN's early investment—in people like Amanpour, and in leasing space on twelve satellites, which relay signals around the clock from nineteen overseas bureaus—affords it another competitive advantage. Catching up is expensive. The annual cost of renting space on twelve satellites averages more than a million dollars apiece. The BBC, which is state-supported, does not have deep pockets. And ABC, which does have cash, has to make some difficult decisions, including whether to pursue its new alliance with the BBC or to build on its 80 percent ownership in Worldwide Television News. The announcement of the BBC-ABC alliance, in March, was made before the partners had defined the nature of their union. "They're in London trying to get out of the mess they got into," a knowledgeable network executive said this June (1993) of the ABC News team of negotiators, which was led by the ABC News president, Roone Arledge. "It was a press release. . . . That's all."

Perhaps the biggest advantage CNN has against upstart competitors is that it is a franchise with instant name recognition. The level of recognition was demonstrated in Somalia in June: while American and United Nations troops were unable to find fugitive Somali warlord Muhammad Farah Aidid, he did let reporters find him, and he declared that to keep abreast of events he had been hiding only in homes that received CNN. Ted Turner's office, in the CNN Center, in Atlanta, also demonstrates the recognition that CNN has enjoyed. The suite outside his office is a veritable shrine: thirty or so framed magazine covers featuring Turner are on the wall behind his secretary's desk; she is surrounded by a forest of plaques, awards, and trophies; everywhere are pictures of Turner with various presidents, of other nations as well as of the United States.

Turner's loud voice beckons a visitor to enter. He paces his giant office in shirtsleeves, clutching a can of Coca-Cola. Then he sinks into a burgundy couch that is nearly a cab ride away from his desk, at the other end of the room. On being asked to describe CNN's strength, Turner says, "We're welcomed everywhere in the world. What's the reason to have another one?" Still, he knows CNN is open to attack from competitors on at least three fronts: it is not yet truly international; it is not local; it has so few overseas partners that its distribution system is stretched thin.

Over the years, Turner warned CNN employees that they would be punished with a fine if they used the word "foreign"; nevertheless, the common perception is that CNN presents news packaged from an American viewpoint. And news from an American viewpoint does not have mass appeal overseas, as the big three networks can attest, since they've never had much success selling their nightly newscasts elsewhere. Parochialism is an issue that the BBC and other competitors regularly use to club CNN with. It becomes a delicate point in a world in which nationalism is resurgent and many nations are already alarmed about American "cultural imperialism."

In describing CNN as "a global, English-language news service," Peter C. Vesey, who runs CNN International, identifies both CNN's niche and its frailty. From the outset, CNN has spoken in one language, and Turner has decreed that most newscasts be assembled in Atlanta. CNN has nineteen overseas bureaus; CNN International has a staff of a hundred in Atlanta and just three in London. While CNN is working to improve these numbers, at present 40 percent of the twenty-four-hour newscast on CNN International from Monday through Friday is recycled from CNN's domestic news; on weekends, 80 percent is recycled. The international news is delivered from anchor desks in Atlanta, London, and Washington, and is produced in Atlanta, where Vesey is based. In March, responding to the BBC's World Service Television linkup with Star TV, the most powerful satellite distribution system in Asia, CNN launched a week of live programming from Asia. But instead of using a local news staff CNN sent three American anchors—Bernard Shaw, Larry King, and Lou Dobbs—to Hong Kong and Tokyo.

John C. Malone, the chief executive officer of Tele-Communications, Inc., the nation's largest cable operator, and one of the part owners and board members of CNN's parent company, Turner Broadcasting, worries that CNN has too strong an American identity. "This is a debate I've had with Ted Turner for years," he says. "Ted's going to have to have some local partners. Ted hates partners." Indeed, before the BBC linked up with Star TV, whose parent company is linked to the Chinese government, Turner rejected an alliance with Star. "They took a deal we turned down," Turner says of the BBC. "They're working for the Chinese."

Tom Johnson says CNN is shoring up its defenses and broadening both its language and its anchor-correspondent base. "We will also serve the world in other languages," he says. "How many of those languages, I don't know." But surely, he goes on, CNN will one day replace the two hours a day of subtitled translations now shown in places like Japan, and, in addition to Spanish and Russian will probably broadcast at least some news in Japanese, Chinese, and French. CNN is making a push to hire more non-American journalists—it recently recruited Riz Khan, who was born in South Yemen; Sonia Ruessler, who has dual Argentine and Dutch citizenship; and Linden J. Soles, a Canadian—to anchor CNN newscasts from Atlanta. "We're not all American-looking and -sounding," says Amanpour, who herself was born in Britain, and whose parents are British and Irani.

CNN's strong American identity reflects a deeper vulnerability: By pursuing its stated goal of being "the world's only twenty-four-hour global television-news network," CNN is obviously determined not to be local. It does not provide abundant local, or even national, weather, sports, or community news in the countries it sends its newscasts to. It does not promote local personalities. "That's where he's truly vulnerable," Herb Granath, the president of Capital Cities/ABC's Video Enterprises, which has many partnerships all over the world, says of Ted Turner. By presenting international news in English, Granath says, Turner bumps into the reality that more nations are demanding news in their own language. "I do believe there will be a national news service in most European countries with twenty-four hours of news in the local language," Granath goes on. Such local services are, or will soon be, operating in France, Hong Kong, England, Germany, and parts of southern China. In this sense, CNN suffers from the same threat the new video democracy poses to any channel that offers a uniform program at a single, prescheduled hour: viewers want, and increasingly enjoy, the freedom to customize what they watch, and to make their own schedules.

A third CNN vulnerability flows from the first two: CNN's lack of local partners. "Ted Turner does not like to be a junior partner to anybody," says Ed Turner, CNN's executive vice president in charge of news-gathering, who has been with CNN since its inception and is not related to Ted. While CNN has affiliate contracts with several hundred broadcasters around the world, who pay the news network to broadcast whatever part of its news report CNN chooses, it has fewer local partners than it would like who distribute CNN as a channel. Nor does it have channels of which CNN can claim at least partial ownership.

Prodded by his board, Ted Turner has in recent months undertaken what one of his executives calls "a strategic shift." In June, he acquired a stake in the German all-news service, but he says that so far this investment is not paying off. "It's losing a lot more money than CNN did in our start-up," he told me in June. In addition to investing in the Russian TV channel, he has made barter deals with other broadcasters. CNN has a contract with TV Asahi, in Japan, which pays CNN ten million dollars annually for the right to unrestricted use of its material in exchange for distributing two CNNI news hours over its cable system. Recently, CNN expanded its relationship with China Central Television (CCTV), the Chinese national network, which can reach as many as six hundred million viewers. In return for CNN news packages, CCTV provides CNN with news footage and with the right to sell two minutes of advertising a day (one minute in prime time) on CCTV. According to Kay Delaney, the executive vice president of advertising sales for Turner Broadcasting, CNN hopes to recruit a single buyer—Coca-Cola, for example—to pay ten million dollars a year for the spots.

"If there's one sea change taking place for Ted, it's to have more partners," Tom Johnson says. However, Turner's strength has been as a buccaneer, an inflexible visionary, and partnerships require accommodation and compromise. The combination of what are still too few local partners and an admittedly weak satellite signal over Asia means that CNN has vulnerabilities in its distribution system. The situation in Asia is tough, Ted Turner says. He notes that little English is spoken there, that the BBC has a link to a stronger satellite signal, and that some parts of Asia have over-the-air networks but no viable cable or other distribution system. "There are people staying up nights trying to get that worked out," Johnson says. Herb Granath says, "Turner's distribution in Europe is primarily in hotels and embassies. And if you watch CNN overseas, you'll discover he has few sponsors."

Turner claims not to be worried by the flurry of competition. The BBC, he knows, is pressed for cash—one reason, he suspects, that the British have sought that still dubious alliance with ABC. "It's not so easy," Turner says of his competitors. "Anybody can do it, but you've got to be prepared to lose a lot of money. The more people that get into it as partners, the less there is on the upside."

But overseas partnering does seem to be the vogue as communication companies seek to share risks and costs and to lower local political barriers. Right now, CNN's stiffest competition is not any single international competitor but multiple competitors who change from continent to continent. In Europe, it comes from, among others, Sky News and Euronews, whose programming includes news from the BBC. In Africa and the Middle East, the Middle East Broadcasting Company, which owns United Press International, is now a CNN rival, and so is the BBC. In South America, the Reuters/Telemundo alliance looms, and so do NBC and Grupo Televisa S.A., the world's largest Spanish-language broadcast company, which now owns half of PanAmSat L.P. and plans to launch three private satellites that will distribute Spanish news and programming. In Asia, the BBC has more viewers than CNN, and there are burgeoning news services in Japan, China, Indonesia, and India. And, as the boundaries between the television set and the computer and the telephone blur, telephone companies may continue to join with one another or with giant communications companies, as they have begun to do both here and abroad.

Knowing just who the players are and who their allies are becomes increasingly difficult. CNN competes with Murdoch's Sky News but has a modest agreement with USA Today Sky Radio to distribute news on United and Delta Airlines flights. John Malone's TCI is a part owner of Turner Broadcasting, yet in Asia TCI is allied with the BBC as a partner of Star TV. "One of the problems one runs into on the transnational stage is that you start running into yourself with partners," Granath observes. ABC, for example, has the new alliance with the BBC and yet sells news and its magazine shows to another sometime BBC rival—Sky News—in Europe. In Asia, Turner is allied with Time Warner's Home Box Office and Capital Cities/ABC's ESPN subsidiary to distribute programs through a common satellite whose chief competitor is the BBC.

This suggests that there are at least two competing future models for world news. One is Ted Turner's. "Nationalism is not growing," he says. "Internationalism is growing. Look at the growth of the United Nations. In Somalia, the United States is part of the United Nations force." The Cold War is over, nations are no longer forced to choose sides, and CNN can be a single, common international carrier of news, Turner says. In the next breath, however, he says, "Localism is strong. I know that." Even so, he thinks he has certain advantages. "The networks have closed down their international operations," he says, somewhat hyperbolically. Murdoch's Fox network, which a year ago threatened to start a news service, has recently scaled back its domestic plans. "We're the only distribution system that goes out to the whole world." And his will remain a low-cost network, Turner declares, noting that "the beauty of one language is that it spreads the costs."

In many ways, the development of global news parallels the development of an interconnected world economy. The world's elites—travelers, government leaders, diplomats, corporate and communications officials—crave a common database, which CNN supplies. They also want to know that a news standard is upheld—that the news service they use strives for fairness and balance. For this niche CNN is now rivaled only by the BBC.

But there is a chasm between the elites and the mass of viewers, whose vision is primarily local. Instead of one channel, there may be hundreds of channels linked to global sources of news. TF-1 and Canal + can provide twenty-four-hour news in France, as they intend to, and then link with ABC or with the Associated Press or Reuters to provide news from the rest of the globe. Thus, one model competing with Turner's might be a consortium of local or national news services linked to other national or to international services.

Like the broadcast networks, CNN is exposed. "The concept of a world news network"—with bureaus everywhere and a single channel broadcast in English—"is vulnerable," Robert Wright, of NBC, says. "A lot of the world is still in the prior-to-television-news developmental phase. What is happening is that groups—language groups, cultural groups, and national groups—want an indication on the air that they are being specifically served. And as privatization increases and more video news is available—distributed either by cable or by satellite or by broadcasting—those units that I just described will be very hungry. I don't think they will accept the national-international news package delivered via NBC or CNN as their bread and butter. They're going to want a much more customized version of that. The same way that WNBC would have no news business here in New York if all we offered were an hour and a half of national and international news delivered by Tom Brokaw."

To reach this larger mass of viewers, CNN would have to dramatically change its vision of a single, English-speaking global network. But to effect that change Turner would need to seek partners and would need to localize. "We don't want to alienate the American traveler who wants to know how his ball team did last night," Turner maintains. "But we are evolving. We will have editions in different countries."

Perhaps Turner will continue to change. Or perhaps he is now as self-satisfied as the networks once were. He claims to be unfazed by the competition. "If I were going to worry, I'd worry about the one-fifth of humanity that doesn't have decent drinking water," he says. "I'd worry about an area the size of Michigan that becomes desert every year. I don't worry too much about CNN. CNN is doing a pretty good job. And each passing day we get stronger."

POSTSCRIPT:

The challenges to Turner would accelerate when Rupert Murdoch acquired Star TV in Asia and also formed a partnership to distribute his products through direct-broadcast satellites in South America. By 1995, the three broadcast networks and Fox were vowing to start twenty-four-hour cable-news services, and NBC even forged an alliance with Microsoft to provide news both on cable and on-line. In the end, ABC and CBS dropped out of the news arms race.

When the Federal Trade Commission approved the merger of Time Warner and Turner (including the minority stake held by TCI's John Malone), it insisted that these cable partners agree to distribute at least one competitor to CNN. John Malone announced that ten million of his cable homes would distribute the new Fox news—in return for a sizable fee and the option to purchase 20 percent of the Fox news network. Bob Wright of NBC protested that Time Warner and Malone were conspiring to keep his stronger news service off their cable channels. Time Warner, which had been negotiating with Murdoch, in the end decided to grant a cable news channel to NBC, which provoked a loud battle and lawsuit from Murdoch.

By early 1997, a judge had ruled in favor of Time Warner and Turner; there were just three all news cable channels—CNN, MSNBC, and Fox; and it was still unclear whether any of this made economic sense, especially since the actual average national audience for either MSNBC or Fox News was probably smaller than the number of viewers watching a local newscast in, say , Phoenix. And this audience could be further cannibalized by fast-growing on-line and television and radio news services like Bloomberg News.

11

THE MAGIC BOX
Interactive TV?
(*The New Yorker*, April 11, 1994)

Kate seemed to be the perfect customer for Time Warner's Full Service Network—the first switched, digitized, fiber-optic, multimedia, interactive TV system. She was just thirty, and so was presumed to be open to change; she was affluent, and therefore willing to pay for what she watched; and on a form she had filled out for a recent Time Warner "viability test" she had expressed a desire to be liberated from network programmers, who told her what she could watch and when. Kate (not her real name) said she viewed about seventeen hours of television a week. When she was handed an unfamiliar air mouse, a remote-control device, she was soon handling it as if she had been using it for years. And then the down button on the remote got stuck. "A pothole on the information highway," said Paul Sagan, the managing editor of Time Inc.'s embryonic News on Demand, one of many menu choices on the proposed Full Service Network.

Sagan, along with his project partner, Walter Isaacson, who was then the editor of new media for Time Inc., and other Time executives, observed Kate from behind a mirrored window at the offices of a testing service in Elmsford, New York. They wanted to see how the Full Service Network would fare in tests. Their laboratory was a makeshift living room with two Danish-modern sofas and a twenty-four-inch TV. They watched as Kate and a succession of other viewers, with the verbal assistance of Beth Blumenthal, a professional tester, toyed with the remote-control device—as they clicked icons on the screen to call up that evening's nightly news, or weather, or sports, or in-depth accounts of President Clinton's health plan. There were, in addition, icons for various other services on FSN: movies on demand that could be paused or rewound; published reviews or highlights to help a viewer choose what to watch; video shopping catalogues; food services; video games; video phone calls.

"Our new electronic superhighway will change the way people use television," Time Warner's chairman and chief executive officer, Gerald M. Levin, had said when he announced the network, in January 1993. "By having the consumer access unlimited services, the Full Service Network will render irrelevant the notion of sequential channels on a TV set."

The executives grew excited as Kate, a composed Geena Davis look-alike, said things that seemed to justify Levin's decision to invest five billion dollars over the next five years to create FSN. She told Blumenthal that she watched a lot of news, including CNN, *60 Minutes,* and *20/20,* and said that she liked the element of control that Time Warner's variety of news choices allowed her. If she got home too late for the regular news hour, she could still see her favorite show when she—not a network—chose to schedule it. Blumenthal asked Kate what she would pay for such a news-on-demand service, and she replied, "I'm a person who would pay for convenience." Blumenthal pressed, asking Kate how much she would pay. "Comparable to whatever I pay for HBO," Kate said. Would she pay fifteen dollars a month for such a service? Certainly. Twenty? Probably.

Time executives who watched Kate and other test subjects from behind the glass witnessed such positive results that even those who had initially been skeptical were soon caught up in the excitement of a possible "killer app," as a breakthrough application is known in computer lingo. After Levin saw tapes of the tests, Isaacson felt that they had pierced whatever clouds of gloom remained. "I now know our organization is firmly committed to it," he said of the news service. "They now feel we have a good product."

Perhaps more than any of the other cable, telephone, computer and consumer-electronics companies that are planning to test such services—these include Viacom, Hearst, Bell Atlantic, Sega, U S West, Microsoft, AT&T, IBM, Comcast, Tele-Communications,

Inc., and the Discovery Channel—Time Warner is a devout believer in interactive video on demand. Last January, Levin told a television-industry convention, "The losers will be those who decide that they can wait and watch, that there is no urgency involved, that they can go on working and producing in splendid isolation from a technology that will transform the way people live, work, entertain and educate themselves." Build it, Levin fervently believes, and they will come.

Levin and Time Warner are convinced that the Full Service Network will transform not just television-viewing but the use of all leisure time. They see people like Kate as typical potential customers, who would feel liberated by the new technology. After a research session of about an hour, Kate stepped outside to sign some forms and I slipped out and asked her if she would really be willing to pay as much as twenty dollars a month for news on demand. Her answer suggests the limitation of anecdotal research, and would probably depress Time Warner's researchers. "I live on a trust fund," she said, nonchalantly. "I just tell my lawyer to pay the bills."

Four assumptions underlie Time Warner's ambitious plans: that customers want to program for themselves, and not merely be passive viewers; that customers are willing to pay extra to be so empowered; that in return for convenience, customers will pay as little attention to the charges as they do to their video-store bills; and that there will be few technical glitches. By the end of this year, Levin plans to have tested these assumptions in about four thousand households in Orlando, Florida.

"There is no way on a conceptual basis to pretest products of this kind without actually putting them into the home and allowing the consumer to respond under natural conditions," says Tully Plesser, who has conducted market research for TCI, which is experimenting with video on demand. "It's the same problem faced by advertising agencies in trying to pretest television commercials. You can't replicate the effect of the surrounding in the home, the consumer's mood, the competition within the house for your time, the noise level, in a laboratory setting."

Nonetheless, a host of communications giants are now conducting their own tests. Each is betting that video on demand won't fail, the way Ford's Edsel, or quadraphonic sound, or solar energy did. Instead, they are hoping that video on demand—the equivalent of a video jukebox that both plays and talks back—will enjoy the kind of explosive success that CDs have had. When CDs were introduced in 1983, skeptics proclaimed that customers would not pay more for a CD even though the sound was better, and would not discard entire record and tape collections and purchase expensive new equipment. A decade later, CDs account for 60 percent of all music sales.

Time Warner's laboratory for these tests is in a three-story mirrored-glass building in the Maitland commercial center, about five miles north of Orlando. Sixteen executives work there, and the brain of the Full Service Network is contained in a twelve-hundred-square-foot operations room. The chilled room is filled with stacks of disk drives, called servers, and is equipped with software from Silicon Graphics and an AT&T switching system referred to as Asynchronous Transfer Mode, or ATM. The servers house a mammoth digitized library. James A. Chiddix, Time Warner Cable's senior vice president for engineering and technology, says that initially the library will contain more than a terabyte of memory—the equivalent of a million megabytes. (Personal computers usually have hard-disk memories of between twenty and eighty megabytes.) The single terabyte will store the equivalent of five hundred motion pictures, and the switches will permit a thousand customers to access the library simultaneously.

To request a selection, a customer will click a remote-control device, sending an infrared signal to a box on top of the TV set. The box will send the signal through the coaxial cable to an aluminum node box, no more than three-quarters of a mile away, which will be located either underground or on a telephone pole. Connected to the other end of the node box is a fiber-optic cable, whose tiny strands of glass fiber can carry vast quantities of information without traffic jams. This wire will speed the customer's request to the operations center, and the center will comply with the request and then route the selection back to the customer. The entire process will take a nanosecond—the time it now takes a remote-control button to change channels.

This will be possible, Time Warner believes, thanks to the emergence of four technologies. The first is digital video compression, which squeezes as many as ten signals into a space previously claimed by one, and thus transforms a narrow cable wire that carries fifty signals into a highway that can carry five hundred. The second is the invention of a fiber-optic superhighway, made up of thousands of strands of glass fiber with virtually unlimited capacity. The third is microprocessing, which allows for the creation of a vastly expanded digitized library. And the fourth is switching technology, which has advanced to the point where it can route and track this information from library to customer and back.

The prototype of the Time Warner system sits in the Maitland office of Thomas Feige, the president of the Full Service Network. When the network is operational, it will have the normal free-TV and regular-cable channels. But in addition, there will be what amounts to a shopping mall on the screen, through which customers can navigate by clicking icons on the screen—for interactive games, movies, shopping, news and sports, education (learning through interactive instruction), concierge services (food, reservations, etc.), mailbox (e-mail and video conferencing), MY TV (customized shows), telephone (videophone, long-distance, or cellular-like services). A click on the movie icon, for instance, summons to the screen a menu of choices, and the customer can either select the desired movie or browse through movie choices by category (the movies of, say, Gregory Peck or Alfred Hitchcock) or watch promotional shorts of each film choice or call up movie reviews from *Time* and other sources or select from a list of new releases. The cost to the customer, Feige expects, will be three or four dollars a movie. "We're going to do an analysis of the video stores and charge close to what they charge," he says. "If we compete with hotels, we will charge their price."

Convenience is obviously a lure. Customers would not have to go to video stores to make their movie selections or to return tapes. Because the video server should theoretically be able to duplicate unlimited copies of each movie, the customer would never have to take a second choice (the top twenty titles are often out of stock in video stores). Unlike most pay-per-view movies, which must be watched at specified times and cannot be paused or rewound, the movie in this system can be turned on and off to fit the viewer's schedule over a given number of hours. The convenience of a movie on demand can be coupled with another convenience—food delivered on demand. I selected *The Fugitive* on Feige's system, and before it appeared on the screen, an advertisement came on. "Hungry?" it asked. "How about trying a delicious pizza from Marcello's?" Click yes, and then select plain cheese, mushrooms, sausage, or the other toppings. With the second click, the food is ordered. Human beings don't speak to consummate a transaction. Rather, the computer in the set-top box sends the order and the name and address of the customer to Marcello's. Some twenty minutes later, when the order arrives, the customer can pause the movie, go to the door, and get the pizza.

One day, Levin promises, this system will offer such additional conveniences as the ability to renew a driver's license without braving long lines. Interactivity could also allow state lotteries to sell tickets more easily, and to save some of the money they now pay to retail outlets. (Of course, such a service would inevitably encourage gambling.) The long-term economic rewards can be stupendous. Shopping by catalogue and other forms of shopping at home now constitute an $80 billion annual business. Entertainment—video stores, movie theaters, theme parks, music, books, video games, theatrical plays, gambling—is a $350 billion business, and is growing twice as fast as overall consumer spending.

The Time Warner test, which was originally scheduled for April 1994, was postponed until the end of the year. Time Warner executives say that they had been rushing to meet an artificial deadline—one they'd imposed upon themselves. "We hyped it to motivate our people to get it done," Edward Adler, a Time Warner spokesman, concedes. The deadline also allowed Levin to demonstrate that he was a visionary in the mold of his late predecessor, Steve Ross. At the beginning of March, however, Time Warner decided that it was better to miss a deadline than to start a system that might baffle the customer, or even crash.

The corporation's institutional memory may also have dictated caution. Back in 1977, Warner Amex Cable, which was a part of Warner Communications, touted a new program called QUBE, an interactive-TV system it planned to offer to thirty thousand customers in Columbus, Ohio, and eventually to the entire nation. QUBE never left Columbus, and was folded in the mid-eighties. Yet Time Warner remains wedded to the idea of video on demand. In late 1991, it launched a relatively simple cable service, called Quantum, for five thousand customers in Queens. Customers were offered a hundred and fifty channel choices, some fifty of which were dedicated to pay-per-view movies. Popular movies were broadcast at fifteen-minute intervals on six to ten channels simultaneously. Less popular titles were broadcast every thirty or sixty minutes on the remaining channels.

Richard Aurelio, the president of Time Warner's New York City Cable Group, says, "What we learned in Queens is: One, people will buy more if you give them more choices. Two, you can get them away from the video store; the thing that holds them is the ability to pause and resume. Three, we found that younger people find it easier to deal with technology than older people do. Four, we learned that the number of channels is immaterial. People don't think of it as a hundred and fifty channels; they think of it as the seventy-five analog channels they had before, and then they enter a new world. Five, we learned you need a really good navigator system, where people feel they have a personal navigator to help them. We didn't have enough of that in Queens. It wasn't friendly enough."

One reason Time Warner is rushing in Orlando is that they know they are in a race. An early test was organized by GTE, the nation's largest Baby Bell. When it launched "Main Street" in 1992, GTE promised a cornucopia of interactive services for its customers in Cerritos, California. The results have been modest. Most of its three hundred and fifty customers were upper-middle-class Asians, re-ports GTE spokesman Larry Cox, and they were more interested in educational uses and sports than they were in movies. Undeterred, GTE is testing elsewhere.

Tele-Communications, Inc., had more success with movies on de-mand when it joined with U S West and AT&T for a test among three hundred homes in Littleton, Colorado, also starting in 1992. It was a primitive system: customers phoned in their order, then waited as an operator loaded the chosen movie in the equivalent of a VCR; but unlike the VCR, the system did not allow customers to press pause or rewind. Tele-Communications' boast that usage "was twelve times greater than the national pay-per-view buy rate" is mis-leading. It does not mean TCI succeeded in siphoning customers from the video store; it means that usage rose among those who al-ready paid to watch movies on TV. Nevertheless, TCI has expanded its test group.

A variation on this theme is the twenty-four-channel test launched by Comcast, the cable operator, and the Discovery Chan-nel's "Your Choice TV," which repeats recent programs the viewer may have missed—yesterday's *60 Minutes* or today's episode of a soap opera. The small test in West Palm Beach proved a few things to Brian Roberts, CEO of Comcast. "Soap operas that replay at night are a real winner," he says. But while Roberts is optimistic, he is cautious. "I believe we've found in the cable business that when you give consumers more choice, some people appreciate it. Not all. The question is: Is that choice enough to justify the cost? That ques-tion is not answered yet."

Discovery has expanded this test to just over two thousand house-holds in eight markets. It licenses programs from the networks in ex-change for the right to charge customers. Viewers are then offered a list of categories to chose from—comedies, news, sports, soaps—

and a simple menu. In the Syracuse test areas, for example, one could order *Saturday Night Live* for one dollar and twenty-nine cents, *Sesame Street* for seventy-nine cents, and *Nightline* for forty-nine cents. The assumption is that some viewers are too passive to tape shows on their VCR but may want to see shows they missed.

Bell Atlantic is trying something else. For the past year, they provided three hundred of their employees in northern Virginia with a free choice of movies, health information, sports, cartoons, and other fare. They started a pay test of one thousand households in a Washington, D.C., suburb in 1993. Eventually, they hope to extend the service to two hundred and fifty thousand customers in northern Virginia. They call it "Stargazer," and it, too, was primitive when it was finally inaugurated. Viewers receive a program guide and about seventy choices for each time period. They then call a programmed answering machine, which asks for their subscriber number, PIN number, and ID number, and then for the four-digit code number of the program they selected. The entire process, according to Larry Plumb, Director of Bell Atlantic's Video Services Company, takes about thirty seconds. Plumb acknowledges that real success will come only when viewers can choose instantly by remote control. "The problem is that it's beyond the state of the art to do this at a price people can afford," he says. "Today's box costs five thousand dollars," which is about what Time Warner is paying per set-top box. Plumb believes the economic model for video on demand is the personal computer, which became accessible to the masses only when prices plunged.

One problem with introducing interactivity, no doubt, is that the public generally doesn't have a clue about what it all means. To ask viewers what they think of a service they can't yet comprehend is like expecting an infant to read. While no one can be sure of consumer acceptance, what is known is that the technology will soon be able to deliver nearly unlimited choices on command. And it is known that consumers want to be empowered to make choices. Cable television and the Fox network would not have appropriated a third of the three networks' nightly audience if this were not so. Nor would shopping malls and catalogues and cash machines and credit cards have become as ubiquitous. It is known that customers will pay for quality—witness the phenomenon of CDs—but whether only larger networks can afford the steep cost is not known. It is known that viewers will reject technology if it is as user-unfriendly as the early VCRs. It is known that providing movies on demand is a killer app, but not whether movies alone will justify the cost of a switched, interactive network. It is known that marketing will be important, for with a plethora of choices it will be harder to learn what's available. It is possible that viewers who are confronted with a blizzard of choices will simply fall back on the familiar brand names—CBS, ABC, NBC.

Those with a commercial stake in interactivity and information or entertainment on demand share several common assumptions. They reject the couch-potato model and believe that viewers want to take an active part in programming. They also assume that the navigator or remote-control device must be user-friendly. People will flock to interactivity, Gerald Levin says, "if it is simple and easy to use, like television."

Levin also assumes that two other sources of revenue await. "First is advertising," he says. Interactive television will allow advertisers to deliver sophisticated graphic messages targeted specifically to current or likely customers. And, second, Levin assumes that he can snare "a piece of the telephone business."

The average family now pays a monthly bill of about fifty dollars for local and long-distance telephone service, and about twenty dollars for basic cable TV. Will people pay more? Can they afford to? W. Patrick Campbell, the executive vice president for corporate strategy and business development for Ameritech, one of the seven Baby Bells, told an industry panel in February that Ameritech estimated that the average monthly household bill for telephone and cable and interactive services would climb to a hundred dollars within five years. Another panelist, Dennis Patrick, who is the president of Time Warner Telecommunications, guessed that the average consumer "will be paying less," largely because companies like his could deliver movies and access to long-distance telephone service more cheaply than the video store and the Baby Bells.

This poses another question: Can the companies putting together pay-per-view services do a better job of camouflaging the true costs? Joseph J. Collins, the chairman and CEO of Time Warner's national cable operations, says of the hundred-dollar-a-month bill, "That's the wrong way to look at it. . . . The tradition in the cable business has been to put everything on your bill. So on your bill you get billed for basic service, you get billed for a second outlet, you get billed for program guides, HBO, et cetera. We've done that because, historically, it's been an efficient way to do it. That's not necessarily how other businesses run their billing practices. In fact, I suspect that as we see the Full Service Network begin to operate we're going to see a very different type of billing, where some things may be billed indirectly. For example, if you go to the video shopping mall and order a new sweater from L. L. Bean, once you order it, it comes from L. L. Bean, and so does the bill." Or, he says, maybe the monthly bill will be reduced, because in return for access to customers, advertisers would subsidize part of the costs.

For Time Warner and the other companies embarked on this venture, there are at least four potential sources of revenue. There is the advertising dollar. There is the charge to consumers for use of the cable/telephone wire. There is the toll to be charged to those program services which want access to this wire, such as L. L. Bean. And there is the license fee paid to the cable company by a software supplier of, say, *Murphy Brown* episodes from Warner Brothers or *The Larry Sanders Show* from HBO.

Because this is uncharted terrain, the economics are murky. Time Inc.'s Walter Isaacson concedes that it is unclear how to structure a price for a service that has yet to be distributed. Also unresolved, he says, is who should pay whom, how to set advertising rates when the customer base isn't known, and how the distributor (the wire) and the manufacturer (the program supplier) should divide advertising revenues. The financial models are built on sand. Still, businessmen dream. "If we can do four to five dollars in additional cash flow per month for twelve months," says Comcast president Brian Roberts, "we can almost completely pay for rebuilding our entire cable system with fiber. That's fifty dollars a year, and if we borrow at the normal ten times cash flow, that's five hundred dollars per customer to spend. If you have one hundred thousand subscribers, that's fifty million dollars, or what it costs to rebuild a fiber system."

Joe Collins of Time Warner challenges Roberts's assumptions. How does a business calculate how much incremental revenue it needs to earn a profit, he asks, when it does not yet know its incremental costs? Until it knows its costs, how can it price? And it can't know costs or prices until it knows viewer preferences.

Still another consideration is what a more assertive federal government will do. Will the Federal Communication Commission's recent decisions to scale back cable prices by 7 percent really take effect, and will this slow the private construction of the electronic highway? Will local broadcasters and the networks be given more channel space, so they can challenge cable? Will cable siphon off telephone business? Will telephone steal cable revenues? Will government allow it? Will the more powerful personal computer—not the television set—be the instrument for interactivity?

The human or social consequences can be viewed in two ways. Some maintain that this video democracy will enhance a sense of community, others that it will fracture communities. "This provides opportunities for more communication," says Levin, who emphasizes that the TV set will still offer free TV. There will be video bulletin boards, he points out, and interactive games, and video telephone calls, and interactive tutorials, and local C-SPANs, and home movies shown to friends and family. "It will be like the Internet," Levin says. "This will create a video family that enables people to share a range of experiences." Critics, on the other hand, say that there will be fewer common experiences. Fewer citizens will be watching the same thing: rarely will two-thirds of American TV viewers be glued to a miniseries like *Roots*, as they were in 1977, and inclined to talk about race relations at the water cooler the next day.

Critics also point out that the news can be personalized; that is, viewers can skip important news they don't want to watch. And there is the danger that executives in the business of providing news on demand will think too much of cost and too little of journalism. Increasingly, there is a tendency in the news business to confuse "coverage" with journalism. The networks substitute pictures they buy from international services and allow a correspondent in London or Washington to supply an authoritative-sounding voice-over to pictures of, say, the latest massacre in Bosnia. It saves money, but it doesn't provide what could be provided by a reporter on the scene who speaks the language, has witnessed the event, or can draw on known and reliable sources.

On the other hand, viewers will be able to skip the Tonya Harding stories and to get more in-depth news. Instead of a lowering of standards, Walter Isaacson says, the opposite can occur as television no longer requires homogenized, entertaining information for a mass audience. "On a service like this, a really good journalist who did a good piece on Sarajevo may have trouble getting it on a twenty-two-minute newscast," Isaacson says. "But we can give that person forty-five minutes, or whatever. You don't have to appeal to a mass market."

It may be that a switched, interactive network is an invention that no one wants. The first of Thomas Edison's 1,093 patented inventions was an electric vote-recording machine to register the ayes and nays of legislators on a big board. When Edison traveled to Washington, congressional leaders told him they didn't want it. "It takes forty-five minutes to call the roll," a congressional sachem told him. "In that time, we can trade votes. Your machine would make that impossible." But it may also be that everyone will want interactive TV. There is no way to find out except to do what Time Warner is doing: build it, and hope they come.

POSTSCRIPT:

Time Warner built it, and so far they haven't come. The Orlando magic box has not proved to be the killer app Time Warner hoped it would be. Nor have any of the other tests altered the media landscape. Time Warner's Full Service Network is functioning in Orlando, but there have been glitches. The servers and switching devices didn't function properly; the system overloaded when burdened with too many requests for the same movie. While those involved in Orlando continue to tout their system, Time Warner chairman Levin has noticeably toned down his evangelical fervor for the magic box.

The modest results from video on demand did not, however, deter the Baby Bells from announcing in 1995 that they planned to provide video services to compete with cable. Spurred on by Michael Ovitz, two different groups of seven telephone companies forged separate consortiums. Each hired experienced programmers. They ordered set-top boxes. They visited Hollywood and announced that they would be a cash-rich customer. They invested millions. And then they froze—worried that the expenses were too vast, that the technology was not there yet, that maybe they should bet on a business they understood, like long-distance or cellular-telephone service. By early 1997, one consortium had closed and the other was barely visible.

No one had yet cracked the core mystery: What do consumers want, and what will they pay for it? The catch-22 is that we cannot know whether a video magic box will be accepted by a mass audience until the price comes down, and yet, as happened with computers, the price won't come down until a mass audience is reached. As Intel's CEO, Andy Grove, said of the Internet in early 1997, to make money, companies must be prepared to lose money: "Columbus didn't have a business plan when he discovered America," said Grove.

12

THE HUMAN FACTOR
Troubles in Disneyland
(*The New Yorker*, September 26, 1994)

My car phone rang. It was August 30, and I was on my way to Kennedy Airport to fly to Los Angeles for a dinner with Jeffrey Katzenberg, the chairman of Walt Disney Studios. He had told me earlier by phone that he would give me the real story of why, just six days before, he and Michael Eisner, the chairman of the Walt Disney Company, had announced their divorce. But he was calling now to cancel dinner. He explained that Eisner, for whom he had worked a total of eighteen years, both at Disney and, before that, at Paramount Pictures, was furious over Steven Spielberg's comment to the *Los Angeles Times:* "Jeffrey Katzenberg's exit will be Michael Eisner's Machiavellian loss—and Corporation X's El Dorado."

"I've got to let the situation calm down," Katzenberg said over the car phone. Eisner, he said, had asked him to quiet friends like Spielberg and David Geffen, the billionaire entertainment impresario. The publicity was hurting the company, Eisner had told him. And, Katzenberg felt, it was hurting his own reputation with prospective employers and possibly weakening the leverage he could exert in extracting a generous contract settlement from Eisner. Both men had a financial interest in quieting the row. Katzenberg said that if Eisner ever found out he and I had met he would, rightly, feel betrayed. A man who had so dependably delivered what Eisner wanted that he had been nicknamed the Golden Retriever was now mistrusted. "Sorry," Katzenberg said. "The rules have changed. I have given Michael Eisner my word that I'm not going to do any more interviews unless he specifically directs me to."

I phoned Eisner when I returned home. "I'm not ready to discuss it," he said, explaining that after at first talking to reporters he had decided that the company should shut up. Having undergone quadruple heart-bypass surgery just four weeks earlier, he said, he needed rest, and he added, "My wife is ready to kill me." The one comment Eisner did make about the breakup was that he had to deal with what he referred to as "this thing."*

The "thing" he meant had come about on the morning of Wednesday, August 24, 1994—a year after he and Katzenberg went for a walk during a stay in Aspen, Colorado, and first discussed Katzenberg's desire to broaden his role at Disney. Ten years before, when they left Paramount together to take over the moribund Disney, its revenues were $1.4 billion. In 1993, its revenues were more than $8.5 billion. The pretax profits of the film and TV studio, which Katzenberg ran, jumped from $2 million in 1984 to about $800 million in 1994. Throughout the past decade, no film studio has reported greater profits.

All this helps account for why the announcement on August 24, 1994, of Katzenberg's departure from Disney has shocked the entertainment industry. "Stability" is a word that is much invoked but little honored in Hollywood. The best-run studios clearly are those with stable leadership, such as that of Eisner and Katzenberg at Disney, and Robert Daly and Terry Semel, the chairman and the president, respectively, of Warner Brothers. Enduring partnerships are so rare in Hollywood that when one dissolves, the town leaps to attention. Haiti and health care took a back seat to Eisner and Katzenberg, with residents of Hollywood asking, "How could such a successful team fail?"

The failure is in part a tale about relationships, a reminder that the path to the so-called information superhighway is pocked with potholes, many of them man-made. Two gifted entertainment executives ended up in positions and reached conclusions that neither wanted. Both were swayed by pride and a stew of other emotions.

---

*In the end, Disney officials chose to speak, fearful that Katzenberg and his allies were talking to the press.

Their breakup was dictated by personality, not performance—by human more than business factors.

In 1976, when Katzenberg and Eisner first met, Katzenberg, at twenty-five, was a low-level assistant to Barry Diller, who was then running Paramount Pictures, with Eisner, then thirty-four, as his second-in-command. By most accounts, during the eight years they spent together at Paramount, Eisner saw Katzenberg as a gofer—a short, brusque assistant who wore dark suits, white shirts, and large square glasses and did what he was told. Eisner admired Katzenberg's machinelike efficiency. In 1984, Diller clashed with Paramount's CEO, Martin Davis, and left the company. Davis, who barely knew Eisner, chose Frank Mancuso to run the studio. Eisner accepted an offer to become the chairman of Disney, with Frank Wells, who had been vice chairman at Warner Brothers, joining him as president and chief operating officer. Katzenberg was invited to join them as a junior partner. "I told him that Eisner doesn't want a partner," recalls a former Paramount executive who was close to Eisner and whom Katzenberg consulted. "Michael will never accept you as a partner."

"If I'm good enough, he will," Katzenberg told the executive.

For the next decade, Katzenberg tried to prove his worth. While esteem for him grew in the Hollywood colony, which saw that he had become an able and creative executive, he never felt fully accepted by Eisner. They had a standing business dinner every Monday when both were in town, and yet, Katzenberg told friends, in fifteen years the only social dinner they had together with their wives was during the week after it was decided that he would leave Disney. David Geffen says that Katzenberg, having had an at times distant relationship with his father, "didn't want to be treated as a golden retriever," and adds, "He wanted Michael's approval." But Eisner, having had an aloof father, and having worked for a scold like Diller, apparently found it difficult to give praise. He never complimented Katzenberg for lashing *The Lion King* into shape, even though it has already generated over a billion dollars for Disney. "As tall as he is, he's a little guy," Geffen says of the six-foot-three-inch Eisner.

Katzenberg wanted to be a sounding board, a partner, the way Frank Wells was. He was enthralled with his job, loved coming to the office in jeans and a Mickey Mouse T-shirt, loved taking his wife and twins to Disney World, as he did each September after joining Disney. When the reviews for the Broadway version of *Beauty and the Beast* were snide last winter, Katzenberg was unfazed. As long as they said the musical was a commercial extravaganza and would attract audiences the way *Cats* did, he was content. He cared about commerce, not art. But as good as he was at his job, Katzenberg never shook Eisner's first impression of him. "A golden retriever fetches," a Disney executive who knows both men says. "He does not think for himself. He goes in the direction his master wants." Katzenberg never succeeded in convincing Eisner that he was a wise, well-rounded executive.

In August of 1993, Eisner, Wells, and Katzenberg went together to Aspen for an industry conference. Within Disney, Wells was known as the diplomat, the man people went to see if they wanted to persuade Eisner to change a decision or if they simply wanted to complain. He also did the unglamorous grunt work—rescheduling corporate debt, negotiating labor contracts, and finding investors to reduce Disney's exposure. Frank Wells was a generous contributor to Democratic candidates, and flirted with the idea of one day running for office—maybe the U.S. Senate. He had already scaled six of the world's seven highest mountain summits, taking two sabbaticals to do so. His climbing feats implicitly announced to everyone in the company, "My life is not just about work. I compete with myself, not with anyone else." In Aspen, Wells and Eisner agreed on the broad outlines of a new seven-year contract.

Eisner then turned his attention to Katzenberg, telling him he hoped that he, too, would agree to a generous multiyear contract. The two men talked about how Katzenberg had a month to decide whether to exercise the option in his contract, which would allow him to leave the company in September of 1994. If Katzenberg did not exercise that option, his employment would continue through September of 1996, and he would receive a significant but as yet undefined bonus and stock payment.

Katzenberg had been thinking more and more about declining to renew, for he was troubled by what he felt was a growing estrangement from Eisner. Disney was now a behemoth, with interests in film, video, television production, music, books, theme parks, plays, interactive games, consumer products, hotels, real estate. The relationship between Katzenberg and Eisner had become more formal. Although Eisner approved the studio's major productions and often read the scripts, more and more of his time and energy were consumed by a controversial theme park near the site of a Civil War battlefield in Virginia and, especially, by Euro Disney, in France. Building Euro Disney had required an investment of $4 billion, much of it from outside sources. In the company's 1993 annual report, Eisner candidly described Euro Disney as "our first real financial disappointment," so "dreadful" that he would grade the effort as "barely a D." Personal strains added to the business strains. Eisner sometimes felt pushed by Katzenberg, while Katzenberg felt that Eisner resented the fact that the son was straying from the nest. And Katzenberg was now a more visible presence around Hollywood, in contrast to the impeccably anonymous Frank Wells.

Eisner invited Katzenberg to take a walk with him through downtown Aspen. Katzenberg later described to a friend the following conversation:

"What do you want?" Eisner asked.

Katzenberg told Eisner that he saw himself as a builder, and he wanted "new mountains to climb." He wanted to be treated as a partner.

Eisner asked if he wanted to be vice chairman and sit on the Disney board.

Katzenberg said, "Does it make sense for Frank to be vice chairman and me to be president?"

"No. That would be perceived as a demotion for Frank," Eisner said.

According to the version Katzenberg has related to colleagues, Eisner then said what Katzenberg longed to hear: "If for any reason Frank Wells is not here—if he decides to run for political office, if he goes off to climb the summit—you are the number two person, and I would want you to have his job."

Subsequent events pivot on whether this conversation occurred in fact or only in Katzenberg's imagination. Eisner has denied it. Close Katzenberg associates—among them Geffen, who served as Katzenberg's chief strategist over the following year—insist that Eisner made the promise. Katzenberg can recite for friends the exact site at which Eisner reportedly uttered those words—in front of Boogies, a diner in downtown Aspen. Although Eisner, like most executives, sometimes says things to stroke his employees' egos, Katzenberg loyalists view him as a liar. "Michael is well known in Hollywood for being careless with the truth," says Geffen, who is one of the few Hollywood executives unafraid to speak their minds. There is anecdotal evidence for such an interpretation. A former deputy who remains friendly with Eisner remembers that he would make commitments to go ahead with a movie, and then, as soon as the producer left, tell the deputy, "Get me out of it."

"But he'll say you lied," the deputy would tell him.

Eisner is reported to have replied, "That's not lying. It's business."

This anecdote had a certain resonance, for I remember interviewing Eisner last year, and at the end of our meeting he mentioned a director whose work and politics he abhorred. "He'll never make a picture for Disney," he confided. Coincidentally, that night at an industry dinner I was standing near this director when Eisner approached and said to him, "I love your work. When are you going to make a movie for Disney?" Still, as Eisner might say, that's business. Or, at any rate, that's Hollywood. (The director has since made a movie for Disney.)

A second interpretation is that Katzenberg is not telling the truth. Eisner would never make that pledge to Katzenberg—so say five men who have talked to the Disney chairman about it. They say that Eisner never wavered in his conviction that Katzenberg lacked the scope to be the chief operating officer—lacked the kind of broad business acumen that Wells had as a business lawyer and as the vice chairman of Warner Brothers. A key Disney board member declares, "It's not consistent with the way Michael would have behaved. You don't make decisions in the abstract that far in advance. I don't believe Michael made a blank commitment of that kind. He never told the board, and I believe he would have. The board has given him total support, not just because we respect his abilities but because we're never surprised. If Jeffrey believes he heard it, I think he misinterpreted it."

This suggests a third interpretation: Both men believe they are telling the truth. Perhaps Eisner, who is a brilliant salesman, "intimated to Jeffrey that he would expand his responsibility," says a student of the two men, and Katzenberg heard more than was said. "Could he have led him on?" asks an Eisner friend. "It's possible." But what wasn't possible, he thinks, was that Eisner would have made such an offer without first discussing it with his board.

This much is certain: at age forty-three, Katzenberg was determined to enlarge his portfolio of responsibility. As was true of Eisner a decade earlier, when they both worked at Paramount, he wanted a bigger job. Katzenberg, who had had only a year of college, worked to expand his range of knowledge. He read, including books on management. Known as a ruthless cost-cutter and micromanager of movie production, he struggled to be more sensitive. He talked freely to reporters and was widely admired for his candor. He socialized with his peers, taking rafting and other outdoor trips with fellow executives. He thought of himself as becoming more relaxed, more social, though that is not always what associates thought. "Jeffrey's never had a social dinner with anyone," says a longtime Disney executive who is aware that Katzenberg sometimes had three different guests at a meal but asked each to wait while he completed his allotted half hour or so with each of the others. "Even when Jeffrey goes rafting, it's business," says a Katzenberg friend. "It's bonding—so that when they get back to town they can do business. This town is not particularly good at real friendships."

By 1993, Katzenberg had become more visible—as an executive with a wide circle of friends, as a fund-raiser for AIDS causes and Democratic politicians, as a political presence in Washington, as a Hollywood potentate. But he craved new challenges. And he craved Michael Eisner's approval. So, to test the relationship, when lawyers notified Katzenberg that if he stayed in his contract through 1996 he would receive $100 million in bonuses and stock options, Katzenberg stunned Eisner by declining.

Katzenberg had decided to call Eisner's hand. If he did not have a new mountain to climb in the remaining year of his contract, he would leave, he said. This refusal, associates say, proved Katzenberg's sincerity and his lack of greed. Disney director Stanley P. Gold, who had helped recruit Eisner and Wells for Disney, says, however, that the refusal is proof that Eisner could not have told Katzenberg in Aspen that he would one day get Wells's job. "I believe that Jeffrey's giving notice shortly after his alleged conversation is proof that it never occurred," Gold says. "If it had occurred, I can't imagine that Jeffrey would, within thirty days thereafter, give Michael and the company notice that he was leaving." Indeed, the chronology of the Aspen walk and Katzenberg's giving notice, a month later, is the most compelling argument that Eisner did not lie.

Katzenberg's retort is that he was hoping not to leave. Between September of 1993 and April of 1994, he became ever more publicly visible. "When Jeffrey could not get recognition inside, he went outside," a Katzenberg aide explains. Eisner, unsuprisingly, did not much like feeling pressured. Nevertheless, during this period the two men spoke of their joint search for new challenges. Eisner asked Katzenberg to spearhead Disney's attempts to become a major Broadway producer by taking over a theater and launching the effort with a stage adaptation of *Beauty and the Beast*. They talked about Disney's buying a television network—a proposal that Katzenberg fervently supported. Both men were hopeful about their chances of reaching an accommodation. In fact, a member of the board's compensation committee recalls that by late winter he was confident that

Katzenberg would sign a new contract. "It was my impression from discussions with Frank Wells that we had just about closed up with Jeffrey on a new contract," the Disney board member says.

Fate intervened on Easter Sunday, when Frank Wells died, in a helicopter accident in Nevada. The next day, Eisner abruptly announced at a staff luncheon in the company's Burbank headquarters that he would assume the duties and the title of president. It was a way of reassuring nervous investors that, as a one-page press release issued on April 4, 1994, said, "we have in place a strong management structure that will carry Disney forward."

Katzenberg had admired Wells, as had almost everyone in the room, but, still, he was focused on himself. He seethed. "I don't think I blinked for the entire lunch," he told an associate. He had been in his office down the hall since six-thirty that morning, yet Eisner had told him nothing. He thought he was Eisner's partner, he told friends, yet, despite eighteen years of working together, despite the fact that they probably spent more time together in those years than either did with his own children, despite their conversation in Aspen, he had been treated as an appendage. During the meeting, he fought to keep his composure. "If he had assured me, privately, that it didn't mean anything, fine," Katzenberg told an unassailable source. "But by his actions Michael assured me that he meant everything. It told me that he wouldn't share with me."

That night the two men had their regularly scheduled Monday business dinner at Locanda Veneta, on Third Street in Beverly Hills. They talked about Wells, and then some about business matters. Both men were shaken by Wells's death. The subject of Katzenberg's role never arose. He told one source that it was Eisner's place to raise it. Perhaps, a close Katzenberg friend suggests, he was afraid to confront his figurative father. Although Katzenberg normally sleeps well, that night he restlessly prowled through his house. An image was building in his mind that Eisner was condescending to him, he told friends.

The next morning, for the first time in ten years at Disney, a visibly angry Katzenberg told Eisner's secretary it was important that he and Eisner lunch that day. Eisner got the message, and they met in a private corporate dining room. Katzenberg wasted no time, as he later told friends, recounting the conversation in detail. "You had promised it to me," he said to Eisner. "I don't understand what you're doing. I don't understand your putting out that press release without talking to me. I don't understand why you said nothing at dinner. I don't understand why, after eighteen years, you wouldn't first talk to me. If you don't want to do what was promised, I'm leaving."

Eisner was furious, and is said to have retorted, "You're putting a gun to my head!"

"No, I'm just holding you to your promise."

"Are you telling me that if I don't do this you'll leave?" Eisner is said to have asked, incredulous at what he saw as immature, insensitive behavior.

"I want you to do what you said you wanted to do," Katzenberg responded. Later, Katzenberg informed his friend David Geffen that at the end of lunch he said to Eisner, "I'm going to leave."

The luncheon meeting left both men enraged. Katzenberg thought that Eisner had reneged on a promise made in Aspen. He felt deeply wounded, and couldn't fathom why Eisner would ignore him. Eisner thought that Katzenberg was audaciously disloyal to Wells. Even Katzenberg's most ardent supporters, though they believed he had reason to be upset, thought that he had pushed too hard too soon—had given ammunition to those who said that he was not fit to suceed Wells. Thus a business virtue—that he pushed hard—became a personal liability.

At the April board meeting, Eisner apprised his board of the conversation. Most of the directors were livid. "I didn't think it was appropriate," one of the more temperate directors says. "We had to digest an enormous loss. Frank's body wasn't off the mountain yet. It exhibited bad taste and judgment." Harsher words were spoken by Roy Disney, the board's vice chairman and the one director who really knew Katzenberg from working with him on animated films, and who could not abide him, and said so. Obviously, Katzenberg told associates, Eisner had never informed the board that he'd been promised the job. And because he wasn't on the board and had failed to court its members as allies, his side went unheard.

The board members began to think they might not be able to satisfy Katzenberg's desire for a bigger job. Meanwhile, Katzenberg was becoming more and more pessimistic. "He said from the beginning that Michael would be unwilling to share, to make him his junior partner," Geffen recalls. "I said, 'No way. He won't risk losing you.' "

Katzenberg, goaded by Geffen—whom many in Hollywood came to blame for reinforcing Katzenberg's combative tendencies—hardened his position: he wanted Frank Wells's title. Eisner, goaded by his board, was determined not to give it to him. Eisner kept Wells's office, which adjoined his own, empty for months, as if it were a shrine. For Eisner, Wells was irreplaceable. While the clock on Katzenberg's contract was running down, the two longtime associates were drifting apart. Katzenberg cared about the title, and wanted confirmation of his own stature. Eisner didn't want to give the title because, he told his board, he didn't see Katzenberg as the kind of astute partner that Wells had been. A key Disney director says of Katzenberg's being named COO, "It would be a waste for him to do that job. In a large sense, Jeffrey's job as chairman of the studio was a more important job than Frank's job. It would be silly to dilute his effectiveness as chairman of the studio. Frank's job involved so much detail—hard, technical kinds of things."

The real plotline here is about two people who are "married," a friend of both Eisner and Katzenberg observes. "And suddenly one party says to the other, 'You're used to me staying home all these years. Now I want to go out.'" On top of this, the friend adds, Eisner was like a jealous spouse, "threatened" by Katzenberg's declaration of independence and enhanced public profile.

Nevertheless, with Wells gone, Eisner did confer more power on Katzenberg—"more trinkets" is how Katzenberg described it to friends. He gave Katzenberg responsibility for Hollywood Records, and for Disney's efforts to promote interactivity with electronic games and its prospective linkups with the Baby Bell telephone companies. More responsibility was accompanied by more anxiety, and produced ambivalence in both men. Katzenberg was enraged that Eisner never raised the issue of his expiring contract, as if he wanted him to leave, which Katzenberg couldn't believe. Eisner continued to be angry about what he saw as Katzenberg's lunge for power. Nevertheless, however embittered their relationship was becoming, at moments both Eisner and Katzenberg still thought that something could be worked out. They finally agreed to resolve this matter in August of 1994, when both could get away from the office.

Then, once again, fate intervened, when, on July 15, specialists told Eisner that his heart was weak, and that he must immediately undergo a heart-bypass operation. The next day, doctors performed a quadruple bypass on Eisner. Katzenberg, according to an aide, learned of the operation when he phoned Eisner at home to give him the weekend movie grosses, and Eisner's wife, Jane, said, "Oh, Jeffrey, I meant to call you." By contrast, Roy Disney had been notified the day before, and had rushed to the hospital from a castle he owns in Ireland. A press release announced that Roy Disney, not Katzenberg, would be Eisner's stand-in. Soon stories appeared in the press that Katzenberg might leave if he didn't get Frank Wells's job—stories that Eisner's champions attributed to Katzenberg. A friend who visited Eisner at the hospital says, "He was lying in a hospital bed, tubes coming out of every part of his body, drifting in and out of consciousness, and when he's awake he sees that he's being pushed and pushed in the press."

Eisner appeared to push back. On August 1, *Newsweek* published an interview that Eisner had given a month before his bypass operation, in which he said that the company would now be "modeled after Rupert Murdoch's News Corporation. In other words: a powerful CEO without a strong second-in-command." Eisner added that the plan was the "suggestion" of Roy Disney. *Time* reported in its August 1 issue that Eisner and the board "may solicit a list of outside candidates as early as this week" to fill Wells's post.

Not surprisingly, Katzenberg was despondent. He took the first week of August off and holed up at his Malibu home. "By this point, Jeffrey was retreating from his role as the perfect corporate citizen," says his close friend Jim Wiatt, the president of International Creative Management, who visited with Katzenberg during this period. "He was starting to say, 'I'm getting tired of not being dealt with.' He felt he had to force a confrontation."

And so he did. The timing was atrocious, so soon after Eisner's operation that he had not returned to the office. Eisner felt pressure to settle up with Katzenberg, and invited him to his home, in Bel Air, one evening during the second week of August. Eisner told Katzenberg that he was agitated by press stories that Katzenberg was pushing for a bigger job and a better title. Katzenberg is said to have told friends that he had cut Eisner off and said, "I never told Frank Wells how much I liked working with him and how much I loved him. I realize I've never told you how much you've meant to me." So he told Eisner how much he admired him and how thankful he was for all he had learned. But Eisner was given reason to question whether the "love" was real when Katzenberg concluded, "Having told you all that, it's also time for me to move on."

"Have you taken a job?" Eisner asked.

"No."

"Is it something we can discuss?" Eisner asked.

"We can talk about it, Michael, but I suspect that the decision is carved in stone by other people." He knew that he was isolated from the board, and instead of blaming himself he blamed Eisner for keeping him on a leash. He continued, "Roy Disney is hostile to me. You think I want to go into a board meeting and know that I don't have the complete support of Roy Disney or Stanley Gold, or know that the only reason I'm there is that you shoved this down their throat?"

Eisner knew he had received an ultimatum. He tried to buy some time. He asked Katzenberg to put on paper what he saw as his role, how he would define the partnership he sought, how he would reorganize the company, what executives he would replace, what corporate acquisitions and strategy he would recommend. In brief, Eisner wanted Katzenberg to crystallize something that they had often discussed: how to reinvent Disney.

Over the next ten days, Katzenberg drafted a four-page memorandum, but even as he did so he was dubious whether a partnership could work. He was torn. Intellectually, he didn't believe Eisner would share power, yet he knew that Eisner did not want to lose someone whose movie and television divisions generated half of Disney's pretax profits. Emotionally, he wanted to escape from Eisner's shadow, yet he believed in the Disney family. He had a love-hate relationship with Eisner and a love-hate relationship with his current job. He was on the ledge, wanting—and not wanting—to jump.

On August 24, Eisner pushed Katzenberg off the ledge. In a calculated show of strength that allowed him to take control of the situation, Eisner summoned Katzenberg to his office and, in what both men have described as a relatively amicable conversation, said that there was no way to reconcile their positions. Katzenberg had brought along his four-page memo but never got to present it. Instead, Eisner handed Katzenberg a draft of a press release, and asked him to review it. The release, to go out that day, would announce not only Katzenberg's departure but also a new, decentralized reorganization, which would split Katzenberg's job responsibilities into three parts—live-action motion pictures, the animation division, and television and telecommunication services.

"I saw the seeds of division between Jeffrey and Michael," observes William M. Mechanic, who left Paramount with both men and who worked at Disney until a year ago, when he was named president of Twentieth Century Fox. "But because of business interests, I didn't think it would come to this." Business logic dictated that they remain a team; human logic dictated something else.

At first, both men felt a measure of relief that their ordeal was over. But the emotion was short-lived. The initial reaction among Katzenberg's many loyalists was that Eisner had been ruthless. "It was about as cold as could be," one Disney staff member says of their parting, not knowing that Katzenberg had been equally cold in pressing Eisner for Wells's job. What is missing in the horror that has been expressed over Katzenberg's termination is a sense of irony. Katzenberg was not shy when it came to terminating employees and slashing costs. "No one could do cold and callous better than Jeffrey," a longtime Disney executive who respected him recalls. "He was as tough and as mean as anyone in business."

At Disney, the initial anger over the rupture was mixed with tears—genuine tears, for Katzenberg was also admired as a leader who was generous with praise and bonuses, who remembered names and birthdays. "In a world where no one has a heart, he is wonderfully protective and humorous, despite his reputation," says Harvey Weinstein, the cochairman of Miramax, whose successful film company Katzenberg had persuaded Eisner to acquire in 1993. When Katzenberg took his family to Disney World during Labor Day week, the entire animation department surprised him with a keg of his favorite drink (Diet Coke) and presented him with a huge drawing showing Katzenberg surrounded by the animated Disney characters that they had created together over the past decade. Then two hundred and fifty members of the department formed a line so that he could sign their *Lion King* books or drawings. Katzenberg's wife, Marilyn, stood to one side and cried.

During the tearful interlude, however, the spin patrols were at work. Eisner talked to reporters. "This is not a Shakespearean tragedy," he told the *Los Angeles Times*. "This is people moving on with their lives, and doing new and interesting things." The press release had said, "With heartfelt thanks and obvious regret, I wish him well in his future endeavors." Eisner played the breakup as the growing pains of a child who was intent on finding his own apartment. Katzenberg wanted his job, Eisner coolly suggested, and it wasn't available. As chief executive, he had made the necessary tough-minded decision. Eisner, having lost a president and undergone a heart-bypass operation, wanted to convey the notion that this was an inevitable split and that Disney remained strong.

Ironically, after a burst of orchestrated outrage from friends, Katzenberg, for different reasons, shared that aim. The audience he had his eye on, he told friends, was the communications executives who attended conferences such as the one the investment banker Herbert Allen, Jr., held annually, and to which he had regularly been invited. Katzenberg did not want to look to this audience like a non–team player, a whiner. He needed to take the high road. "The way I handle this speaks to what my character is," he told friends. A second business reason that Katzenberg sought to mute criticism of

Eisner or Disney was that he hoped to get what he told friends was "a king's ransom" from Disney as a contract settlement. Although his contract was to expire at the end of September, Disney, as he interpreted it, owed him a generous exit payment, because he enjoyed profit participation. A number of his projects, including *The Lion King*, swelled his line of credit.

Which is where I came in. The day after I asked Disney spokesman John Dryer for an interview with Eisner, Eisner phoned me, late in the afternoon of August 30, to say that he did not want to cooperate on a story about the split. But he promised to grant me exclusive access to Disney if I waited a few months. "It's too early. I am so tired of the stories," he told me. A few hours earlier, after discussing it with Eisner, Katzenberg had called to cancel our dinner while I was en route to the airport. Now Katzenberg chimed in that he thought Eisner's idea that I wait two months or so and in return win exclusive access to Eisner and Disney was a good one. The next day, when I phoned Eisner to say that I was proceeding with this story, he told me that my request was the final "straw" and said, "We are not going to cooperate with anyone else." (That night he dined with a *Vanity Fair* reporter, Kim Masters, who had been at work on a Disney story for several weeks.) A week later, Eisner faxed me from his home, saying that the reason he was not talking was that "I simply promised Jeffrey that for the time being interviews from me about him are over."

The contract-settlement discussions got ugly. Eisner did not think he owed Katzenberg any money, he told me in a telephone conversation when I was in California in early September. He did not budge from his insistence that he would not treat this parting as a "gossipy, tabloid event. I'm not going to give you the opposite view from David Geffen's. What does he know about this?" But one view he would share was that Disney owed no payment to Katzenberg. "His contract is over," he said. "When people's contracts are over, they're over."

Did he owe Katzenberg money?

"Not in my mind," he replied.

Because Katzenberg felt otherwise, he quietly retained a noted Hollywood lawyer, Bert Fields. Katzenberg set a September 9, 1994, deadline for reaching an accord, and then amended it. While Katzenberg was at Disney World, a Disney lawyer asked Katzenberg's staff how fast he could vacate his offices—an inquiry that was blown up into "They asked him to leave!" So exclaimed a Katzenberg friend on September 8. Geffen warned that if Eisner didn't work out a settlement, "he's going to have to tell the truth under oath about everything." He added, "Eisner's lack of kindness, lack of generosity, and inability to give credit were simply shameful." Katzenberg, although he did not believe that Eisner had known of the lawyer's call about vacating his office, felt "debilitated," an associate says.

On Saturday, September 10, Eisner softened the hurt by visiting Katzenberg at home. Katzenberg told friends that he was determined to stop playing the role of supplicant; he told Eisner that the ball was in his court and that he would wait for a contract proposal, to which he would then give one of two responses—either "Yes" or "Talk to my lawyer."

What should have remained a private business disagreement instead escalated into a messy public divorce. The Disney split has unsettled Hollywood. "This is the biggest Hollywood story I've seen since I took over *Premiere,*" says Susan Lyne, who has been the editor of the magazine ever since she launched it, in 1987. "It is the end of the most successful twenty-year partnership in Hollywood. And it sent out ripples throughout the industry. Wherever Jeffrey goes, he will replace someone." And his departure will remind all those in the entertainment colony of the fickleness of their lives.

"There wasn't a person in this town who didn't look in the mirror and say, 'If it can happen to Katzenberg, it can happen to me,' " a Hollywood agent says.

There are also possible consequences for Disney. "It's a major loss," says Stanley Gold, who has been harshly critical of Katzenberg. "He's a talented fellow who helped drive that department"—the Disney Studios—"to new heights. He will be replaced, as all managers are replaced. But to suggest there is no loss? You won't get that from me."

While Joe Roth, Katzenberg's successor as studio chief, is respected in Hollywood, a longtime Disney producer who admires both Katzenberg and Eisner says, the loss of Katzenberg will be felt by producers and directors and writers, for they will now be confused about what projects to take to Disney. "Jeffrey imparted a worldview," he says. "It's what made the place work. If you couldn't put into a fifteen-second tease why people should see this, we shouldn't make it. It's another way of saying 'high concept.' "

The departure of Katzenberg could also have an impact on Disney's strategy. Eisner has been criticized for being too cautious about acquisitions, and certainly Katzenberg tried to get him to open Disney's wallet and buy a network to guarantee the distribution of Disney's television products. But Eisner has argued publicly that as long as access to the so-called information superhighway is available to all, Disney needn't invest in a delivery system. Whether the delivery system of the future is a phone wire or a cable wire, over the air or over direct-broadcast satellites, a computer screen or a wireless system, he insists, none will dare shut out Disney as long as it stays a software factory.

Perhaps as a way of demonstrating Disney's stature—of telling the world that his heart is strong and that Katzenberg's departure has been taken in stride—Eisner has made a major move in the last two weeks; according to three executives who say they know, Disney has entered negotiations with General Electric to acquire NBC. In urging me to wait with this story, Eisner suggested that if I did I would glimpse Disney's true strategy, because there were deals pending. "There are a lot of things changing in our company that will be helpful to you," he declared. But Rupert Murdoch, the Fox chairman, observed, "The problem for Michael is this: he has to keep making the same contribution to the company he personally made over the last ten years. And not try to do Frank Wells's job. Not that he'd be bad at it. But he can get people to do that."

It is also possible that Jeffrey Katzenberg will return as a competitor. But what studio job is open to him? Right now, there doesn't seem to be a studio vacancy. Besides, Katzenberg insists that he wants to run more than just a studio—an ambition that could limit his options. Murdoch, on being asked if he would hire Katzenberg for Fox, said, "He wouldn't come here except for my job!" But he added, "Anyone would love to have Jeffrey working for him." His loyal friend David Geffen said of Katzenberg, "I would trade fifty percent of my future for Jeffrey's in a second. I've offered to go into business with Jeffrey and be partners. It's one of the many things he will consider."

Nevertheless, as the recent aspirations of other entertainment executives have demonstrated, one needs deep pockets to be a mogul. If Barry Diller were a billionaire rather than a millionaire, he would probably own CBS or another network. Since Katzenberg is without really deep pockets, a friend of his said, "the problem is that he has to wait for the job that he wants, and then has to wait to be asked."

This Hollywood tale is another reminder that intangibles often matter more than tangibles like profit and loss or business strategy or so-called synergies. Jeffrey Katzenberg left a big job for reasons having to do more with psychology and personal chemistry than with performance. He wanted a bigger job and a better title, and those desires might have been finessed by Eisner. The head of a major studio remarked, "I would have said, 'Jeffrey, you can have the title. You are my designated number two. But don't stop making movies.'" But then Katzenberg would have had to finesse his rage. Another Hollywood figure said, "If I had been asked for advice, I would have told Jeffrey not to play hardball when Michael was vulnerable"—after Wells's death and Eisner's heart operation. "Jeffrey should have said, 'I'm here for you.' I would have told him that patience pays off."

"What's interesting to me about them is that as a dealmaker my job is to help preserve relationships," said a major Hollywood agent. "This one was a no-brainer. Easy to preserve. Yet I look at the train wreck and I say, 'How did it happen?' There was no mediating influence. There should have been, because these people genuinely like each other in work. This was not about business. This was about needs."

In any case, the die had been cast by midsummer. Katzenberg and Eisner had taken positions they couldn't back away from. As a result, Katzenberg left a job that he loved and Disney lost a talented executive that it didn't want to lose. Primal forces were at work, which could not be controlled by the mind. "Ambition is a force of nature," an admirer of both men comments. "It goes back to Richard III."

POSTSCRIPT:

I was mistaken in thinking that Katzenberg would have difficulty finding a mogul's job. Not long after this story appeared, Katzenberg, Geffen, and their friend Steven Spielberg announced the formation of a new Hollywood studio, Dreamworks SKG. Katzenberg was mistaken in his hope that a financial settlement could be amicably reached, and he announced that he would drag Eisner and Disney to court, seeking more than a hundred million dollars. Eisner hired Michael Ovitz to fill the job Katzenberg wanted, and asked him to try to reach an accord with Katzenberg. He could not.

Eisner tried hard to block this *New Yorker* article. After trying to convince me not to report it, he phoned Tina Brown, the editor of *The New Yorker*, in an attempt to convince her not to publish it. She told him that the decision as to whether or not there was a story here was my call. When the piece appeared, Eisner was furious. He wailed to associates that I was anti-Disney, a phrase he has used to describe other journalists as well. He complained that I had violated an off-the-record pledge, which, of course, is a journalistic felony. When I pressed him some months later to amplify this charge, he mumbled a general complaint but did not cite an example. As best I can figure, he was probably referring to the mention of the director whom he had privately denounced to me and then that very night publicly praised. Since I didn't name the director, I thought Eisner's confidence was not betrayed.

Katzenberg was also upset by this piece, though he was gallant when he phoned. He said that he was honestly trying to square his self-perception with the perception that he had been insensitive in pressing for the number two job right after Frank Wells had died and soon after Michael Eisner had endured a major heart operation. As a journalist, I have been able to conduct a fair number of subsequent interviews with Katzenberg, who enjoys a better press than Eisner, in part because he is a straight shooter. Eisner rarely returns my calls.

A few colleagues at *The New Republic* seized on the lead paragraph of this piece to poke fun at my propensity to drive around with a cellular phone and talk to people like Katzenberg on it. Actually, I didn't have a car or a cellular phone. The reason I began this piece with "My car phone rang" was that I thought in an article about Hollywood it might be funny to mimic Hollywood. I was on vacation when this story broke, and *The New Yorker* arranged for a car to drive me to the airport. My wife phoned this car to say the dinner was canceled, and then I called Katzenberg.

In the absence of Katzenberg, Disney has continued to thrive under Joe Roth. And in August 1995, Eisner did make the bold move Katzenberg had pushed: he purchased Capital Cities/ABC. He also surprised the entertainment world by luring Michael Ovitz to Disney as his number two. But in little more than a year, Ovitz was gone. Whatever Ovitz's failings, which I chronicled in an August 1996 *New Yorker* article, this failure was also Eisner's, for once again the Disney chairman did not—perhaps could not—share power with someone who also sought to take a bow.

13

THE BEHEADINGS
Successful and Unemployed

(The New Yorker, February 12, 1996)

On the afternoon of Tuesday, January 16, 1996, Frank Biondi's secretary told him that Sumner Redstone, Viacom's chairman, wanted to see him at three-thirty the next day. This was an unusual request, since Redstone usually popped unannounced into Biondi's office, which was just a few steps away from his own. "I knew then that something was going on," says Biondi, who at the time was Viacom's president and chief executive officer.

Redstone, who is seventy-two years old and is the principal owner of Viacom, recalls being extremely nervous as he entered Biondi's office and said, "Frank, there isn't any easy way of saying this, but I've reached the conclusion that our business relationship should be terminated. I just came to the conclusion that your vision and mine as to how the company should be led are incompatible." With that, Redstone did what Biondi had done when he fired Richard Snyder, the publisher of Simon & Schuster, in June of 1994: he handed him a draft of a press release announcing his dismissal.

Although Biondi was on the board of Viacom, he did not know that Redstone had conducted a secret board meeting that morning at the law offices of Shearman & Sterling. Nor did he know that since Christmas Redstone had been discussing firing Biondi with two executive vice presidents, Philippe Dauman and Tom Dooley, and the company's outside communications counsel, Ken Lerer. Biondi had been at Viacom for nine years, and Redstone had credited him with building it into the fourth-largest entertainment company in the world, behind Time Warner, Disney, and Bertelsmann. But some tension had crept into their relationship in the last few years. As Biondi says, "the dynamics" between them had shifted slightly in 1993, when Redstone selected Dauman, his longtime counsel, to succeed him as chairman. He knew that Redstone had been upset when Nickelodeon's president, Geraldine Laybourne, announced last December that she had been lured away by Disney, to become the president of Disney/ABC Cable Networks. Redstone hated losing her, and Biondi's cool detachment about the loss had annoyed him. Still, Biondi could not imagine that this was a factor, since Laybourne had told them both that Viacom could not match the broader opportunity Disney was offering.

Biondi also knew that it annoyed the work-obsessed Redstone whenever he left around six for the gym or a game of tennis, and when he took regular vacations. Dauman and Dooley and others would often go to dinner with Redstone, but Biondi felt secure enough to believe that he didn't need to schmooze with the boss, and didn't need to work all the time. Although they weren't pals, he believed that he and Redstone were a good fit: the disciplined manager and the brash entrepreneur.

On January 17, as Biondi sat across from Redstone, he only glanced at the seven-page draft press release, then responded in a typically subdued fashion. He recalls saying, "That's a surprise. Why?"

"You should know," Redstone replied.

"I don't," Biondi said.

"Talk to Tom. He should know," Redstone responded, referring to Dooley. Biondi said that he hadn't been able to find Dooley all day and had been told he was out of the office. Redstone insisted that he was there, and Biondi replied that he was not. "I'll go find him," Redstone said, and he then left and did not return. From four-thirty until midnight, Redstone recalls, he was on the telephone, trying to lessen the shock for dozens of people. One of the first calls he made was to Gordon Crawford, a senior vice president at Capital Research Company. Capital's clients own 8 percent of Viacom's stock.

As for Biondi, he phoned his wife, Carol. "I have good news and bad news," he remembers saying.

"What's the bad news?" Carol Biondi asked.

"We're not going to China," he said, referring to a trip that was to have been part business and part pleasure. "The good news is that I'm going to be able to try some new things, because I'm going to be leaving. Sumner just walked in and said he wants to take my job."

Frank Biondi is merely the latest in a lengthening line of entertainment and communications executives who have been forced out in recent months or years: Warner Music's chairman, Robert Morgado, last May, and his successor, HBO's chairman, Michael Fuchs, last November; the Disney Studios' chief, Jeffrey Katzenberg, in September of 1994; the Sony Corporation's American president, Michael Schulhof, in December of last year, and the Sony studio head, Peter Guber, in September of 1994; Simon & Schuster's publisher Richard Snyder in June of 1994; and the Paramount Communications president, Stanley Jaffe, earlier in 1994. Recent provoked or forced resignations include those of CNBC's president, Roger Ailes, who left in January of this year; Hearst Magazines' chief, Claeys Bahrenburg, who stepped down last November; and Fox's Barry Diller, who left the company in 1992. Biondi, like most of the others, was by objective measures a success. (Schulhof and Guber, whose financial performance was dismal, were exceptions.)

Under Biondi's leadership, Viacom had grown from a $1 billion company to a nearly $12 billion colossus. He had created a strong management team and had smoothly steered both Paramount and Blockbuster into the Viacom tent. Earnings were slightly lower than analysts anticipated in 1995, but Viacom's operating cash flow (before interest, taxes, and depreciation) was $2.3 billion, up $500 million from the previous year. Last year was "the best year any entertainment company had on the numbers. We were up twenty-five percent," Biondi says. He predicts that the company, including Blockbuster, will continue to enjoy double-digit rates of growth.

Morgado could claim a similar success. When he took over Warner Music, in 1985, the company's revenues were just under $1 billion. At the end of 1994, five months before he was dismissed, he was presiding over a $4 billion global music division that generated $720 million in cash flow for Time Warner. Yet Gerald Levin, Time Warner's chairman and CEO, told his board that Morgado had to go, because he had alienated many of the company's music executives, some of whom had departed. Michael Fuchs had overseen the growth of HBO, which went from six hundred thousand subscribers in 1976 to thirty million by 1995. A few months after Fuchs was promoted to manage the music division as well, Levin told the Time Warner board that he did not march in step with management. Barry Diller was the chief architect behind the successful building of the Fox network, in 1986. Jeffrey Katzenberg played a major part in the resurgence of the Disney Studios. Richard Snyder transformed a sleepy publishing house into a giant. In just a little more than two years, Roger Ailes helped transform CNBC from a start-up cable network that lost money into one that produced operating profits of some $50 million last year.

The reason Biondi was let go was not that he wasn't a team player, as was said of Richard Snyder. He was not accused of disloyalty, as Fuchs was, or of an inability to get along with some underlings, as Morgado was. No one believed that Biondi had pushed too hard to be promoted, as was said of Katzenberg, or that he was hapless, as was said of Schulhof. Biondi was fired for other reasons. "Frank is not an aggressive person," Redstone says. "He is not a person who seizes the day. That doesn't mean he's not a good manager." It just means, Redstone continues, that his skills were not appropriate for what the chairman calls "the new Viacom"—a company that has to analyze less and act more quickly than it did in the past. While expressing respect for Biondi's talents and integrity, Redstone criticizes him for lacking passion. "I think passion is important," Redstone says. "My passion is rationally based. It plays a role in leadership—to be able to instill excitement."

Common themes can be found in the firings of Biondi and many of his brethren, and they may have to do less with job performance than with the nature of the entertainment business (and human nature). The business has changed because of its "awesome and escalating concentration of power," Redstone says. And Levin, who in the past year fired Morgado and then Fuchs, says, "We're not in the furniture business. We're an idiosyncratic, ego-driven business. To spend two years turning out an album is idiosyncratic. The management of this business does not lend itself to traditional management practices."

Biondi, too, believes that the entertainment industry, where decisions are often made by people who operate on instincts and hunches, is unlike any other. "If there is a single unifying element, it probably has to do with the size of the stage and number of people on it," Biondi says. "It's an ego thing. . . . At Coca-Cola or Ford, executives are part of the fabric. They're a team. They do things the Coca-Cola way. In communications, you don't have this tradition and history. There's high turnover. There's not a lot of feeling of permanency, partly because the business has been reinvented the last fifteen years. Ted Turner was running a TV station in Atlanta fifteen years ago."

Howard Stringer, who was formerly the president of the CBS Broadcast Group and is now the chairman and CEO of Tele-TV, an effort by a consortium of telephone companies to provide TV programming, believes that the "fabric" in entertainment and communications has been replaced—in part, at least—by the will to power. "We are now entering the Napoleonic period of the entertainment business," Stringer says. "Everyone wants to be Napoleon—'I'm in charge.' Everyone wants to be perceived as uniquely powerful." Biondi, too, says, "You're seeing the emergence of the quasi-owner executive. Someone like Michael Eisner, much to his credit, became the embodiment of Disney."

Redstone, however, shrugs off the idea that ego plays a role. "I have passed the stage where I have to bolster my ego with news stories," he told me, adding, "I am not a stargazer." As he said this, I studied the photographs that line his office wall. They are all of celebrities, albeit of another era: Jerry Lewis, Gregory Peck, Jimmy Stewart, Edmund Muskie.

Geraldine Laybourne, who may be the most powerful woman in the television industry, says she never saw any sign of a Redstone ego. "Sumner never had trouble with me getting publicity, or with [Paramount Studio chief] Sherry Lansing or [MTV chairman] Tom Freston getting publicity." Yet there is some evidence that Viacom's public-relations department sought to be sure that Redstone received at least equal time with Biondi in all future articles; when the magazine *Wired* planned to put a photograph of Biondi on its cover, according to a top editor there, "the flacks at Viacom" insisted that Redstone be included. Viacom's spokesman, Carl Folta, does not deny this, he merely falls on his own sword. "Sumner Redstone knew nothing about this. It was our decision," he says. It is clear that the attention paid to other mergers rankles Redstone. "People forget—they're caught up with Disney and ABC or Time Warner and Turner, and they forget how much Viacom has changed," he told me. It is also hard to miss Redstone's own changed appearance. In recent months, he has replaced the rather casual look he favored with a new, coiffed hair style.

Geraldine Laybourne suggests that gender, too, may play a role in all the industry shifts. "I think men measure their achievements— and this is a gross generalization—by who's on top," she says. "You see a phenomenon like, 'OK, Rupert bought this. What am I going to buy?' Women measure their success differently. . . . It's not that women can't compete; it's just that we compete in different ways." Perhaps.

Gerald Levin believes that the speed of changes in technology, the publicity, and the nervousness of Wall Street all have conspired to create a new climate for the executives at the top, and make the executives more anxious to find associates who can make them feel comfortable. "In every case, it's a question of performance, if you define performance broadly enough," Levin says. "Performance can mean psychological tension. It can be like a marriage where people grow out of each other. Many of these conflicts involve a CEO's comfort level. It's not always black and white when someone should be fired. . . . The most important thing is the feeling of a relationship. Words like 'family' and 'marriage' are important. In doing deals, you need trust or they come apart. Wall Street often misses the point. . . . Anytime someone makes a logical business decision, go look at the emotions underneath."

This response surprised Robert Morgado, in the light of his firing by Levin last year. Since that happened, Morgado has used his severance pay, of roughly $60 million, to fund Maroley Communications, a media-investment company. When I passed along Levin's thoughts, I didn't tell him whose thoughts they were. "Comfort level with what?" Morgado replied. "What is comfort? Does it mean that someone wants to be comfortable so he doesn't feel threatened?" Wasn't it reasonable to expect, say, loyalty, from an underling? "I won't dispel that as part of the equation," he answered. "But if it's loyalty that springs from insecurity, then it breeds bad decisions. It breeds the plumbers of Watergate. It breeds survival at all costs." And, no doubt, it bred the classic case of insecurity that drove Gulf & Western's former CEO, Martin Davis, to drive out Paramount Studio executives like Barry Diller and Michael Eisner in 1984 because they were featured on the cover of *New York* and he was not.

Of course, the line between a neurotic and a justified action can be a fine one. Michael Eisner was, even his friends agree, jealous of the media attention Jeffrey Katzenberg received. But if he also honestly believed that Katzenberg lacked the financial and Wall Street skills to serve as COO, the position Katzenberg sought, then Eisner acted rationally. Although they had been colleagues for twenty years, Michael Fuchs did not hide his disagreements with Gerald Levin. Unlike many at Time Warner, he stabbed him in the chest, not the back. If Levin fired Fuchs because he felt threatened by a strong dissenter, then he behaved neurotically. If he sacked Fuchs because he was loudly undermining the company and its proposed merger with Turner Broadcasting, as Levin told his board, then he had to act. If Redstone couldn't bear sharing credit with Biondi, he might seek counseling. If he honestly believed Biondi was too cautious and careful to be CEO, he was justified.

Has a new era of beheadings arrived? "I don't think the percentage of turnover is any higher," the investment banker Herbert A. Allen says. "What's new is the intense publicity." Redstone, who since Biondi's dismissal has added the title of CEO to his letterhead, thinks that the new nature of the communications business inevitably speeds the turnover rate. "There are diverse themes within the various companies," Redstone says. "But there is probably a general, common theme, and that is, with . . . companies becoming bigger and bigger, with their operations becoming more and more diverse and more and more complex, and with the world in which they live becoming more and more complex, the issue arises in the minds of those who control these companies: Are the managers they have chosen equipped to face the challenges that are arising?"

Wall Street and the press also have altered the landscape. "I think we've compressed the period of time when things occur," Levin says. "Consolidation has brought more competition. And there always needs to be a next step. Most of these companies are financed through the stock market. They grow through the support of the market, and it's a much more unforgiving market. The press can bring someone down now, because perceptions are more important. They can affect the way a board feels or the way investors feel."

Even so, the entertainment business could not survive without risk—and risks hinge on hunches, not certitudes. Levin invested millions in a video-on-demand experiment in Orlando, Florida, maintaining that it would become the next rage. It didn't. The Internet did. Redstone complained that Biondi let Paramount make movies from screenplays that both he and Biondi didn't like, yet Redstone admits that he "loved" the studio's script for its $55 million Christmas offering, *Sabrina*, which was a box-office dud. In the end, CEOs struggle to guess right: Should they invest in direct-broadcast satellite or cable, long-distance or video services? Interactivity? High-definition TV or more channels? A studio or computer software? Global media companies must decide: Do they pour money into Western or Eastern Europe, Asia, South America, Africa, the Middle East? Do they gamble that local governments will allow them to act alone, or only with indigenous partners? And which partners?

Biondi, who worked at HBO and at Columbia before joining Viacom, says that what has changed is the level of uncertainty: "There's much more of a sense that 'I've got to do something.' There wasn't this sense in the eighties." And scared companies often act scared. In terms of growth and profitability, "these are not industries that are threatened," Jorge Reina Schement, professor of communications and information policy at Penn State, observes. "More fundamentally, these industries don't know where they're going, so there is a tendency to change people."

A source of their terror is the more pronounced role played by Wall Street. That's an audience that AT&T was playing to when it rushed to announce the layoffs of forty thousand employees. Wall Street is the audience Redstone had in mind when he made his quick call to Gordon Crawford, among others. And it is why CEOs spend much more time today then they did ten years ago making presentations to major institutional investors, or courting Wall Street analysts as if they were political delegates. Levin knows that if he doesn't get the price of Time Warner stock to rise—a price that has stalled despite two bull markets in six years—he may not be able to finance the expansion he wants, and he could be ousted. That creates insecurity, but it also encourages a tendency to treat the market as a more intelligent institution than it is.

The press reinforces this short-term preoccupation. CNN, for example, now frequently billboards its version of stock-market "Winners and Losers." Viacom is said to be a loser, because Blockbuster dipped for a couple of months and is threatened by video on demand. Video stores may one day be threatened by electronic delivery of movies, but that bleak threat is not an imminent one. Over the next four years, Blockbuster expects to triple its stores overseas, from 1,333 to 4,000.

On the Monday after Biondi was dismissed, his office, on the twenty-eighth floor at Viacom's headquarters, on Broadway, was vacant. He had been moved to another Viacom office, on Eighth Avenue; there, on the fifteenth floor, Biondi and his secretary were the only visible employees.

Biondi is too disciplined—and too intent on negotiating a comfortable severance agreement—to show any public anger at the way Redstone treated him. In his bare new office, he was a picture of calm, no doubt aided by the knowledge that he will walk away with a rich settlement. Nevertheless, this is the second time that Biondi has been fired. The first was in 1983, when he was the president of

HBO: his boss, Nicholas Nicholas, dismissed him because of strategic differences, and chose to replace him with Michael Fuchs, who was then Biondi's closest friend. "Having been through this thing once before, I view this somewhat less competitively," Biondi says. "When I left HBO, I felt I had to prove them wrong. This time, I don't feel that way. It was a great run." But it must be humiliating for a man who grew up in a home where his father, a scientist, and his mother, a strict housewife, instilled in him the belief that merit wins. By every measurement he knew, Biondi had twice proved his merit, yet he was twice treated as if he had failed.

There are some small consolations. When Fuchs replaced Biondi at HBO, the move nearly wrecked their friendship. That both men have now been fired, just months apart, appears to have cemented a new bond between them. One of the first calls that Biondi got was from Fuchs, to say that when he heard the news on the radio he almost crashed his car, because "I was so shocked and distracted that I coasted past a stop sign and into the middle of an intersection." Fuchs invited the Biondis to dinner at his home, in Bedford, New York, the Saturday after Biondi was dismissed, and raised a glass to toast him. "This completes the cycle," Fuchs says. "We'll never be rivals again. Who knows? Maybe we'll be partners."

There's also a chance Biondi will be asked to become CEO of a company like MCA. On January 26, a week after he was fired, he had lunch with Edgar Bronfman, Jr., the CEO of Seagram, which acquired MCA last year. Biondi, a top Seagram executive says, is a prime candidate for that post.

Reflecting on all the carnage he has witnessed over the past several years, the second-ranking executive of an entertainment giant, who is nervous that he may be next, searches for a historical comparison. Finally, he blurts, "The entertainment business is like the French Revolution. All the beheadings are public. You get paraded through, and then you get beheaded. It's embarrassing. But the fact that you're paid so much money and there are so many others who are being beheaded has to make you feel a little better."

POSTSCRIPT:

Frank Biondi was quickly hired as CEO of MCA, which competes with Viacom. Michael Fuchs, as of early 1997, had not yet landed the kind of top entertainment position he seeks, no doubt because his blunt manner is perceived as evidence that he is not a team player. Barry Diller let me know he was annoyed to be included among successful but beheaded executives.

Diller insisted he was not forced out. In one sense, this is true: he was not terminated, as Biondi, Fuchs, and Snyder were. But he was forced out. Diller asked for a bigger piece of the candy store, and Murdoch refused. Murdoch, as was later true of Redstone, decided he wanted to run the candy store himself.

Since becoming CEO, Redstone has not been noticeably more aggressive than Biondi was, at least measured by the scant number of deals Redstone has made. In truth, a reason Redstone executed Biondi is that he was modeling himself after Murdoch, a man everyone in the industry has their eyes glued on. At a July 1996 industry panel at Allen & Company's annual summer camp for media moguls, Redstone appeared on stage with Ted Turner, John Malone, and Robert Wright. The subject of their panel was advertised as the future of communications, and a moderator began the discussion by asking: What is the number one issue communications companies should worry about over the next twelve months?

Redstone, Turner, and Malone weighed in with solemn or rehearsed answers. However, when the moderator asked NBC's Wright this question, Wright paused dramatically for effect, scanned the two hundred and fifty or so familiar faces, then declared, "That's not the question! The question is: Where's Rupert?"

14

THE PIRATE
Rupert Murdoch

(*The New Yorker*, November 13, 1995)

When Rupert Murdoch arrives at his office on the Twentieth Century Fox lot, in Los Angeles, the single television is set to CNN, and the sound is off. At about eight-thirty, Dot Wyndoe, Murdoch's assistant of thirty-three years, spreads newspapers out on a shelf across from his desk: "TOP COP NICKED MY WIFE," blares one of his London tabloids, the *Sun;* "LOVE JUDGE COMES HOME," screams the *New York Post.* These are the only noisy elements in the office. Even the phones are quiet. Numerous calls stack up and are announced on an electronic monitor at his right elbow. Pastel paintings by Australian artists hang on snow-white walls over white couches and armchairs.

Murdoch's hair is turning white. He is sixty-four, and his shoulders stoop slightly. His voice is soft, and his manner is unfailingly courteous, as he sits with one leg tucked under the other. Only the hard brown eyes suggest that he is a predator.

He spends most of his time on the phone. He phones while driving his BMW to work, and he starts phoning from his desk as soon as he arrives, about seven A.M. He always apologizes for disturbing an employee at an odd hour or on vacation, but the apology is more a ritual than a sign of genuine contrition. He hardly ever says hello or goodbye.

Murdoch lives in Beverly Hills, in a Spanish-style house, on six acres, formerly owned by Jules Stein, the founder of MCA. He's often away, doing business out of a Gulfstream jet he owns, or from one or another of the offices he keeps in several cities, including New York, where he works from a high-rise on Sixth Avenue near Times Square. This office is also white, and it has seven silent TV sets.

"Eric, sorry to wake you," Murdoch was saying from the New York office one recent afternoon. It was before dawn in Australia, and he had reached Eric Walsh, a lobbyist based in Canberra, who has been advising him on how to get the Chinese government to sanction his Star TV satellite system. "I was going to go to Hong Kong for a week, but I've been invited for a one-day conference in Beijing," he told Walsh. Murdoch wanted to know if he should reach out to the Chinese. And, if so, when? "Do I need an alibi?" he asked Walsh. "Or can I just write and say, 'I need a half hour of your time'?"

Walsh cautioned Murdoch to stay in the background. "Nobody understands what's going on in there," Murdoch mused in response, slipping down further in his soft leather chair, until his head was not much higher than his desk. "Everyone has different readings. You just never bloody know."

Without the support of the Chinese government, Star TV can have no paid subscribers, and advertisers stay away. That situation translates into big losses—projected at eighty million dollars in the current fiscal year, Murdoch says. He owns a satellite service that can potentially reach two-thirds of the world's population, yet, because of widespread concern in Asia about "cultural imperialism" and the impact of uncensored images and information, he has had to curb his aggressive tendencies. "My Chinese friends tell me, 'Just go there every month,'" he says. "'Knock on doors. It may take ten years.'" This is not the message that Murdoch wants to hear.

Murdoch is the chairman, CEO, and principal shareholder of a company, the News Corporation, that produced nearly nine billion dollars in revenue this year and more than a billion in profits, but he feels frustrated. He is frustrated by China. He is still frustrated because he has no international news network to supply his Fox network here, or his Sky or Star satellite services in Europe and Asia, while Ted Turner has CNN. And until last month he was frustrated because he owned the rights to televise sporting events all over the world but didn't own a sports network, like ESPN.

Overall, however, these frustrations present mere skirmishes in a global war that Murdoch is winning. All the media deals and maneuvers of the past few months have come about, in part, because Disney and Time Warner felt that they had to catch up to Murdoch. "He basically wants to conquer the world," says Sumner Redstone, the chairman of Viacom. "And he seems to be doing it." Former press lords, like Lord Beaverbrook, William Randolph Hearst, and Henry Luce, were dominant figures in a single medium on a single continent. Rupert Murdoch's empire spreads across six continents and nine different media: newspapers (his company owns or has an interest in a hundred and thirty-two); magazines (he owns or has an interest in twenty-five, including *TV Guide*, which has the largest circulation of any weekly magazine in the United States); books (HarperCollins and Zondervan, the dominant publisher of religious books); broadcasting (the Fox network; twelve TV stations in the United States; 15 percent of the Seven network, in Australia; and Sky Radio, in Britain); direct-broadcast satellite television (Star TV, in Asia; 40 percent of BSkyB, in Europe; half ownership of Vox, in Germany; and a yet to be named joint venture with Globo, in South America); cable (the fX network, in the United States; Canal Fox, in Latin America; and half ownership of Foxtel, in Australia); a movie studio (Twentieth Century Fox); home video (Fox Video); and on-line access to the worldwide Internet (Delphi). Murdoch not only reaches readers, he also holds the electronic keys to their homes.

Murdoch moves more swiftly than most rivals, takes bigger risks, and never gives up. In fact, despite public denials he has made, in late October 1995 he was still contemplating ways to reverse the deal that Time Warner announced in September to buy Turner Broadcasting System—a deal that blocked some of Murdoch's own expansionist plans. One scheme that he discussed internally and ordered his bankers and lawyers to dissect carefully was to attempt a takeover of Time Warner, then valued at more than forty billion dollars. "We're working hard at it," a central figure in the News Corporation said in late October, days before Murdoch became convinced that the effort would fail. The impediment, two participants say, was not finding partners but figuring out how to avoid the steep capital-gains taxes on the sale of Time Warner's various pieces to eager buyers. The idea was to make a bid for Time Warner in the next few months, before the merger with Turner was consummated. There were internal discussions about such potential partners as the Bronfmans, whose Seagram already owns just under 15 percent of Time Warner; U S West, which owns 25 percent of Time Warner's entertainment assets and opposes the terms of the Turner merger; and John Malone, the president and CEO of Tele-Communications, Inc., the world's largest cable company.

Murdoch created the first global media network by investing in both software (movies, TV shows, sports franchises, publishing) and the distribution platforms (the Fox network, cable, and TV satellite systems) that disseminate the software. Within the next few years, the News Corporation's satellite system will blanket South America, in addition to Asia and Europe and parts of the Middle East and Africa. "Basically, we want to establish satellite platforms in major parts of the world," Murdoch explains. "And that gives us leverage here." If a cable-box owner or a programmer—John Malone or Time Warner, for instance—wants to reach a foreign market covered by one of Murdoch's satellite systems, Murdoch can extract favors from the programmer in the markets it controls in the United States. "What we're trying to do is put ourselves in a position in other countries that some of these cable companies are in in this country," he candidly says. He wants to be the gatekeeper.

To advance his grand plan, Murdoch arranged a summit meeting in suburban Denver on August 10 with John Malone, of TCI. Murdoch called the meeting because he believed that Malone was necessary for capturing two of the missing pieces in his empire—the sports network and the news network. Malone owns fifteen regional cable sports channels and is a partner with Charles Dolan's Cablevision Systems in other regional sports channels around the country; these, when joined with Murdoch's and Malone's overseas sports holdings, could become the foundation of an international sports network. In addition, Malone was Ted Turner's most influential shareholder and could link Murdoch with CNN. Murdoch was assuming that Turner was a possible partner, for he knew that Malone was openly dismissive of Time Warner's management.

The day before the meeting, Murdoch summoned to his California office Chase Carey, the chairman and CEO of Fox Television, and Preston Padden, the president of network distribution for Fox and the president of telecommunications and television for the News Corporation. Murdoch, from behind the oak table he uses as a desk, began by noting that at that moment Malone was trying to help Turner finance a bid to acquire CBS, which would compete with Fox. "I think he intends to screw us," Murdoch said.

However, Murdoch knew that Turner and Malone needed cash to finance a network acquisition, and this might prove to be his opening. "The first thing we do," Murdoch told his colleagues, "is ask: 'Do you have CBS in place?' If not, we ask: 'Is there an asset you are willing to sell to us?' "

Murdoch wanted CNN, and he wanted a sports link with Malone and Charles Dolan, Malone's partner in six regional sports networks.

"If you could convince Turner to partner with us and TCI, the game is over worldwide!" said Padden.

The three men held another caucus the next day, huddling in green upholstered armchairs on Murdoch's Gulfstream as it headed toward Denver for the meeting, at 2:00 P.M. They knew that Malone and Turner needed cash to take over CBS, and they talked about how Turner might be encouraged to sell something.

"The only asset we'd be interested in is CNN," Carey said.

Padden asked what might happen if Fox became a one-third owner in CNN.

"You don't want to hand over all your news efforts to Ted," said Murdoch, who is as wary of Turner's being too liberal as Turner is of Murdoch's being too conservative.

A car waited at the Denver airport to take Murdoch, Carey, and Padden to TCI headquarters. The meeting was held in a stark, glass-walled conference room dominated by a black granite oval table. The only refreshments were a few cans of diet soda and a thermos of coffee on a granite sideboard; each corner of the room was occupied by a rubber plant. Malone, attired in a red-and-white-checked short-sleeved button-down shirt and chinos, was careful not to sit at the head of the table. Flanked by two TCI executives, he took a seat across from the Murdoch trio. The table was bare except for a single folded sheet of paper in front of Malone.

The six men, who did not rise from their seats for the next four hours, began with industry gossip, which soon bored Murdoch. He changed the subject: "Are you getting it together for Ted and CBS?"

"We know where the money will come from," Malone answered. "The question is Time Warner. I think their strategy is to drive Ted nuts." Time Warner, which, like TCI, then owned about 20 percent of Turner Broadcasting, had the right to veto all acquisitions, and it was thwarting the takeover of CBS. Malone said, contemptuously, that he thought Gerald Levin, the Time Warner chairman and president, was seesawing, because his management was engaged in an internal war.

Rather offhandedly, Murdoch asked, "Does Ted want to sell any assets?"

"No," Malone replied, with equal aplomb. "I don't know of any asset Ted wants to part with." He knew that Murdoch was fishing, hoping he could hook CNN.

Privately, Murdoch describes Malone as "the most brilliant strategist" he knows. Malone says the same thing about Murdoch, and over the years they have been both adversaries and allies, depending on the venture. It's difficult to keep track of partnerships in the communications business these days, because the players change sides depending on the country or on the deal.

Malone had several goals in this meeting. He wanted to see if there were areas where he and Murdoch could do business together, and he wanted to avoid conflicts. Both were interested in creating a sports network that could compete with ESPN, and he thought he could get Murdoch to invest in the regional sports channels that TCI shared with Charles Dolan's Cablevision.

"The fact is," Malone told Murdoch during their summit, "we tried to merge our sports with Chuck's maybe fifty times." Each time, the plan unraveled because Malone and Dolan could never agree on who would run the partnership. But Malone was convinced that if a third party came in—someone "who may be more objective than we can be with each other," perhaps Murdoch—a successful merger could be achieved. "Between us, we have all the regional sports networks in the country controlled—except Minneapolis," he said, and that came to thirty-three million subscribers. "The real value for us," he went on, "is putting these together," thereby creating a powerful entity for buying worldwide sports rights, as ESPN does.

Malone and Murdoch then began a *tour d'horizon*, starting with a review of what each of them was doing to reach the estimated seventy million television households in Central and South America, and how they could combine their own efforts and those of their separate satellite partners. "If we could get the two interests aligned in separate spheres"—one aimed at Mexico and Central America, the other at Brazil and much of South America—"and you guys ran it, I got to think that's the trump card in Latin America," said Malone. He proposed to be the outside link in South and Central

America—"the rubber joint," as he referred to it—that he wanted Murdoch to be between Malone and Charles Dolan.

The conversation soon switched to Europe. "That's the next area where we have to see if we can avoid conflict," said Malone, aware that he was dependent on Murdoch's thriving BSkyB satellite delivery system and the movies and other programming that Murdoch had locked up through long-term deals with the Hollywood studios that needed to pass his tollgate to reach TV sets. Malone did most of the talking. "I guess at this point Europe is pretty stable," he said, adding, "The only other thing I could think of in Europe was the Travel Channel. Do you guys have the Travel Channel on Sky?"

"Yes, but we don't own it," said Murdoch.

"We don't own it either," said Malone, who would like to. "We see it as an opportunity for interactive stuff. We want to get involved. Gates, in particular, is very interested in it," he said, referring to Bill Gates, the chairman and CEO of Microsoft.

"How are you doing in that partnership?" asked Murdoch.

"About like ours!" shot back Malone, and they both laughed. "Some good days, some bad days. The area of conflict is something called, 'At Home,' "—a high-speed cable connection to the Internet that will allow the user to instantly download information. He added, "Bill was [also] a little taken back that we were an investor in Netscape"—the software company that developed a product that simplifies access to the Internet. But Malone remained confident of a solid relationship with Microsoft. "The real question of our relationship with Microsoft," he said, "is what the government will allow without triggering an antitrust action."

A short time later, Malone said, "The thing we haven't talked about is Japan." He barely moved as he spoke, keeping two fingers pressed to his temples. "We have a deal there in sports, and we have to discuss how that relates to Star."

"Star isn't there at the moment," Murdoch said. "But we are pretty far down the road." He asked how Time Warner was doing in Japan.

"They're running way behind us," Malone said. "They came to us and asked, 'Why don't we do a joint venture?' " He went on to say that he had rejected this for two reasons: first, it was dumb politics, since most nations carefully limit foreign ownership; second, "Time Warner is brain-damaged in terms of making decisions. You've got to move fast." Malone then returned to sports: "All these pieces fit, Rupert. We ought to be able to arrive at this"—a sports merger—"in painless fashion." He added, "We think Asia will take a while."

"Japan is now," Murdoch said.

"Either you guys get in or ESPN will own sports," Malone warned.

"We got to go in and kick ass," Murdoch agreed.

Dolan was difficult to negotiate with, Malone warned.

"If we have to go without him, we'll go without him," Murdoch said. They could always fold in Cablevision later. What Murdoch needed was some numbers from Malone.

The single sheet of paper slid across the table. It contained, Malone explained, what TCI thought each of its fifteen regional sports channels was worth.

Afterward, I asked Murdoch and Malone separately if there was anything they had discussed alone. Only one thing, Murdoch replied: in a private moment, Malone had "tried to talk me into buying five percent of Time Warner to put it in play." Malone also conceded that they had talked about Time Warner.

Murdoch thought that he had accomplished half his mission. He was confident that he and Malone would make a sports deal, but he felt that CNN was slipping away, not least because Turner would probably spurn him. Turner, like a fair number of communications-company CEOs, thinks that Murdoch is a pirate.

CNN became even further out of reach the following month, when Time Warner announced a merger with Turner. This was a real blow. Murdoch and Malone had been outmaneuvered. Malone no longer sat in the cockpit with Turner. His role had been reduced to that of a passenger and an investor. "The picture is confusing," a subdued Murdoch told me in late September. "Strategies are changing day to day." He partly blamed Jane Fonda, who is Turner's wife. "I surmise Jane had a great hand in it," he said. " 'We can change the world together.' "

Despite the setback, Murdoch and Malone's meeting did produce some concrete results. Murdoch chased Charles Dolan, and when, as Malone had anticipated, they could not reach an agreement, the News Corporation and TCI negotiated, and on October 31, 1995, announced that they had agreed to become partners in a new, world-wide Fox sports network to compete with ESPN. They also have tentatively agreed to become partners in a direct satellite system in South America.

Malone sees Murdoch's strategy as much broader than that of his foes. He is not just trying to "get big," Malone observes. "He sees the nexus between programming and platform." Unlike Disney and Viacom, which have both concentrated on programming, Murdoch owns satellite distribution systems that span the globe. And, unlike his competitors in the satellite business, he owns a programming factory—Twentieth Century Fox. Competitors like NBC own neither. And TCI, Malone admits, has cable platforms but is "too weak in programming." Furthermore, unlike Time Warner and Viacom, Murdoch doesn't have a lot of debt: he has several billion dollars of investment capital available. And, unlike any other communications giants except perhaps Sumner Redstone and Bill Gates, he has a controlling interest in his company. "Rupert is a bit like a painter with a canvas, but in his mind the canvas has no perimeters," Arthur Siskind, the general counsel for the News Corporation, observes. "He's going to keep painting and painting."

Murdoch is a pirate; he will cunningly circumvent rules, and sometimes principles, to get his way, as his recent adventures in China demonstrate.

Sometime in 1992, Murdoch took his initial look at Star TV, which was then a five-channel satellite service operating out of Hong Kong and reaching fifty-three countries. He wanted to enter into a partnership with the owner—Li Ka-shing, a Hong Kong billionaire, who was on good terms with the Chinese government, and with Li's son, Richard, who ran the company on behalf of his father—but only if the News Corporation could manage Star.

Murdoch instantly saw that Asia—particularly China and India—was like the American West in the last century: the next frontier. The size of the potential market was measured in billions, not millions. He calculated another advantage. "If we learned anything from Sky," says Arthur Siskind, "it's that there's a tremendous advantage in being first in a market. It means everyone is playing catch up." The first in gets to establish a brand name. Within two weeks, and without going to his board, former Fox executive vice president and counsel George Vradenburg remembers, Murdoch made a half-billion-dollar decision. In July of 1993, Murdoch offered $525 million for 64 percent of Star, and that led to an agreement.

Murdoch was confident that Li and his son could run political interference from China. Others were less certain. Robert Wright, the president of NBC, recalls that one reason NBC had not pursued a deal with Star was the fear of opposition from the Chinese government. Perhaps Murdoch, too, should have worried more about this. The Chinese were still smarting from an abortive attempt he had made earlier that year to buy 30 percent of Hong Kong's largest broadcaster, TVB, despite a law stipulating that no foreign entity could own more than 20 percent of a communications company if it also owned another media outlet, and Murdoch did—the *South China Morning Post.*

Murdoch seems to have thought that he could get around the Chinese authorities' hostility toward his ownership of Star because technically he did not need the sanction of *any* government to deliver pictures from space. Besides, Murdoch is supremely confident—to the point of arrogance, perhaps—that he can get what he wants. "Rupert figured he could find a way to deal with China," George Vradenburg recalls.

If it had not been for the Speech, Murdoch might have found a way. On September 1, 1993, he invited hundreds of advertisers to Whitehall Palace, in London, and gave a speech explaining why the News Corporation was at the cutting edge of the communications revolution. He declared that George Orwell was wrong. "Advances in the technology of telecommunications have proved an unambiguous threat to totalitarian regimes everywhere," he said. "Fax machines enable dissidents to bypass state-controlled print media; direct-dial telephony makes it difficult for a state to control interpersonal voice communications. And satellite broadcasting makes it possible for information-hungry residents of many closed societies to bypass state-controlled television channels."

A month after the Whitehall speech, the Chinese prime minister signed into law a virtual ban on individual ownership of satellite dishes, and a suddenly chastened Murdoch was forced to show solicitude toward a totalitarian regime. He consulted many experts. He moved to Hong Kong for six weeks, and he reached out to Chinese officials. "He analyzed the problem objectively," his public-relations adviser, Howard Rubenstein, recalls. "He didn't make any excuses. It was an error that he made."

Murdoch had spoken like the libertarian he has professed to be, but now he chose to be reeducated, and among the lessons he learned was how deeply the Chinese government detested the BBC, whose World Service news was carried on a Star channel. The regime was especially angered by a BBC documentary that investigated Chairman Mao's unorthodox sexual habits. And, since the BBC was British, it was seen as the ally of forces that sought to keep Hong Kong independent of the mainland.

In April of 1994, Murdoch removed the BBC from the Star network in China and replaced it with Chinese-language films. "The

BBC was driving them nuts," Murdoch says. "It's not worth it." The Chinese government is "scared to death of what happened in Tiananmen Square," he says. "The truth is—and we Americans don't like to admit it—that authoritarian countries can work. There may have been human-rights abuses in Chile. But that country under Pinochet raised living standards. And now it has a democracy. The best thing you can do in China is engage the Chinese and wait."

How does Murdoch explain his new tolerance for dictatorships? "I'm not saying they're right," he replies. "You don't go in there and run a controlled press. You just stay out of the press."

As he did with the BBC?

"Yes," he replies. "We're not proud of that decision. It was the only way."

Murdoch was not the only media potentate to grovel to a repressive regime. In Asia at about this time, Ted Turner, after meeting with Chinese broadcasting officials, criticized the United States for trying to "tell so many other countries in the world what to do" about human rights. And, when the former prime minister of Singapore and his son, the deputy prime minister, threatened to sue for libel because of an article in the *International Herald Tribune* suggesting that the deputy prime minister had been appointed to his post in an act of nepotism, the paper, which is owned by *The New York Times* and *The Washington Post*, issued an abject public apology.

Murdoch's campaign to win over the Chinese authorities was multifaceted. It included replacing Star's English-only format with programming in Chinese; buying sports rights to badminton and other sports popular with the Chinese; joining with four international music companies to create a music channel for local Chinese talent; becoming a partner with the Tianjin Sports Development Company in the construction of four television studios and post-production centers; starting a pay-TV channel in Mandarin; and putting up $5.4 million for a joint investment with the Communist Party organ, *People's Daily*, to provide, among other things, an on-line version of the kind of dull newspaper that he would never publish in the West.

Last winter, Basic Books, a division of the News Corporation's HarperCollins, brought out a hagiography of Deng Xiaoping by his youngest daughter, Deng Rong, who wrote under her nickname, Deng Maomao. Ms. Deng has long served as her father's personal secretary, and is married to He Ping, the head of Poly Technologies, one of the country's largest military conglomerates. The Harper-Collins catalogue announced that the book by Ms. Deng, who Murdoch said received an advance of ten or twenty thousand dollars, would benefit from promotional efforts that would be coordinated with News Corporation media outlets. Despite the meager advance, Murdoch fêted Ms. Deng as if she were Tom Clancy. Last February, he attended a book party on the top floor of the Waldorf-Astoria sponsored by HarperCollins. He gave a private lunch for her at his ranch in Carmel. And he and his wife were hosts at a dinner for her in a private dining room at Le Cirque. There were six tables of eight, and among those who attended were Cyrus Vance and the Chinese ambassadors to Washington and the United Nations. After a four-course meal, Murdoch rose and, a guest recalls, toasted Deng Xiaoping as "a man who had brought China into the modern world." This summer, Murdoch was confident enough of victory in China to have the News Corporation invest $300 million more to buy the 36 percent of Star TV that it did not already own. In a further development that perhaps reflects Murdoch's solicitousness toward the Chinese government, in September HarperCollins dropped out of the bidding for a book by the Chinese-American human-rights activist Harry Wu, who is despised by the Chinese government.

Murdoch's decision to boot the BBC out of China was condemned as "the most seedy of betrayals" by Christopher Patten, the governor of Hong Kong. To espouse freedom of speech at home "but to curtail it elsewhere for reasons of inevitably short-term commercial expediency" was, he implied, immoral—or, at least, amoral. Some businessmen have defended Murdoch. "Being tough is sometimes confused with being amoral," Sumner Redstone says. "They are not the same thing. He's doing what he has to do to solve his problems with the Chinese government. He's being smart to do that." It's just business, in other words.

But some of Murdoch's colleagues demur. "When matters of principle and expediency clash with Rupert, expediency wins every time," says Frank Barlow, who, as the managing director of the London-based media conglomerate Pearson, is chairman of the *Financial Times* and oversees the Penguin publishing group. "I don't think we would have dumped the BBC. From time to time, the *Financial Times* gets banned in some countries. But that doesn't alter our approach. We published the Salman Rushdie book. And stood behind Salman Rushdie." Joe Roth, who successfully ran the Fox studio for Murdoch and has called him a "visionary," nevertheless suggests that he can be coldly amoral. "I think of him in business as a guy who will do whatever he needs to do to get it done."

. . .

Murdoch's detachment can be traced to his childhood, in Australia. His father, Keith Murdoch, was a national hero, a war correspondent in World War I who first reported the slaughter of Australians ambushed by Turkish soldiers at Gallipoli. The massacre, which many Australians blamed on inept British generals, bolstered the anti-British sentiments of the Murdoch clan, which had fled Scotland in the nineteenth century to settle in Australia. After the war, Keith Murdoch was appointed an emissary to London by the Australian government. He became a confidant of British press baron Lord Northcliffe, the first successful British proponent of shrill headlines, fervent reader contests, and giving customers news they wanted rather than news some editor thought they needed to know. Keith Murdoch became a protégé, and returned home to edit the *Herald*, a daily broadsheet. With success, the *Herald* expanded to other properties, and Keith Murdoch, now chairman of its board, was soon running Australia's first newspaper chain. He would battle, as his son would decades later, the Packer and Fairfax newspaper families.

Sir Keith and Lady Elisabeth Murdoch (he was knighted in 1933) had four children, Rupert being the second. He and his three sisters were raised in a large colonial house with sunken gardens, tea parties, ponies, and a tennis court. Their parents tended to be withholding of praise, particularly their mother, who dominated the household while Keith Murdoch was out building an empire. "We've gotten closer as she gets older," her son says today. "He was the soft one. My mother was the stern one. She felt she had to balance his softness." The boy's life was one of leisure and discipline, of rebellion and rectitude. Young Rupert had traits associated with his mother's Irish and his father's Scottish heritage. "He contains within his character both an extraordinary gambling instinct and a certain dour puritanism," a biographer, William Shawcross, has written. "Perhaps that is not so surprising. One of his grandfathers, Rupert Greene, was a roistering, charming half-Irish gambler. The other, Patrick Murdoch, was a stern pillar of the Free Church of Scotland. Traces of each man are found in Rupert Murdoch's conflicting nature." Young Rupert liked to gamble on coin flipping—called "two-up"—and on horses. "I enjoyed the danger," he remembers.

As a boy, young Rupert felt apart. He was sent to a boarding school, Geelong Grammar, near Melbourne, which featured military discipline, cadet parades, and occasional canings, and he rebelled against all of them. He was teased a lot at school, he recalls.

"I was always a bit of an outsider. That undoubtedly came from my father and his position." The sons of the landed gentry at Geelong considered journalism a low-rent business, populated by people who snooped on their families. He and classmate Richard Searby thought of themselves as Marxists, dedicated soldiers in what they expected would be a lifelong struggle against poverty and oppression.

Young Rupert's battles with the establishment continued at Oxford's Worcester College, where he and Searby enrolled in 1950. Sir Keith was now sixty-three, and had a frail heart. Uneasy that his son was a lackluster student and foaming radical, Sir Keith asked a gifted young Oxford Fellow in social history, Asa Briggs, if he would serve as Rupert's "moral tutor." Although separated by ten years, the two men had much in common. Rupert's attraction to leftist politics and aversion to the stratified British establishment was shared by Briggs, who was the first in his working-class family to attend college and was close to several Labour Party figures. Murdoch was an average student. "He could be quite lazy," recalls Briggs. Murdoch focused more on politics. A bust of Lenin glared from his mantle.

Briggs believed Murdoch had a promising future. "I think I'm the only person in Oxford who saw it. He had a sort of vision," recalls Briggs, who became one of the world's foremost authorities on broadcasting and would remain one of Murdoch's mentors. Murdoch spoke often to Briggs of a career in journalism, in broadcasting as well as print. He struck his tutor as daring, opinionated, as someone who, in Briggs's words, "always had to have an enemy."

In early October of 1952, Keith Murdoch died in his sleep. Rupert was "shattered," Briggs reports. The journey home took three days, and Elisabeth Murdoch did not wait for her son to arrive before burying her husband.

Rupert returned to Oxford at his mother's urging to get his degree and to gather journalistic experience as an editor on Lord Beaverbrook's *Daily Express*, but within a year he rushed home. The son had always thought of his father as a press lord, but in fact he had been employed by a newspaper chain; Sir Keith had owned only two newspapers, the *Adelaide News* and the *Brisbane Courier-Mail*, and the Brisbane newspaper was sold to pay taxes. So in 1953, the Murdoch empire consisted of a single newspaper with a daily circulation of under a hundred thousand, and its competitors were intent on putting it out of business. Rupert had "an enemy" and went to war, which even then he perversely enjoyed. Though he was only twenty-two, his ability to focus was a major weapon. "He had few social graces, but when required or when it suited him, he could turn on the charm," wrote Sir Norman Young, a lifelong admirer, in a memoir. "He had little or no small talk and had virtually nothing in common with his peers." He revealed nothing. "I learned early that getting very close to people can be dangerous if you're going to be in a position of public responsibility," Murdoch says today. By twenty-four, he ruled the company.

He married an airline stewardess, had a daughter, and spent most of his waking hours scheming to make his newspapers more popular, more sensational. By the end of the sixties, Murdoch's newspaper operation in Australia was at least the equal of the Packer and Fairfax chains, and exceeded them in one respect. In 1964 he started *The Australian*, the first national newspaper. Unlike many of his other properties, this broadsheet aimed to elevate journalism. Few thought Murdoch would succeed. The country was too vast, too sparsely populated, too little interested in fine journalism. Even fewer thought Murdoch would persist. He surprised detractors by running editorials that railed against the Vietnam War, and by waiting twenty years to earn his first profit on the paper. By the end of the sixties, Murdoch had divorced his wife and married a cadet reporter on his *Daily Mirror*, Anna Torv.

Before Murdoch was thirty he had acquired a number of newspapers and a TV station. He became a British press lord in early 1969, when he bested Robert Maxwell for the *News of the World*, a London Sunday tabloid with a circulation of approximately six million. Murdoch employed his charm to induce the Carr family, which had controlled the paper for nearly eighty years, to spurn a richer offer from Maxwell, whom the Carrs, like others, thought an odious man. Murdoch was seen as the safe choice, and he pledged to run the paper in partnership with the Carrs. But, as with so much of Murdoch's life, his acts are open to multiple interpretations. He argues that his pledge to the Carrs conflicted with his responsibility to shareholders. "It didn't take me long to realize that it was a total wreck of a company," he explains.

His treatment of the Carr family intensified an impression that would grow and forever shadow Murdoch: that his word is counterfeit, and he can't be trusted. The impression is widespread that people in journalism and business help him at a critical point and are then discarded. Eight years after he bought the *News of the World*, he started courting Dorothy Schiff, the owner of the *New York Post*. They had been brought together by Murdoch's friend Clay Felker, the editor and founder of *New York*. Murdoch induced her to sell the

Post to him, and he transformed it from a dull liberal paper into a racy conservative one. Around that time, Felker confided to Murdoch that he was having difficulties with his board. Weeks later, Murdoch betrayed Felker by going behind his back and acquiring *New York*. Murdoch said at the time that the board wanted to sell the company because it thought that Felker was profligate. Most of the editorial staff—including this writer—sided with Felker and quit.

Murdoch's sudden prominence in America attracted notice. *Time* and *Newsweek* featured him on their covers. He was widely reviled. *MORE*, the journalism magazine, headlined a story "Killer Bee Reaches New York," accompanied by a David Levine illustration of Killer Bee Murdoch. *The Columbia Journalism Review*, in an editorial, denounced him as "a force for evil." The fear journalists harbored—that Murdoch would tart up *New York* and *The Village Voice* and use them as weapons—proved false, in this case. Still, the noise was deafening. Murdoch's principal investment banker, Stanley Shuman, says of his client's purchase of *New York* magazine, "It was a sixteen-million-dollar transaction. You would have thought he had bought General Motors! It's just another indication of the narcissism of the press. The emotions ran high. That's when the image of Rupert dropping out of the sky came up."

Actually, several events conspired to blacken Murdoch's reputation. Nineteen seventy-seven was the year of a mayoral election in New York, and as he had done in Australia and England, Murdoch used his newspaper to flex political muscle. Not only did his *Post* editorially endorse Edward I. Koch, the mayoral candidate most opposed to the liberal establishment that dominated the city, but its news columns also tilted in Koch's favor. There were five candidates, but one couldn't tell that from the Koch-dominated headlines and pictures. "He used the *Post* aggressively in that campaign, to put it mildly," recalls Mario Cuomo, who was one of the candidates for mayor that year. Soon after the election, a petition was sent to Murdoch and signed by fifty of the sixty reporters on the *Post* protesting "slanted" coverage. Koch admits, "I could not have been elected if Rupert Murdoch had not endorsed me." Murdoch, in an interview at the time with ABC's Barbara Walters, stated his case baldly. "I'm entitled to support whom I like." That, of course, is not the journalistic issue. Any publisher is free to editorially support a candidate. The issue is whether Murdoch's editorial policy twisted his news coverage, as *Post* reporters said it had. Similar charges were lodged against Murdoch's *Post* in subsequent New York and national elections.

The feeling that Murdoch is a betrayer has been heightened by shifts in his political position. The purchase of the London *Sun*, in 1969, for instance, roughly coincided with Murdoch's transformation from an Oxford leftist into a passionate capitalist. The political epiphany came, he recalls, when he grappled with England's trade unions. "This sounds very subjective and selfish," he told me, "but living in Britain and having to handle fifteen print unions every night and wondering if your papers were going to come out—if anything could make you conservative, it was handling the British print unions as they were in those days." Not long after Murdoch bought the *Sun* and tarted it up with topless women and gossip posing as news, compelling the other tabloids to follow his lead, the editorial voice of the paper veered to the right, and by the end of the decade it had become a fight-to-the-death defender of Margaret Thatcher.

A Conservative had rarely had the editorial backing of a working-class English newspaper before the *Sun* supported Thatcher in the 1979 elections. "I believe we were right to support Thatcher in critical times when she had no other supporters," Murdoch says now. "I think what people don't understand about me is that I'm not just a businessman working in a very interesting industry. I am someone who's interested in ideas." Throughout the eighties Murdoch was protected by the Thatcher government in ways that were crucial to his business, most notably in the broadening of his base as a newspaper owner.

Despite his disdain for the establishment, Murdoch always wanted to own an influential newspaper—one read by the establishment. First, in 1976, he tried to buy the London *Observer*, but was rebuffed by journalists fearful of his reputation for political interference. Then, in 1981, he pulled off a spectacular coup, gaining control of the biggest quality weekly, the profitable *Sunday Times*, and also of the loss-making daily *Times*. Since he already owned the *Sun* and the weekly *News of the World*, it was expected that his bid would get mired in a protracted review by the Monopolies and Mergers Commission, but a recommendation of a referral to the commission was overruled by the government on the ground that the papers would die if Murdoch wasn't allowed to save them.

The staffs of both the *Times* papers were seduced by a pledge that an independent board would approve the hiring and firing of editors and that Murdoch would not interfere in the editorial operations of the papers. In a year, however, the editor of *The Times*, Harold Evans (now publisher of the Random House Trade Group), was ousted by Murdoch. The paper became an editorial partisan for Thatcher.

Encouraged by the political climate that Thatcher had created, Murdoch audaciously schemed to break the stranglehold of the newspaper unions—whose contracts called for eighteen men on a printing press when only five or six were needed—by secretly building what came to be called Fortress Wapping, a modern printing plant and headquarters surrounded by tall fences topped with coils of razor wire. He carried out the Wapping gamble with military precision. While he was negotiating, he ordered computers and new printing presses to be installed and tested in an abandoned warehouse on the Thames, and when negotiations collapsed he stunned the unions by transporting this equipment to Wapping, along with nonunion workers to operate it. When mayhem threatened and thousands of angry workers paraded outside, more than a thousand police were on hand to preserve law and order. Murdoch concedes that he could not have succeeded if Labour had been in power.

Although the Conservatives remain in power in England, Murdoch worries that Prime Minister John Major represents the traditional establishment and that his government will "screw" him—an expression he often uses.

Might his papers support the British Labour leader, Tony Blair?

Murdoch laughs. "We're not tied to any party," he says. "The big question with Tony Blair, who's very impressive, is what he will be allowed to do by his own party if he ever achieves power." In July, Murdoch flew Blair to Australia to speak at the News Corporation's weeklong management retreat, on Hayman Island. Unlike previous Labour leaders, Blair is prepared to entertain the deregulation of much of the communications business. Over cappuccino a few blocks from Parliament, one of Blair's associates whispered to me that although many Labour backbenchers had made a commitment to limit the percentage of broadcast outlets and newspapers that Murdoch could own, "I would be extremely surprised if that commitment was honored." What the party is trying to do, the associate admits, is fudge the issue. Though this associate is privately critical of Murdoch's power, he says, "It would be absolutely mad for me to talk about Murdoch publicly." Blair's goal, he admits, is to get Murdoch's support, but the party would be happy just to keep him neutral.

Murdoch's News Corporation is a large company run like a small one. A single individual makes big decisions, quickly, and for reasons he sometimes does not explain. For example, members of Murdoch's board have wondered about the dollar drain imposed by a favorite

Murdoch toy, the *New York Post*. Murdoch himself says that this year the *Post* will lose "close to twenty million dollars"—a figure that will raise the net total losses Murdoch has endured in the thirteen years he has owned the paper to more than a hundred million dollars. Yet he wouldn't think of closing it. It offers a powerful political platform. And, besides, at heart he thinks of himself as a newspaperman.

As a manager, Murdoch employs an instinctual and hands-on style. He will plunge into one of his businesses for a week or two, focusing all his attention on it until he masters it. He has a minimal staff of experts and shuns formal meetings, preferring the telephone or one-on-one encounters. In most cases, all that Murdoch requires from those who run his various properties is a one-page weekly financial report containing no narrative—just a recitation of expenditures and revenues, budgeted versus actual totals, and this year's revenues versus last year's. Employees feel that Murdoch is always watching. "You have a sense," says the *Post*'s editorial-page editor, Eric Breindel, "that it really doesn't matter if he's in Australia or on his boat. He's going to get that paper and that editorial."

There is no chief operating officer at the News Corporation. The board is composed of thirteen people, only three of whom are non-employees, non-consultants, or non–family members. George Vradenburg says, "He operates the company as a private, independent family company. And it has the strengths and the weaknesses of that. The strength is quick decisions. The weakness is that no one else has independent knowledge of what is going on in all of the company. Everyone is vulnerable to his health."

Murdoch delegates—except when he doesn't want to. As the chairman of a worldwide company, he does often get ensnarled in excruciating amounts of detail. David Corvo, who was hired from CBS News to become the executive producer of a new Fox newsmagazine show, was startled at Murdoch's level of involvement. "He saw all graphics," recalls Corvo, who is today a vice president at NBC News. When Murdoch's trusted outside public-relations counsel, Howard Rubenstein, sought permission to make available to *The New Yorker* certain executives or Fox promotional footage, he went to Murdoch, which was not unusual. What was unusual was that Murdoch personally screened the footage and reviewed the list of executives, adding and subtracting the names of people I might speak to.

On the other hand, he is clearly not as deeply engaged in the movies as he is in, say, *Post* headlines. "The long lead times frustrate him," says Peter Chernin, the chairman of Twentieth Century Fox.

"It's hard to get him involved." Joe Roth, who preceded Chernin as the studio's chief, guesses that Murdoch reads maybe one script a year.

Murdoch's fabled do-it-alone impatience was one of several things that led the News Corporation to the precipice of bankruptcy in 1990. For example, Murdoch negotiated the nearly three-billion-dollar purchase of *TV Guide* from Walter Annenberg without calling on his lawyer, Howard Squadron, or his investment banker, Stanley Shuman.

Murdoch has had only one chief operating officer, Gus Fischer, and Fischer lasted only four years in that position. "He never told people in New York, 'You are now reporting to Gus.' It was difficult," recalls Fischer, who left the News Corporation in March 1995 and is now an independent entrepreneur. "I think he feels much more comfortable having someone who calls him if he wants to go to the bathroom. Don't get me wrong. I'm not saying this in a negative way, against Rupert. I'm just saying it's his style." While many executives have been with Murdoch for decades, there has been carnage near the top of the company. Murdoch, observes Joe Roth, who now runs the Disney studios, "makes you feel fungible." Even Barry Diller, whom Murdoch himself has credited with building the Fox network, was eased out—despite denials at the time—when he asked for a larger slice of the company. Other Fox executives who "were feeling too proprietary," as Howard Squadron phrases it, also have gone.

Personally, Murdoch is a gentleman. He treats executives as part of his extended family; he invited them to his son Lachlan's engagement party, he remembers spouses' names, he rarely raises his voice. "He's not abusive," a senior News Corporation executive says. "But when he turns on you, it's"—he snaps his fingers—"like that." In the past decade, Murdoch has had four deputies who appeared to be his number two—Donald Kummerfeld, Gus Fischer, Andrew Knight, and Richard Searby, the chairman of the board for ten years and a friend of fifty years. In 1992, Searby learned that he was terminated when Murdoch dispatched another executive to visit him carrying a terse note.

Did the firing of such an old friend cause him pain? "You get terribly pained," Murdoch told me, wincing and lowering his voice. "It's the pain leading up to it. Those things don't come overnight. Things like that involve a year or two of pain. And a lot of sadness afterwards." The "sadness" can't be too intense, however, since Murdoch admits that he has not tried to communicate with Searby since.

. . .

Rupert Murdoch is a bit of a prude—"I guess it's my Scottish blood," he says—and a conventional family man. The only photographs in his New York and California offices are of his family. He has four children: a daughter, Prudence, from his first marriage, and two sons and a daughter with his second wife, the former Anna Torv. Elisabeth, twenty-seven, is married to Elkin Pianim, a Ghanaian whom she met in college. Until recently, the couple owned two TV stations in northern California. Lachlan, twenty-four, joined the News Corporation last year and is based in Australia. And James, twenty-two, has completed his junior year at Harvard and has dropped out to start a record company with two friends.

With his children, Murdoch is openly affectionate, hugging them and calling each "darling." "Rupert has two basic attachments—family and business," suggests investment banker Stanley Shuman, who met Murdoch when their sons were classmates at Dalton and who serves on his board. "He doesn't have intimate personal friends." Although he traveled incessantly, Murdoch's offspring claim they had no sense that he was absent. "I don't remember him being away a lot," says Lachlan Murdoch, a handsome former philosophy major and wrestler at Princeton. "For instance, around Christmas we always went to Australia as a family. We went to company events. We spent weekends in Old Chatham, New York. On holidays we'd lease a sailboat. We bought a house in Aspen and skied in winter and hiked in summer."

Anna Murdoch, who is thirteen years younger than Rupert, is a novelist, and serves on the News Corporation board. Their son James says, "I don't know of any confidant other than my mother. I couldn't imagine him going to anyone but my mother. She's as tough as nails." She is also more conservative than her husband. A devout Catholic, she is implacably opposed to abortion and thinks it should be illegal; he does not (though he would eliminate federal funding). "She doesn't like *The Simpsons*," Murdoch says. "I think it's brilliant." She voted for George Bush in 1992, he recalls, while he voted for Ross Perot, as a protest. Today, she favors Bob Dole for president, while Murdoch likes Colin Powell. "He appears to be a man of very fine character," Murdoch says. "I agree with a lot of what he says."

Murdoch's family values and taste have rarely interfered with what goes on in the pages of his papers or in his television programming. The sexually raunchy shows are manufactured rather cynically. "I wouldn't let a thirteen-year-old watch *Melrose Place*," Murdoch admits. Yet Fox not only broadcasts *Melrose Place* but has shifted it from 9:00 P.M. to 8:00 P.M., the traditional children's hour. Murdoch introduced tabloid TV newsmagazines to America in 1986, with *A Current Affair*, on Fox, encouraging others to follow him and to push the envelope, as they say, of bad taste. "He's confirmed the suspicion that shit sells," a former network president observes.

Salaciousness, whether manifested by the bare-breasted women who appear daily on page 3 of the *Sun*, and who are referred to as "*Sun* lovelies," or the sexual escapades of politicians or the Royal Family, is a staple of Murdoch's newspapers. This summer, for example, the *New York Post* got a lot of mileage out of pillorying Judge Kimba Wood for allegedly having a romance with a married investment banker. The Kimba Wood story passed three Murdoch tabloid tests. First, he says, "it's a soap opera," an entertaining spectacle. Second, it fuels envy and resentment—and thus circulation—because it's about "high society," as he calls it. It tells the little people how the other half lives—repressed-schoolmarm judge likes sex! Third, it displays what Anthea Disney, who has worked on Murdoch's London tabloids and now serves as a top Murdoch executive, calls a "visceral" Fleet Street attitude of rebellion against social pretense. She also edited *TV Guide* for four years, and she recalls that Murdoch would always say to his editors, "Do you think this is too upmarket? Do you think it should be more mass?" Kimba Wood was "mass."

There was another reason the Kimba Wood tale pleased Murdoch, he confesses: "She put my friend in jail for ten years." Wood presided over the trial of the investment banker Michael Milken, sentencing him to a stiff ten years in jail for securities fraud (but allowing him to serve only two in return for cooperating with authorities). Murdoch has been loyal to Milken. Last spring, Milken was retained as a consultant in a deal in which MCI committed two billion dollars to the News Corporation; and *Payback: The Conspiracy to Destroy Michael Milken and His Financial Revolution*, by Daniel Fischel, was published recently by HarperCollins and excerpted in the *Post*.

Murdoch's journalism follows a pattern. When he plunged into editing the *News of the World*, in 1969, he immediately applied lessons he had learned as a subeditor at Lord Beaverbrook's *Daily Express* the year after he graduated from Oxford. Murdoch believes that a paper—particularly a working-class paper—has to be "fun," a form of entertainment. The true public trust is to give the public what it wants. "Populism" and commerce drove Murdoch's papers to invade the bedrooms of the Royal Family and to print jingoistic headlines, and populism combined with commerce accounts for the introduction of softer features in both *The Sunday Times* and the *Times* after he took them over.

But is what Murdoch's tabloids do so different from what tabloids have traditionally done in England? "I think it's fair to say that the British don't have the same sense of dignity about the press that Americans have," says Roger Laughton, who spent much of his career at the BBC and is today the CEO of Meridian Broadcasting. "The British press is far more competitive, and very professional, but I wouldn't say it has a range of ethical considerations that it takes into account. That predates Murdoch. He didn't behave out of character with the way the press operated here, or in Australia. Others have tried to do the same many times in the past. That's what being a press baron implied."

Certainly Murdoch's products do not speak with one voice. *The Australian*—Australia's first national daily newspaper, which he started and nurtured for twenty years, until it turned profitable—was used as a political weapon in the 1972 and 1975 national elections, but it is a respectable publication today. As are *The Times* and *The Sunday Times*. Nor are Murdoch's politics as simple as the conventional view of him suggests. He is, for example, a fiery advocate of law and order, yet a quiet opponent of the death penalty. "It brutalizes society," he explains, echoing an objection often lodged against his newspapers.

Murdoch has not tried to insert his conservative views into his movies or television, according to Hollywood associates. This is not true of his approach to news. "He never called me and said, 'Don't do a story.' Or 'Do a story,' " recalls the former Fox News executive producer David Corvo. "But once, when I went over the list of network contracts coming up—Bryant Gumbel, Leslie Stahl, Steve Kroft, and so on—he just brushed them off. I felt that he wanted a conservative. I don't know how you build a credible news division with an agenda."

Murdoch believes that his critics do what he does, the difference being that they're hypocrites and he's not. He complains that reporters for *The New York Times* and *The Wall Street Journal*, for instance, "write more subjectively" than reporters do in England. He seems suspicious of anyone who proclaims journalism a calling. "Journalism has been mistaken over the past two decades for some kind of profession," says one of Murdoch's favorite journalists, Steve Dunleavy, who has worked for him on many continents since 1967. "We're not doctors. We're not lawyers. We're not architects. It's not a profession. It's a craft. We're the same as carpenters or mechanics."

But Murdoch is not a traditional press baron. Frank Barlow points out the difference: "I don't think there was ever any suggestion that the old press barons received any benefit for their businesses. Maybe because they weren't in any other businesses. Now it is not unusual to read that Murdoch gets benefits." This goes to the heart of the critique of Murdoch's journalism: too often, it becomes the servant of his political or commercial interests. What gives him the most pleasure in all his empire, Murdoch told me, is this: "Being involved with the editor of a paper in a day-to-day campaign. Trying to influence people." Barry Diller says, "Rupert is a pure conservative, except for purposes of manipulation."

Murdoch's papers do influence people, and he has used this influence to support political figures, who may then be in his debt. *The Sun* made Thatcher acceptable to its working-class readers. His Australian newspapers have influenced elections. Reporters on both *The Australian* and the *New York Post* protested in the seventies that Murdoch used their news pages to reward favored candidates and to savage those he disliked. The *Post* helped elect the current Republican mayor of New York City and governor of New York State, and the state's Republican and Democratic senators. When the new governor, George Pataki, wanted to meet with corporate leaders after his election last year, he asked Murdoch to act as host of the meeting. When Mayor Rudolph Giuliani's wife, Donna Hanover, was rehired as a TV reporter, it was Murdoch's Fox-owned station—WNYW—that recruited her.

Murdoch nearly lost his empire in 1990, when he was on the edge of bankruptcy, and another threat surfaced in late 1993. The New York chapter of the NAACP filed a petition charging that Murdoch had misled the FCC in 1985, when he founded the Fox network. Federal law stipulates that a foreign citizen cannot own more than one-quarter of a broadcast station's capital stock, and it was then that Murdoch changed his nationality, becoming an American citizen. But a tenacious volunteer lawyer for the NAACP, David Honig, discovered while digging through Fox's applications for station licenses that although Murdoch himself had 76 percent voting control over Fox, his Australian holding company, the News Corporation, indirectly owned more than 99 percent of the equity of its stations.

Thus, the NAACP claimed, Fox had exceeded the foreign-owner-ship limit, thereby depriving minority Americans of an opportunity to bid for a broadcast license.

The issues were obscure. One could argue that Murdoch's original application misled the FCC, or that the government agency was incompetent. The FCC's policy in 1985, one of the five commissioners griped, was "We don't ask, you don't tell." What is clear, as Stuart Taylor, Jr., wrote in a careful probe in *The American Lawyer* magazine, is this: "The Fox legal team put the information from which Fox's foreign equity could be inferred just deep enough in its 1985 application and subsequent filings to escape notice, while disclosing just enough to negate any claim of overt deception. As for the FCC, 'they winked—or they blinked—and it's hard to know which,' in the words of a lawyer familiar with the case."

Murdoch at first viewed the NAACP allegations, Preston Padden notes, "as a minor irritation." It wasn't until the fall of 1994, when NBC filed a similar petition, that he began to fret that it was more than a nuisance. NBC charged that all of Murdoch's 1985 station-license applications were false. This was war. Murdoch directed strategy meetings, edited press releases, and camped with Padden in Washington and visited members of the key congressional committees. "We were not asking anybody to do anything," says Padden. "Our pitch was 'This outburst by NBC was because they didn't like competition in the marketplace.' " It was the O.J. defense: us versus them.

Although Murdoch has never been a major financial contributor to politicians, he bet heavily on a Republican victory in 1994. In the months after the election, and so far without attracting much notice, a subsidiary of Murdoch's News Corporation—an entity that had never made a soft-money contribution—donated a total of two hundred thousand dollars to the Republican House and Senate Campaign Committees. According to Fred Wertheimer, the former president of Common Cause, this dwarfed contributions made previously by any of Murdoch's companies. "These guys haven't played the Washington game that much," Wertheimer says of Murdoch and his people. "But they know it's a candy store. There's an awful lot to be purchased." Immediately after the election, Padden arranged for Murdoch to visit a total of seventeen public officials, including eight Democrats. One of the first visits was with Speaker-elect Newt Gingrich. "It was a ten-minute meeting—maximum," Murdoch says. "We met in the hall, because there were too many people in his office. It was just chitchat. We talked about the chances of his getting his Contract with America passed." Padden, who was also present, says, "None of their conversation had anything to do with this business." Then he adds, "I piped up at the end to say that NBC was going after us." However brief this part of the conversation was, it was initially denied by a Gingrich spokesman, who said that neither NBC nor any other matter before the FCC or before Congress was discussed.

One month after this hallway encounter, it was revealed that Gingrich had signed a two-book contract with HarperCollins worth $4.5 million, likely the largest book advance ever received by a current officeholder and the third-largest advance ever received by an American public figure. Murdoch denied that he knew anything about Gingrich's book contract. "I was telephoned in Beijing on Christmas Eve and told that it had happened," he says. "Howard Rubenstein called me. I went crazy. I knew critics would explode." Gingrich denied knowing that Murdoch was connected to Harper-Collins. The denials did not quiet the Democrats or the editorials. It was unseemly for a public official to take such an advance, and Gingrich implicitly acknowledged this when he belatedly announced that he would accept no advance and would take only royalties that the book actually earned. The most charitable interpretation is that there was an appearance of conflict for Gingrich to receive such a large amount of money from a company with business before Congress; the least charitable is that it was a bribe.

HarperCollins has a history of signing book deals with leaders of countries that the News Corporation has a commercial interest in—from Margaret Thatcher to Deng's daughter, from Mikhail and Raisa Gorbachev to Boris Yeltsin, from a Saudi prince to Dan Quayle. And in some cases (those of Thatcher and Gorbachev) Murdoch actually negotiated the deals. But it is possible that Murdoch was not told ahead of time of the Gingrich book deal. William Shinker, a former publisher of HarperCollins, says the CEO of HarperCollins, George Craig, had the authority to approve payments of that size without clearing the decision with Murdoch. Murdoch says he has since amended the policy; now he must personally sign off on any advance above a million dollars. In a way, Shinker implies, it really doesn't matter whether Murdoch knew ahead of time or not. Murdoch, he says, constantly pushed the company to publish more conservative books. He always told Shinker to make publishing decisions on the merits—and he did not overrule his editors when they told him they didn't think a memoir proposal Pat Buchanan had sent Murdoch should be accepted. However, all the princes at the publishing house knew what the king wanted. "Rupert would accuse me on several occasions of not publishing enough conservative books," recalls Shinker. "He'd joke, 'You're all a bunch of pinkos.' But he never told me to buy one of them. Or how much to pay." Still, if Murdoch didn't know, how come executives at HarperCollins were so deaf to appearances?

By late 1994, Murdoch felt embattled. He wrote a letter to the FCC chairman, Reed Hundt, in December in which he expressed "personal anguish" about the storm surrounding his motives and declared that he was an American citizen and exercised "de facto control of the News Corporation and all its businesses." He felt mistreated by the establishment, and in his deposition to the FCC he said he felt that the agency would be playing a game of "semantics" if it counted News Corporation equity as foreign ownership but did not count Murdoch's voting control of the stock as American ownership. It was, he said, "a witch hunt." Republicans agreed. Larry Pressler, the chairman of the Senate Commerce Committee, and Jack Fields, the chairman of the House telecommunications subcommittee, began making calls to the FCC lambasting the Democrats for picking on Murdoch and Fox. Fields threatened a "top to bottom" review of the FCC if Murdoch was persecuted further. The conservative Heritage Foundation issued a paper on rolling back regulations in which it urged the abolition of the FCC.

Murdoch was crafty, but he may have been lucky as well. In January, Apstar-2, the satellite that NBC had counted on to distribute its programs in Asia, blew up shortly after liftoff. A month later, NBC, which had accused Murdoch of making craven deals to advance his interests, made a craven deal with Murdoch: NBC withdrew its FCC petition in return for a lease of two channels on Murdoch's Star satellite system.

By spring, the tide had turned in Murdoch's favor, and he knew it. On the morning in May when the FCC was to rule, he arrived forty minutes early at the hearing room where the decision was to be announced, flanked by his photogenic wife, Anna, and his daughter Elisabeth and her husband. Murdoch had in his hand a statement thanking the commissioners for exonerating him and Fox from the charge that they had "lacked candor." When the Murdochs took seats in the second row, Preston Padden leaned over to them. "The commissioner has arranged for you to sit in the first row," he whispered. A phalanx of photographers snapped pictures of Anna

Murdoch patting the shoulder of her husband's navy-blue double-breasted suit and straightening his red-and-blue polka-dot tie. The five commission members filed in and listened as the FCC general counsel, William Kennard, announced that although the foreign ownership of Fox did exceed 25 percent, the evidence did not support a conclusion that Fox had misled the FCC; it also confirmed that an American citizen—Rupert Murdoch—did indeed control the parent News Corporation. Moreover, Murdoch was invited to apply for a "public interest" exemption from the regulations requiring Fox to restructure its equity, and he promptly did so.

Murdoch was jubilant. David Honig, of the NAACP, was not. His view of the ruling was this: "The process was tainted. I think they threw the Communications Act in the garbage. Murdoch's Republican cohorts blackmailed the FCC by threatening its existence." Others marveled at Murdoch's political dexterity. "I wish we could be as successful as he's been," Sumner Redstone says.

Murdoch did not rest. A few days after the FCC ruling, MCI announced that it was investing two billion dollars in a 13 percent stake in the News Corporation, and not long before Murdoch had revealed that he would finance a new conservative magazine, the *Weekly Standard*, based in Washington and edited by William Kristol. He acquired a 100 percent ownership of Star TV, and he anointed his son Lachlan its deputy chairman. In August, he combined Delphi, his Internet operation, with MCI's Internet operation. The same month, he raised fifty thousand dollars at a "21" fund-raising lunch for the Senate Commerce Committee chairman, Larry Pressler. Last week, in addition to concluding a sports link with Malone, Fox planned to announce that it had acquired the rights to televise major-league baseball and started a children's-television venture with Saban Entertainment.

In recent days, Murdoch has concluded that he could probably not bring off a takeover of Time Warner at this point, but the mere fact that he contemplated it is testimony to his extraordinary bravado. He would enjoy claiming Warner Brothers, with its huge film library, and HBO, and Time's various magazines.

The Time Warner–Turner merger—like Murdoch's expansion—has fanned the debate over whether media behemoths like the News Corporation and Time Warner will monopolize the production and distribution of information and entertainment. Murdoch would argue that, in the long run, no one will be able to monopolize information. Eventually, even in China, customers will be able to bypass middlemen like Murdoch or the government and summon to their TVs, computers, or telephone screens any news source, any channel, any desired program. It would be difficult for any entity to screen e-mail or books or newspapers downloaded from one computer to another. Technology makes it possible for every citizen (who can afford a computer and a high-speed modem) to become a publisher. Those who fret about monopolies are seen as being trapped in a time warp, railing against past dangers. If government regulators try to impose new anti-monopoly rules, they risk suffocating the world's fastest-growing industry.

On the other side of the argument are arrayed those who believe that Murdoch approaches a monopoly, certainly in Australia and in England. News Corporation newspapers control more than 50 percent of the daily and Sunday circulation in Australia and a third of the United Kingdom's national newspaper circulation. As part of a deal with Murdoch's BSkyB, TCI entered into a partnership with U S West to carry Murdoch's programming on cable in the United Kingdom. By making long-term deals with all the major Hollywood studios for movies, and by using Fox programming, BSkyB dominates most of the product and the chief means of reaching pay-TV customers in the United Kingdom and Europe.

A traditional objection to a monopoly is that it restrains trade in order to control markets. When the distributor also owns the product distributed, conflicts arise. For example, it has been charged that before American cable companies like TCI and Time Warner would give channel space to CNBC, they insisted that it provide only business news, so as not to compete directly with CNN. Media companies, which are becoming entangled in local politics and multiple partnerships, are heading inexorably toward their own version of global *keiretsu*, the informal back-scratching system used by Japanese companies.

The issue of monopolies and what to do about telecommunications legislation will be discussed this week in Washington as nine House and eleven Senate members confer on how to reconcile bills passed in each chamber. On the agenda are matters of momentous significance for communications companies like Murdoch's News Corporation. Among them: Should broadcasters like Murdoch be allowed to own stations reaching 35 percent of American viewers, or should the limit stay at 25 percent? Should broadcasters like Murdoch have to bid on extra channels that technology will make available, or should this extra spectrum space be a gift to existing broadcasters? Should cross-ownership restrictions be lifted, allowing a broadcaster like Murdoch to own a newspaper in the same market where he owns a TV station?

Whether the News Corporation is a monopoly or not, Murdoch wants to keep it a family company. "Lachlan is only twenty-four," Murdoch points out. "He's a young man. Certainly he is conducting himself well. His sister Lis is very keen on coming back into the company. Soon, I hope. I think James is undecided. He's determined to do his own thing. I took all three out to our management conference." To turn over the News Corporation to his young, untested children presupposes three things: that Murdoch will remain in charge perhaps another ten years, allowing them to gain valuable experience; that if he cannot stay another decade Anna Murdoch might be installed as CEO for a period of time; and, finally, that a public company board would deem a family member a suitable replacement. John Malone observes, "They have only one real vulnerability, and that is that Rupert is mortal. They have an Oliver Cromwell problem: Where do they find a successor?"

Does he tire of the constant competition to get bigger, to win each war? When does Murdoch say *enough*? I asked him this question at the end of a long night that started with a drink and a stroll over his six-acre property in Beverly Hills, past the fig trees and manicured English gardens that Anna Murdoch tends. After dinner in the dining room of his comfortable Spanish-style home, Murdoch sipped a glass of California Chardonnay and treated the question as something so alien as to be incomprehensible. "You go on," he responded, opaquely. "There is a global village in some sense. You are competing everywhere. . . . And I just enjoy it."

It was eleven o'clock, and he was tired, his white shirt collar unbuttoned. We talked about an early mentor, Lord Beaverbrook, and he said, "He had a bit of the devil in him. He took pleasure surrounding himself with a lot of hangers-on and corrupting them. He was pretty cynical about everything."

Murdoch would be surprised to learn that his "moral tutor" at Oxford, Asa Briggs, has come to think of him as a cynic. Briggs believes his former pupil will be remembered as "a very important figure in history. He has done something no one else has ever done, which is to reach a global set of interests." But while Briggs praises Murdoch for his boldness as an entrepreneur, he deplores the cynicism that animates many of his products. "He has not elevated taste. No one could ever accuse him of that," Briggs says, shaking his head, hesitating to go on because in the nearly half a century they have known each other he has never said this to Murdoch. If he were to have a serious conversation with his former pupil about journalism, he says, "I'd be bound to talk of subjects like taste and responsibility. It would be tricky for me." Therefore, Briggs sadly concludes, "I couldn't find myself talking to Rupert about one part of his empire, which seems to me doesn't show much moral scruple."

Murdoch believes he has built things that will endure. Still, while the Fox network and Sky and Star TV provide viewers with more choices, they are rarely better choices. Unlike the Sulzbergers or the Grahams, William Paley, or Henry Luce—no matter their many flaws and sometimes outrageous vices—Murdoch will leave no monument except a successful corporation. He has boldly built a worldwide company, but he has rarely elevated taste or journalism. Competitors envy Murdoch's financial success and his bold vision, and politicians fear his power, but the News Corporation is a business that relies on a singular man who is now in his sixty-fifth year. Rupert Murdoch has created a much bigger empire than his father did, but his success may be just as fleeting. "I believe," Roger Laughton says, "that we're talking about Genghis Khan, not the Roman Empire."

POSTSCRIPT:

Murdoch did not want to be interviewed for this profile. For two years I tried to persuade him; each time I asked, he flatly said no. Finally, in the fall of 1994, when I was interviewing him for a profile of someone else, I told him, truthfully, that I had written about many of the leading figures in the communications wars but there was this huge hole: I had not profiled him. Talk to him in six months, he said. I did, and in the spring of 1995 he agreed to cooperate.

Then he got cold feet and reneged. I told Howard Rubenstein, his public-relations adviser, that I was proceeding, with or without his cooperation. I believe that he was probably moved to see me by several factors, including this bit of hardball. Murdoch also thinks he has a story to tell. He thought I would be fair. And he undoubtedly thought I would be charmed.

He can be charming, and effortlessly candid, as only a man without fear can be. But he was unhappy with the outcome. As is often true when a subject opens himself to a reporter—and Murdoch told me that the fifteen or so hours of taped interviews that we did and the access to all his meetings and schedule for a week (the sole exception being a meeting with Michael Milken) represented the most access he had ever granted a writer—the subject feels betrayed. Rejected. Angry. Vulnerable.

This profile runs about 20 percent longer than the version that appeared in *The New Yorker*, since I added biographical and other material excised for space reasons from the original version. The worst, and funniest moment,

I experienced while reporting this profile came during the summit between Murdoch and Malone. For four hours I sat there desperate to pee. Yet I was convinced that if I left the room they'd never let me back in. So I waited.

Murdoch did not wait to lure two of his children, Elisabeth and James, to join the company. She is now vice chairman of BSkyB in England; he now runs News Corporation's Internet ventures outside the United States. Also after the appearance of this profile, the bad blood between Murdoch and Ted Turner spilled over into a public brouhaha when Time Warner and Turner decided, in the fall of 1996, to grant a cable channel to MSNBC, but not to Murdoch's new twenty-four-hour Fox News. Time Warner claimed that the only available cable-channel space was that held by NBC's America's Talking, and since they had a contract to air this channel and since NBC was merely replacing it with MSNBC, Time Warner had no choice but to give NBC rather than Fox News the one available slot. Fox, like Bloomberg News and other program suppliers, protested. But none protested quite the way Murdoch did. Demonstrating his political muscle, Murdoch induced the mayor of New York City, as well as the governor and attorney general of New York State, to treat this esoteric matter as if it were a free-speech cause, for in press conferences and a lawsuit they championed Murdoch as a victim of a monopolistic Time Warner and a vindictive Ted Turner. Meanwhile, Turner felt victimized by this onslaught and by attacks on him in Murdoch's *New York Post*. He denounces Murdoch as a merchant of "sleaze."

While Murdoch's products win few awards for taste, he has company. As the next piece demonstrates, garbage is produced elsewhere. And the reason is the same: Garbage is produced by businessmen who worship the bottom line and care too little about other values.

15

THE POWER OF SHAME
Bill Bennett Takes On
Gangsta Rap
(The New Yorker, June 12, 1995)

It was the kind of snarling, screaming confrontation the producers of *The Jenny Jones Show* hope for. The stage was the eighth-floor conference room at HBO headquarters on West Forty-second Street. The time was 4:00 P.M. on Thursday, May 18, 1995. The actors were William J. Bennett, a codirector of the conservative advocacy organization Empower America, and his Democratic ally C. DeLores Tucker, chairwoman of the National Political Congress of Black Women. They sat on one side of a long table, flanked by five associates. Across from them sat the opposition—six executives from Time Warner, the world's largest media company, and the corporate parent of HBO as well as *The Jenny Jones Show*. They had gathered to discuss an ad campaign mounted by Bennett and Tucker against some of the gangsta-rap music produced by Time Warner music companies. The music, Tucker says on a sixty-second television commercial featuring her and Bennett, "celebrates the rape, torture, and murder of women."

Richard Parsons, president of Time Warner, who like Tucker is black, invited her to speak first, according to detailed notes taken by Bennett and the recollection of seven of the other participants.

Tucker took care to praise Time Warner's accomplishments but said she was distressed by the effect on youngsters of the misogynist lyrics of some of their rap music. "If corporate responsibility dictates that we protect the whales, protect the rivers, protect the environment," she had said earlier that day at the annual Time Warner meeting of shareholders, "then the most important of all earth's resources must be protected . . . our children."

Parsons looked to Michael Fuchs to be Time Warner's spokesman at this meeting. Fuchs, chairman of HBO, had recently been given an additional responsibility, chairman of the Warner Music Group, and he began by speaking generally of music as art, and as a voice of protest since the days of slavery. Time Warner, he said, took pride in the fact that it did not act as a censor. Besides, he said, if the Supreme Court could not define pornography—you know it when you see it, a justice commented—how could Time Warner define violence? Fuchs said he hoped they could enter a real dialogue, that they could address the root causes of crime.

Tucker had something more concrete in mind, and she pulled from a folder copies of the lyrics of a song recently recorded by the group Nine Inch Nails on the Interscope label, which is half owned by Warner Music–U.S. She passed the copies across the table to Fuchs and the other executives. Three times, Tucker recalls, she asked Fuchs to read the lyrics aloud. Fuchs declined, as did the other executives. "I did not think it was the right way to begin a meeting constructively," says Fuchs. "He couldn't read it because he knows it belongs in a porno sewer!" says Tucker. She asked Peter Wehner, director of policy for Empower America, to read aloud these lyrics:

> *I am a big man*
> *(yes I am)*
> *and I have a big gun*
> *got me a big old dick and I*
> *like to have fun*
> *held against your forehead*
> *I'll make you suck it*
> *maybe I'll put a hole in your head*
> *you know, just for the fuck of it*
> *I can reduce you if I want*
> *I can devour*
> *I'm hard as fucking steel, and I've got the power*
> *I'm every inch a man, and I'll show you somehow*
> *me and my fucking gun*
> *nothing can stop me now*

shoot shoot shoot shoot shoot
I'm going to come all over you
me and my fucking gun
me and my fucking gun

If the Time Warner executives were embarrassed, they tried not to show it. Their aim was to stall the ad campaign, to start a dialogue, and to explore improving the voluntary labeling system for violent lyrics. The executives knew that Bennett had a Ph.D. in philosophy and had been a Republican Cabinet official, and they expected him to be respectful, polite. Instead, they encountered a former Brooklyn street tough, an impatient man with a tart tongue. With undisguised disdain, Bennett asked, "Do you consider the words vulgar and potentially damaging to children?"

"Art is hard to interpret," said Fuchs, making a point that would be echoed by colleagues. If they censored Nine Inch Nails today, why hadn't the Beatles been censored yesterday for extolling drugs in "Lucy in the Sky with Diamonds"?

Bennett suggested Time Warner appoint a panel of citizens to certify all its lyrics. Time Warner was not about to cede its decision-making to outsiders, said Fuchs. Besides, he noted, community standards change—witness a 1965 *New York Times* editorial that accused Elvis Presley of selfishly "exploiting the youth of America." At that point Gerald Levin, chairman and CEO of Time Warner, who had arrived late, spoke up. He said that Bennett and Tucker were dealing with a symptom, not a "root cause," of the violence sweeping America. Earlier that day, at a sometimes contentious shareholder meeting, Levin had said: "In the case of music, where there are albums that might offend or disturb certain audiences, those products are, in fact, labeled. I understand that there are voices raised in concern about certain lyrics and imagery that are contained in certain songs published by every worldwide music company." Now he told Bennett and Tucker that the issue was one for the entire music industry, not just for Time Warner. They had to talk, not shout at one another, he said, and to that end, he said, he had appointed Michael Fuchs and Doug Morris, the chairman and CEO of Warner Music–U.S., to help launch "a constructive dialogue" and to "develop standards for the distribution and labeling of musical material."

It was, Bennett would later say, all "pseudo talk." To him, the labeling of records was a public-relations ploy. The problem was the manufacture of garbage, not whether the garbage was appropriately labeled with a sticker that says, "Parental Advisory. Explicit Lyrics." (Fifteen of Warner Music's 1,250 albums were so labeled at the time.) Unlike the movie-rating system, which granted theaters the right to deny children a seat at, say, an R-rated movie, the labeling of records was strictly at the discretion of the record companies, not the store. At the meeting Bennett, who knew this, exploded. "You guys are the bottom of the heap."

"How dare you talk about us this way!" screamed the burly Morris at the burly Bennett. Michael Fuchs said later of Bennett, "He came to town as the visiting marshal. If he said it once, he said it six times: 'This is a no-brainer. You guys are missing the point. I can't believe you think this is complicated. Read this lyric!' "

Bennett refused to discuss guidelines for determining what would be acceptable in lyrics and instead pointed to just a few rap songs. He focused on just one record company, Fuchs said, adding, "This was a hit-and-run act. Personally, I have a lot of trouble with a guy who has conservative credentials and who could probably give me an argument about why we should have assault weapons. And he's telling me how much harm lyrics are doing? . . . He does not come into this situation—if you're talking about youthful violence and family problems, the thing that is tearing at the social fabric—he does not come in with clean hands."

Bennett, who opposes lifting the ban on assault weapons, does talk about social problems, but as a conservative Republican he differs from liberal Democrats like Fuchs when it comes to solutions. And they differed about something else. Although blacks like Tucker and members of the Congressional Black Caucus have led the opposition to gangsta rap, some of the Time Warner executives nonetheless saw race as a motive. "A lot of the attack on rap smacks of being racist," says Fuchs. "It's a fact that white kids are buying black music and are being influenced by it, and that frightens their parents. It's not very different than the feeling my parents had thirty years ago when rock and roll came out—about the influence of black music."

The discussion in HBO's conference room spun around and around. Twice during this meeting, Tucker stomped from the room. Then Levin stormed out.

At one point, Bennett said angrily that the only reason Time Warner agreed to meet was because he and Tucker had produced a TV ad attacking them. Wrong, said Levin. We believe in dialogue.

"Are you telling us that you would have met with us if we had not made the ad?" Bennett shouted.

"Yes," said Levin.

"Baloney!" Bennett exclaimed.

With that the usually mild-mannered Levin gathered his papers and said, "There's nothing more for me to discuss. Mr. Bennett, you clearly are not here to discuss serious issues. Mr. Fuchs and Mr. Morris will continue this dialogue."

Bennett left at 5:45 to catch a shuttle to Washington. Tucker and the Time Warner executives remained and had what Fuchs calls "a heartfelt" discussion. Before the session ended, they asked Tucker if she would draw up a list of suggestions.

For the next week, Time Warner heard nothing. Then, on May 25, Bennett dispatched a letter to Levin that read like a Nike ad: *Just Do It!* "My recommendation is fairly straightforward: Time Warner should stop its involvement and support of gross, violent, offensive and misogynistic lyrics," he wrote. "Anything short of that is, I think, an abdication of corporate responsibility." He continued:

> Unfortunately, warning labels do not prevent these kinds of lyrics from reaching young people. As was mentioned in our meeting, it is not right to hold you responsible for the entire music industry. What you are responsible for—and what you can have an immediate impact on—is the conduct of Time Warner. . . . One of the clear impressions that I and my colleagues came away with is the recognition that you and your colleagues are not willing to make any normative judgments about this music. Without that willingness to make normative judgments, it is simply impossible for you to develop standards. . . . There are some music lyrics on which reasonable people might differ. There are obviously some close calls. But the lyrics which Dr. Tucker and I object to, and which our ad targets, do not fall in that category. . . . Let me reiterate: we are not talking about censorship. Our appeal is to a sense of corporate responsibility, to basic decency, and to citizenship.

The issue of violence in the media had been building for years. In 1993, an agitated Congress threatened the television industry with censorship if it did not curb violent programming. To pacify the Congress and parents, the broadcast and cable industry eventually agreed to provide guidance to parents by voluntarily labeling programs with violent content or language. Industry executives now accept labeling as an act of corporate responsibility, just as the movie studios accept a rating system they once opposed. The loudest burst of applause President Bill Clinton received during his marathon 1995 State of the Union address was when he scorned Hollywood's "mindless violence and irresponsible conduct." Then, in March

1995, a John Leo column in *U.S. News & World Report* provoked a new storm. Leo wrote:

> The schlocky *Jenny Jones Show*, the first show on which a guest who was humiliated later was charged with murdering his humiliator, is a Time Warner product. The most degrading commercial picture book about human sexuality may be Madonna's $49.95 porn book, which, I am told, pictorially indicates that she is game to have sex with everything but babies and folding chairs. It was published by Time Warner. . . . In the movies, the all-time low for cynicism and historical lies (Oliver Stone's *JFK*) and for graphic, wholesale serial killing presented as fun (Oliver Stone's *Natural Born Killers*) were both produced by Warner. . . . But it's in the music field that Time Warner does most of its damage. C. DeLores Tucker, chair of the National Political Congress of Black Women, says Time Warner is "one of the greatest perpetrators of this cultural garbage." She may be understating the case. From the rise of 2 Live Crew and Metallica, through the national uproar over Ice-T's cop-killing lyrics, down to Snoop Doggy Dogg, Nine Inch Nails and Tupac Shakur, the sprawling Time Warner musical empire has been associated one way or another with most of the high-profile, high-profit acts, black and white, that are pumping nihilism into the culture. . . . Last week Time Warner bought another chunk of Interscope, the hottest record company around, and now owns 50 percent. This is the cultural equivalent of owning half the world's mustard gas factories.

Soon after the column appeared, Leo says, Levin, furious, telephoned him and berated him for questioning the company's integrity.

"If integrity is the question, why not sell Interscope?" asked Leo. (Interscope produces many gangsta-rap recordings.)

"Why don't you attack Rupert Murdoch?" countered Levin. "He's worse than us." Levin and his colleagues were particularly irked because they believe that Leo, a former *Time* employee, saves his bile for what was once his home.

Those who wrote letters to the magazine sided with Leo. Democratic senator Bill Bradley wrote, in part: "Too often those who point to government as an enemy of freedom and a destroyer of families are strangely silent about the market's corrosive effects on those same values in civil society." Leo's column moved Bennett. "My wake-up call was John Leo's article," he says. Bennett phoned Leo and soon got in touch with C. DeLores Tucker, whose views on this matter he gleaned from the column. They would forge a potentially powerful alliance—bipartisan, biracial, and bipolar in that it united many liberals, feminists, and blacks with the new right.

Time Warner should have anticipated it. They had been subjected to withering criticism in 1992 for publishing the rapper Ice-T's song "Cop Killer," and Levin had mounted a much-ridiculed public defense of Ice-T's right to free speech. Without fanfare, the following year Time Warner severed its ties with the rapper, obviously deciding that while Ice-T was free to speak, companies were also free to say they did not wish to sponsor songs that glorified cop-killing. However, the issue of violent lyrics did not die, and so Levin announced at Time Warner's May 18, 1995, annual meeting that he had asked Michael Fuchs and Doug Morris to "take the lead" in developing industrywide standards. Levin spoke feelingly of his son, who used rap music to help reach his students at Taft High School in the Bronx, and of his nineteen-year-old daughter, who was engaged in social work and who spent her summers working in shelters that housed "battered wives and abused children." Like his own children, Levin in effect pleaded that he and his executives were good citizens. They cared. Trust us, he told his shareholders. The industry need not resort to censoring music but instead should participate in a voluntary labeling system.

When he defended Ice-T, Levin had confronted his accusers directly on First Amendment grounds, however specious. Today, Levin and Time Warner have abandoned this argument. Instead of meeting the opposition head-on and defending specific songs or performers, they say, essentially, Let's change the subject. Let's talk about Bennett's political motives. They claim that as the dominant American company in the music business, they are being picked on because Time Warner is a much easier target than a Japanese company like Sony, or Germany's Bertelsmann. They insist that Time Warner is not the worst offender—it is only the fourth-largest publisher of rap music, and gangsta rap accounts for only 3 percent of all music sold in America. They assert—much as the National Rifle Association does when it insists that people, not guns, kill people— that rap music is not responsible for the prevalence of violence in society. They speak of Time Warner's many good deeds, including some visually arresting public-service ads, developed jointly by HBO and the Warner Music Group, which feature rap performers and are meant to deglamorize guns and violence. They produce lists of black academics and other leaders who extol gangsta rap as meaningful art.

And they resort to guilt: Who are we to judge? Asked if some lyrics troubled him, Fuchs responds, "Yes, but I'm not sure I should be the translator. I'm a forty-nine-year-old white guy raised in a middle-class Jewish environment. There's clearly another society growing up within this country and for various reasons it has a different language, different values, and a different approach to life in America than what a typical middle-class white American has. Let's remember, this is only language. It's not screaming 'Fire in the theater!' And it's not guns." What Fuchs and his colleagues in the music industry struggle not to do is talk about specifics—like the poison often contained in the songs of Nine Inch Nails or Tupac Shakur, who sat in prison for an act of violence while his Interscope album flew near the top of the charts. Beleaguered on several fronts, Time Warner strives not to admit error.

In a country where the most outrageous crimes and salacious imagery are now so common that they have lost their power to shock, Time Warner obviously doesn't deserve to stand alone as a target for opprobrium. The crime-drenched local TV news broadcasts, and Dennis Rodman's bizarre, and brutal, behavior on the basketball court, spring to mind, as do Howard Stern, or Calvin Klein ads. Then there is verbal violence. G. Gordon Liddy, who was convicted in the Watergate break-in and is today a popular talk-radio host, has boasted on air of using pictures of Hillary and Bill Clinton for target practice. Recently he advised listeners that if surrounded by federal firearms agents they should fire at the agents' head or groin, avoiding the bulletproof vest. "Head shots, head shots. . . . Kill the sons of bitches." Bob Mohan of KFYI in Phoenix said of gun-control advocate Sarah Brady: "She ought to be put down. A humane shot at a veterinarian's would be an easy way to do it."

What does Bennett have to say about such verbal violence? To his credit, Bennett does not hesitate to criticize his allies on the political right. He has phoned a fellow conservative Republican, Rupert Murdoch, who owns the Fox network, to complain of the "trashy" promotions for *Beverly Hills 90210* that affronted him and his eleven-year-old son while they watched the National Football Conference playoffs. "In speeches, I mention Fox more than Time Warner," Bennett says. He has denounced the National Rifle Association's harsh attacks on federal law-enforcement officers, charging that their cop-bashing sounded very much like the "cops are pigs" talk of the Weathermen or the Students for a Democratic Society in the sixties. As for talk radio, he is critical—"I don't think the Liddy stuff is helpful, and I've said so publicly"—but he believes music is a more powerful influence on children. Though Liddy "is on the edge," he says, nevertheless he is "part of the broad political debate, which ought to be pretty robust. The music is about making money and selling stuff to kids."

The core issue is the nature of corporate responsibility. Why was it OK, for example, for former liberal activists like Warner Music's Ken Sunshine to accuse Dow Chemical, the maker of napalm, of irresponsibility during the Vietnam War, but not OK for Tucker and Bennett to lodge a similar complaint against Time Warner? "In the context of what gets produced out there, we may not be doing better," admits Sunshine, vice president of Warner Music–U.S. "But I know there are things Warner Music will not produce. Anything that talks about killing, we won't produce." He called back to add the following to his list: "Any song that promotes harm against a group, that says, 'Kill the Jews! Kill the blacks!' We wouldn't put that out now." But they will put out a Nine Inch Nails song that promotes harm against women. "For those who would pretend not to understand," wrote *Daily News* columnist Stanley Crouch, who is black, "try this: put on some gangsta rap and remove the words 'nigger,' 'bitch' and 'ho.' Replace them with 'kike' or 'Jew.' Everything will suddenly become clear."

Are Time Warner executives deaf to this issue? Asked if this song and several others were trash, a senior Time Warner executive says, "Now I'm talking off the record, right? The answer is yes. I'm a father too."

So why do fathers lend their companies' imprimatur to trash? There are, to be sure, matters of principle involved—artists need a climate of creative freedom, and consumers have a right to make their own choices. There are reasons to worry about the increasing demand for censorship, from those who seek to ban books (the religious right), from those who wish to punish speech on campus (the politically correct left), from those who introduce legislation to criminalize any "obscene, lewd, lascivious, filthy, or indecent" materials sent by one computer to another (Democratic senator James Exon and Republican Slade Gorton). But the battle over censorship, as Time Warner executives privately concede, has more to do with profits than principle. Music is a twelve-billion-dollar business, with stupendous profit margins because production costs are cheap, particularly when it comes to the new performers who record rap music. Asked why they produced rap, a senior Time Warner music executive whispers, "The first issue is profit."

But he cites a second reason that is linked to a principle. If Time

Warner sold its 50 percent stake in Interscope—as some senior executives at the company would like to do—the company would lose not only cash but perhaps talent as well. "A Madonna or Neil Young might leave Time Warner if they thought we were buckling under to Bill Bennett," confides a Time Warner executive. A prominent music talent agent, who says he would not represent gangsta-rap performers because "their music desensitizes people," nevertheless cautions that Time Warner would be hurt if it sold Interscope or banished certain performers in a way that appeared antiblack.

All of this poses a problem for free marketeers like Bennett and others who trumpet "family values": How can they square their faith in free enterprise with such sordid behavior? "That's why I believe in two cheers for capitalism, not three," says Peter Wehner of Empower America. "Capitalism doesn't have a conscience. The argument that because it's free enterprise, that somehow this makes it right, is not compelling to me. I don't think a company like Time Warner ought to talk about being a good corporate citizen if they put this stuff out. Just because it sells doesn't mean it's good." As Marx said—and conservatives like Wehner will concede—private greed vies with the common good.

Admittedly, one wouldn't want publishers, say, to publish only books whose viewpoints they supported, nor would one want studios to just make art movies. Business leaders do sometimes sacrifice profit for social reasons. Fearful that a madman had inserted poison into Tylenol bottles, Johnson & Johnson CEO James Burke decided a decade ago that the company's good name was worth more than the cost of recalling the product from shelves. When Grant Tinker was chairman of NBC in the mid-eighties, he listened to Howard Stern one day on WNBC radio and was appalled by his foul language. "Why do we air Howard Stern?" he asked Bob Walsh, who oversaw the radio division.

"Howard Stern is the difference between profit and loss at WNBC," answered Walsh.

"Look," responded Tinker, "if that's the only way to make money we shouldn't be in business." Tinker canceled the Stern broadcast.

Similarly, Bill McNulty, the general manager of KCKC-AM, canceled the popular Gordon Liddy program as too incendiary.

And sometimes public protests compel a company to sacrifice profits. In the spring of 1989, NBC aired a tasteless Aaron Spelling series called *Nightingales*, which featured nurses in wet T-shirts. Although the one-hour drama was a ratings success, the network canceled it after nurses' groups and others protested that it stereotyped nurses as sex objects. (Perhaps more crucially, advertisers like Chrysler said they did not want their products featured in such an unfriendly environment.)

The solution most often advocated by companies like Time Warner—placing a warning label on certain music—is a palliative. "The labeling program that exists now is voluntary," explains Hilary B. Rosen, president of the Recording Industry Association of America, the trade group for two hundred and fifty record companies. "The decision whether or not to sticker is a record company's alone." And, as Rosen concedes, companies will not sticker if it hurts business. She mentions the music of Prince, which teenagers find sensual and sexually suggestive. "If it meant that a sixteen-year-old wouldn't buy Prince, the company wouldn't sticker Prince," says Rosen.

Companies, like politicians, respond to noise. Bennett and Tucker are planning more ads. Tucker says that "this is going to be a major subject" at the convention of the National Political Congress of Black Women on July 9, 1995. She is writing letters to the members of Time Warner's board. "I just believe that Levin is not going to continue this," she concludes. There is a precedent for her optimism. Two years ago, Michael Fuchs, like other television executives, at first resisted pressure to label violent programming. Today, Fuchs admits: "The jawboning has caused a de-emphasis on violence on television. And although I was not in favor of that initially, I've been on the public record as saying I do believe that there is gratuitous violence. All of that [pressure] has had some results. I do know from inside the programming departments in the networks that clearly there has been an emphasis on getting rid of some violence. It is clearly on the radar. So the government jawboned us, and it worked to some extent."

The task at hand is to use the club of public opprobrium—to link reputable executives to the sometimes disreputable things their companies do—in order to alter the corporate culture. As Bennett puts it: "Madison said that if every member of the Athenian Assembly had been a Socrates, the Assembly still would have been a mob. Things take over. People sometimes subordinate the best that's in them."

POSTSCRIPT:
This piece was written for *The New Yorker.* Because it was delivered late and space was tight, an eight-hundred-word version ran in the Talk of the

Town section. Ironically, several people at the magazine objected to the printing of the misogynistic lyrics of the Nine Inch Nails song, on the grounds that they were offensive. Precisely. The lyrics were published. Time Warner did sell its stake in Interscope early in 1996—to Edgar Bronfman's music division at MCA. As a Canadian-owned company, it is less susceptible to political pressure—but just as susceptible to public pressure.

I recently witnessed one vivid expression of the clash of values at an international conference in Davos, Switzerland, on the uses of the Internet, moderated by Esther Dyson, who chairs the Electronic Frontier Foundation in Washington and is an Internet pioneer. Dyson is famous for her ardent defense of an unfettered Web, but at Davos she found herself surrounded by high-powered foreign officials who just as ardently believe that the Internet must be made to serve society, not the individual. Chan Heng Chee, Singapore's ambassador to the United States, explained that her government licenses Web sites. "By licensing you are asking for responible use," she said, acknowledging that transgressors would be punished.

Dyson was as offended by this assault on freedom as William Bennett was by gangsta rap, but others refused to cede her the high moral ground. Hisham El Sharif, an Egyptian official educated in the United States, sided with Singapore, saying that he favored regulating "human dignity things"—such as expression that might be construed as racist—and urged the adoption of international standards, to be negotiated by "a United States of the World." Dyson and others from the West, including Uffe Ellemann–Jensen, a former foreign minister of Denmark, denounced this as a menace to individual liberties and liberal values.

"I am not a liberal," Mohammed Larijani, a member of the Iranian parliament, instantly declared. He revealed that the Internet was a fact of life in Iran, one he had helped expand to fifteen Web providers. But he and his government opposed "pollution"—a term he defined as embracing attempts by people like Dyson to spread Western democratic values on the Web. "A nonliberal system does not equal intolerance," he added, explaining that his country favored "community" over "individual" values.

If we can't agree on a precept of the World Wide Web— "the free flow of information"— asked Paul Sagan, an American who had supervised Time Warner's new media efforts, does this not mean that instead of one World Wide Web we will be seeing "a First World, a Second World, and a Third World Web?"

"Our censorship is an enlightened one," said Ambassador Chee of Singapore, expressing a sentiment most Americans would find unenlightened. This exchange is a reminder that discussion does not always result in common understanding, which means there can be no power of shame when there are no common values.

Beyond these prickly questions about values loom the dominant mysteries every company confronts: What does the consumer want? And is the product user-friendly? To answer these questions, the Highwaymen spend, and often lose, vast sums. Among the biggest gamblers is Microsoft, as the final chapter in this book demonstrates.

16

Jumping off a Bridge
Microsoft, and Michael Kinsley, Enter Cyberspace

(*The New Yorker*, May 13, 1996)

When Michael Kinsley announced, last November, that he was leaving Washington, D.C., to edit an on-line magazine in Seattle, colleagues and friends reacted as if he were moving to another planet. And, in a sense, he was. In Washington, Kinsley was celebrated as an editor and writer, and he had won fame and large lecture fees as the cohost of CNN's *Crossfire*. Telephones rang; attention was paid.

On Kinsley's new planet—Microsoft's corporate headquarters, in Redmond, just outside Seattle—the phones rarely ring. There are nearly nine thousand employees on this collegelike campus, and they converse, silently, by e-mail. Ties and jackets are shunned, jeans and uncombed hair are "cool," and the denizens speak a language of their own. A person might say, for instance, that he needed to locate the "owner" (person in charge) and then "drill down" (learn more) in order to "achieve granularity" (grasp the complexities) and then get to "2.0" (the next level) while avoiding "a bandwidth problem" (an overloaded schedule) or a "communications disconnect."

Kinsley has always thought of himself as a contrarian, an impression he fostered as the editor of *The New Republic* and as the writer of the magazine's "TRB" column. At Microsoft, he maintains that role because he has become the observant anthropologist. The Microsoft natives "have their own inside-the-Beltway culture here, too," Kinsley says. "They're not following primaries. They're following the new release of Netscape." The young men and women who work in each of the thirty Microsoft buildings receive the same free sodas and coffee, eat the same subsidized meals, and inhabit the same one- or two-"module" windowed offices. "Corporate socialism," he marvels. He also notes, "I'm the oldest person I run into all day." At forty-five, he is eleven years older than the average employee, and is older than every senior executive but one at Microsoft.

Even Kinsley's appearance sets him apart. He wears L. L. Bean chinos and cotton shirts, white socks, and the round clear-plastic-framed eyeglasses he first purchased from England's National Health Service—the kind of austere "war-criminal glasses," as he calls them, that were worn by former Defense Secretary Robert McNamara. But, since few of his new colleagues are concerned with appearances, he blends in. "Someone said to me that going from CNN to here is like going from a company where men wear makeup to a place where women don't," he says.

The relevant question about Kinsley, though, is not sartorial; it is whether the electronic magazine he is conceiving will be real. He is, metaphorically, jumping off a bridge—only he doesn't know whether water or cement awaits below. He does, however, know this: more than a third of all American households have computers, according to the national research firm Odyssey, and a rapidly expanding quarter of these households are connected to the Internet. Kinsley also knows that an on-line magazine is dirt cheap, since there are no costs for paper, printing, or postage. And he knows a larger truth: traditional middlemen—the Post Office; billing services; travel agents; department stores; video, record, and book stores; movie theaters; bank tellers—are threatened by new delivery systems, from cyberspace to wireless technology. Those middlemen—who get between the product and the customer—will become "roadkill" on the information superhighway, in the phrase of Microsoft's technology guru, group vice president Nathan Myhrvold. Myhrvold, a thirty-six-year-old Ph.D., writes lively, term-paper-length missives every couple of weeks, and circulates them among fellow executives. Myhrvold believes that cyberspace is to America in the late twentieth century what Frederick Jackson Turner said the frontier was to America in the nineteenth century: the engine driving an economic democracy. A Myhrvold memo from September of 1993 declared:

One of the most general and dramatic aspects of the information highway is to virtualize space and time. Put another way, the highway will break *the tyranny of geography*—the stranglehold of location, access and transportation that has governed human societies from their inception. . . . Assuming that the network is there, a person in Redmond or Manhattan or nearly anywhere else will have *equal* access to goods and services presented on the network.

But this is stargazing. Back on planet Earth, none of the hundreds of electronic magazines started in the past year or so—not *HotWired*, or *Word*, or *Salon*, or *Feed*, or *Mr. Showbiz*, to name a few—have satisfactorily solved the puzzle of how to bill and collect from subscribers, or how to justify rates for advertisers. As for the audience, emerging technologies could make lots of information available, but it is not clear how aggressive—and invasive—these electronic publications are prepared to be. Nielsen Media Research can't even agree with itself about the number of people using the Internet: it reported last October that there were twenty-two million adult users and this April that there were just over nineteen million. One of the most widely trafficked entertainment magazines on the Internet is the year-old *Mr. Showbiz*, but its mixture of entertainment-industry reporting and gossip attracts only twenty-three thousand daily visitors, according to its deputy editor, Jim Albrecht. "Like every other site on the Internet, it's a money loser," Albrecht, who is twenty-eight and is based in Seattle, says. "I've never heard of a site that makes money. The gamble is that this is going to be a big advertising industry." Build it, and they will come.

What Kinsley has set out to build is a weekly public-affairs magazine somewhat to the left of *The New Republic* but much quirkier than the pigeonhole he found himself imprisoned in as the "liberal" on *Crossfire*. The new weekly magazine, which will be called *Slate*, will run to about twenty-five thousand words (*The New Republic* runs to about forty thousand words), and the most optimistic goal, according to Russell Siegelman, who until recently was the general manager of on-line services for Microsoft, is a paid circulation of about a hundred thousand. Siegelman, who is thirty-three, is Kinsley's immediate superior.

To retrieve *Slate*, those who subscribe—Microsoft is discussing rates of between twenty and thirty dollars a year, and about three dollars for a single issue—will connect to the Internet and type "http://www.slate.com," and it will appear. Other magazines are "read" in the same way—sometimes to the consternation of World

Wide Web users, who find that text and pictures are occasionally retrieved with excruciating slowness, or sometimes, when traffic is
very heavy, not at all. As we've seen with bold promises to provide
video-on-demand and five hundred channels and handheld computers, technology sometimes moves more slowly than the human
mouth.

What will set *Slate* apart from most electronic magazines, Kinsley
hopes, will be a traditional, linear sensibility. He will run pieces
longer than seven hundred words, which is generally thought to be
the maximum attention span of those who scroll cathode-ray screens
and skate from site to site with a mouse. He will try to navigate
between becoming a data bank, as are many Web sites, and a hip,
dripping-with-irony magazine where people write faster than they
think. Kinsley is an unabashed elitist, dismissive of the tell-me-
what-you-think Zeitgeist of the Web. He does not apologize: "I'm
too old to go whoring after twenty-somethings." Later, he adds,
"I'm operating on the assumption that you can give people a meal."

But there are those who think this is the wrong approach to the
Internet. Esther Dyson, the president of EDventure Holdings, who
was a Harvard classmate of Kinsley's and later an Internet pioneer,
believes that an electronic editor must be more "an intellectual bartender than a chef." She says, "Michael doesn't suffer fools gladly.
Generally, you want someone who's a little kinder. You need to understand: you're not in control. In the interactive world, the bartender doesn't do all the talking. The bartender knows who should
talk to whom." If Dyson's guess—and it can be only that—is correct,
then Kinsley the editor is destined to become another middleman,
another roadkill in the abyss of cyberspace.

"I was your classic nonathlete" is how Kinsley describes being
brought up as the only son in a Jewish home in Birmingham, Michigan. His younger sister, Susan, is a lawyer and writer who is now
working with refugees in Sarajevo. Michael's father, George, who
was a surgeon, died of a heart attack in 1974, at age sixty-seven; he
was by most accounts, including his son's, a modest man who
worked far too hard. His mother, Lillian, a former high-school
debating champion who now lives in West Palm Beach, pushed
Michael to excel in school and to take piano and art lessons. He
starred as a student at Harvard; as an editor at the *Crimson;* as a
Rhodes scholar at Oxford, where, he says, he "turned into an An-
glophile"; and then at Harvard Law School. In 1976, he was a
twenty-five-year-old third-year law student when Martin Peretz,

who had been a college teacher of his and had bought *The New Republic* two years earlier, proposed that he become its managing editor. He accepted and moved to Washington. Two years later, he ostentatiously quit *The New Republic* after a spat with Peretz, who almost instantly lured him back with a raise and a grander title, editor. For the next dozen years, he alternated in that job in three-to-four-year stints with Hendrik Hertzberg (now the editorial director of *The New Yorker*). Kinsley had a kid-in-a-candy-store sort of journalistic experience: in 1981, he moved to New York to become editor of *Harper's* (for a troubled twenty months); returned to Washington and *The New Republic* in 1983 to write "TRB" for the next eleven years; again became editor of *The New Republic* in 1985; went to England in 1989 as the "American Survey" editor of *The Economist*; and crossed the Atlantic again seven months later, to sign up with CNN. By contrast with his hot, snarling "liberal" posture on *Crossfire*, his "TRB" columns, op-ed pieces in *The Wall Street Journal*, and articles in *Time* and *The New Yorker* were models of clinical precision.

Somehow, Kinsley's journalistic reputation survived the *Crossfire* food fights—six and a half years of shouted battle with the "conservative" bunker. Kinsley made no secret of a yearning to run a magazine again. Yet when an offer came to edit *New York* magazine, in January of 1994, he agonized for weeks. Then he said yes. And then, seven hours later, he said no. A man who had grown up in a leafy Detroit suburb and owned a house in Chevy Chase, Maryland, he dreaded the idea of living in New York City. He also dreaded making a decision. The *New York* magazine episode, he admits now, "was a midlife crisis."

What the decision did do was put Kinsley in a frame of mind to change his life. He grew a beard. He gave up his "TRB" column. "I spent a year and a half sulking over my stupidity in turning it down," he says. He saw himself, correctly, as skilled at spotting and nurturing talent, as someone who knew how to make a home for writers, treat them as peers, serve them potluck dinners in Chevy Chase. By the spring of 1995, he recalls, "I was determined to be on the next train to pull out of the station no matter where it was going—provided that I was the engineer."

He talked with the Time Warner, Inc., editor in chief, Norman Pearlstine, and others about editing either what he calls "a highbrow newsmagazine, roughly like *The Economist*, or an overtly liberal magazine to occupy the space *The New Republic* used to occupy." But the conversations were proceeding slowly, partly because such publications have high fixed costs and low circulation. What ignited his imagination was a chance conversation with Jeffrey Dearth, then the president of *The New Republic*. Kinsley recalls, "He mentioned to me that there are now storefronts in shopping malls in suburban Virginia that will set you up with a Web page for a magazine and that it could be like *The New Republic*." Only a lot cheaper. Kinsley was not entirely computer illiterate. He had been exchanging e-mail for years with the writer Nicholas Lemann, his closest friend, and had just begun to explore the Internet. "I would say I have computer-nerd tendencies but keep them suppressed," he explains. The conversation with Dearth, however, galvanized Kinsley to write Pearlstine and ask, "Would you be any more receptive to a magazine on the Web?"

Pearlstine arranged to meet him on August 7. In preparation, on July 27 Kinsley sent a five-page, single-spaced memorandum sketching what he, interestingly, called "a traditional magazine, or something fairly close to it, distributed on the Web." Kinsley calibrated his "motivation" this way: only 1 percent was "because of the new technical possibilities of *presentation* on the Web (e.g., hyperlinks, sound, video)"; 50 percent was "because it miraculously sidesteps all the printing and distribution problems/costs of a traditional magazine"; and the remaining 49 percent was based on his sense that this was "a new medium, it's almost certainly the wave of the future, and . . . we want in." Kinsley's focus was more on the message than on the medium, and he vowed that those who worked on the magazine would not get distracted by technological "bells and whistles."

In his memo Kinsley contrasted his approach with that of other electronic publications by saying that his would have "an attitude," but it would differ from the voguish "free-floating cynicism and snideness" that is the "cheap attitude" prevalent on the Internet. He proposed to charge subscribers for the magazine rather than offer it free (as most on-line publications do). He proposed a weekly print version as well.

The presentation to Pearlstine went well, and further discussion followed. But, as days and then weeks passed, Kinsley grew impatient. Then he saw the September 4 issue of *Newsweek*, which mentioned that the Microsoft chairman, Bill Gates, was "looking to hire a big-name editor" for a new news division, the Microsoft Network. Kinsley knew the Microsoft executive vice president, Steven Ballmer, slightly from a Harvard connection. Kinsley wrote to Ballmer and asked, "Do I qualify as a big-name editor?"

He did, even though Ballmer said *Newsweek* was wrong—Microsoft was not then searching for an editor. But, as *Newsweek* had reported, the software giant was eager to expand into the content business, to develop new products as well as new technology, thus avoiding the fate that Nathan Myhrvold had predicted for most middlemen. In fact, by the summer of 1995 Microsoft had concluded that investing in new technology alone would not permit the company to maintain its astonishing growth—net revenues soared to nearly $6 billion last year, and net income was $1.453 billion. The company was belatedly beginning to focus on the Internet and three areas of content: an on-line content provider like the Microsoft Network; a small number of "branded" properties, which could include a magazine like Kinsley's; and consumer software. Ballmer telephoned Kinsley and told him he was interested, and said he'd get back to him.

When Ballmer got off the phone, he e-mailed several colleagues. All expressed eagerness to explore teaming with Kinsley. The most passionate proponent was probably on-line-services chief Russell Siegelman, who read *The New Republic* cover to cover each week and calls himself "a huge fan" of Kinsley's. Another executive drawn to Kinsley was Myhrvold. In a memo that Myhrvold distributed internally last July 16, "Principles for Online Content and Brands," he noted that records, most books, and movies were " 'published' by companies which play essentially no role in the actual creation of the material," and that for most entertainment-information companies, with the exception of entertainment companies like Disney, "a strong brand is not a large source of value. People do not buy records because they were published by Sony Records, or Geffen Records—instead they buy Madonna, Aerosmith or Zubin Mehta. . . . The best model for online content in the existing media world is the magazine business, where editorial brands can become the primary enduring sources of value." Such magazines, Myhrvold argued, would be cheap to produce and distribute on-line, and didn't require a vast audience to succeed.

In Ballmer's view, however, there were two drawbacks to Microsoft's aligning with Kinsley. As Ballmer, a prematurely bald man of forty who wears bright crew-neck sweaters and whose booming voice is often enlisted to fire up the troops at corporate rallies, spelled out the drawbacks for me, the first was "Is this a business?" and the second was political: "Does this company have to worry about having him in terms of having a point of view?" Kinsley describes himself as "a perennial malcontent." He relishes skewering public figures and, just as frequently, public pieties. "I am honestly bewildered by the frequent complaint that American culture is hostile to religion—that religious beliefs are routinely belittled and held up to scorn," he once wrote. "Does anybody really think it's harder to stand up in public, in 1993 America, and say, 'I believe in God,' than it is to stand up and say, 'I don't'?" He once deflated Vice President Al Gore by describing him as "an old person's idea of a young person." As a thirty-year-old, Kinsley eviscerated the very Washington talk-show circuit he later became part of in a savagely funny piece, "Jerkofsky & Company," with sendups of George ("George III") Will reciting from Bartlett's quotations and Hugh ("Sidewall") Sidey boasting of "my major lunches around Washington" and Marvin ("Jerkofsky") Agronsky lowering his voice to say, "Concern mounts in Washington about any number of things. Life on earth continues, but doubts arise about its purpose or justification."

Russell Siegelman invited Kinsley to visit Microsoft in mid-September. Siegelman and Peter Neupert, the vice president for strategic partnerships, took him to dinner, and the three discussed Kinsley's ideas and the possibility of a marriage with Microsoft. Kinsley welcomed their thoughts. What he said he wouldn't welcome was any limitation on his editorial freedom. The magazine had to be "lively," he recalls saying to them, and "the only way to be lively is to make trouble. You understand?" They said they understood. "In my experience," he continued, "trouble comes not when you attack a company but when you attack the owner's friends. For example, Warren Buffett"—a friend of Bill Gates. Understood, they said. Were they prepared for the possibility that he might attack Microsoft? No problem.

On September 18, a Monday, as is the Microsoft custom when recruiting an executive, Kinsley was scheduled to be interviewed individually by most of Microsoft's hierarchy, except Gates. Kinsley pressed each about editorial independence but also talked at length about his vision for an electronic magazine. In turn, they pressed him on politics. The group vice president Pete Higgins, who, along with Myhrvold, presides over the Applications and Content Group, remembers worrying about whether Kinsley was too "left" and wondering aloud whether his magazine might be paired with a conservative counterpart. The Microsoft executives, despite their belief that geography was becoming irrelevant, told Kinsley that, "for synergy reasons," he should edit the magazine on their campus, and not from Washington, D.C., or New York.

On this first visit, there was "synergy" between Kinsley and Microsoft. "I thought he was a kindred spirit," recalls Patty Stonesifer, who, at thirty-nine, is the senior vice president of the Interactive

Media Division. "He has a kind of intellect that likes to break down issues and consumer needs and marketplaces into an understandable and manageable and parcelable challenge." At the end of the visit, Kinsley told Siegelman, "If you guys are interested in doing something with me, I'm interested."

They were interested. That very night, in an e-mail sent to seven Microsoft managers at 10:13, Siegelman wrote:

> Having him be an MS employee would legitimize MS as a content company to a large degree. . . . I think we can get him, but it will take the right offer and selling. His one big issue is church and state: everyone today gave him the right message about senior mgmt not mucking with editorial decisions, but I think he will need constant reassurance of this. My bottom line: I think that Mike is super and would be a fantastic catch.

The following Friday, Kinsley e-mailed Siegelman:

> A few more thoughts as you guys decide whether/how you want to proceed. . . . Pete Higgins seemed to have some concern about my being "on the left." I hope you can reassure him that my politics are pretty eclectic, averaging out I suppose as a moderate-liberal with a libertarian streak. One thing I would NOT be interested in would be being "paired" with a conservative product. That would be unfair to me, and unfair to the real left. It's one of the things I'm trying to get away from at Crossfire.
>
> But it did occur to me that, even if whatever-it-is-we're-talking-about starts out as a single "magazine" product, it might contain (let's call them) "modules" or mini-magazines from different political viewpoints. Not just left/right, but mainstream conservative; mainstream liberal; progressive left; libertarian; agrarian/luddite/unabomber (!); etc.

Even before Microsoft made Kinsley an offer, these exchanges with the technology-oriented company had planted an idea that would flower. "I like your idea of 'mini-magazines,' " Siegelman e-mailed to Kinsley. He went on:

> This makes sense. In the world of the Internet it is not always clear when the "boundaries" between "publications" start and end because everything is hyper-linked. We definitely want to create an online magazine that has brand, and everything hangs together as "one" magazine. But one way to think of this is as a conglomeration of "mini-modules" that can be linked together in several ways.

"Glad you like the 'modules' notion," Kinsley e-mailed back, revealing that this new medium was beginning to alter his message.

> I've been thinking further about that and a variation (or addition) might be a sort of on-line McLaughlin Group, only more sophisticated and intellectually honest: really smart people . . . discussing some issue (eg, Medicare reform) in a dialogue supervised by me, purged of sound bites and cheap shots, and with some hope of dialectical progress toward agreement/solution.

The internal Microsoft debate persisted for six weeks, largely because the debaters were vexed by the "political" issue. But they knew that an association with Kinsley would "teach us a lot," Patty Stonesifer recalls. The Microsoft people also knew that they were late to see the potential of the Internet and needed content for it.

By late October, Kinsley had heard nothing concrete from Microsoft or Time Warner, and was becoming anxious. "I then called Microsoft and Time Warner and said I needed a decision," Kinsley says.

The ultimatum led Siegelman to go see Nathan Myhrvold and Pete Higgins and say, he recalls, "Look, we've got to nail this down. If we wait much longer he's going to work for Time Warner. . . . Can I send an e-mail to Gates that you will back?" Gates had been briefed by Ballmer and Myhrvold and was in agreement. Although Gates had yet to meet Kinsley or even exchange an e-mail with him, he was telling people that Kinsley's initial electronic-magazine memo was "brilliant." Gates gave his assent.

Siegelman thereupon called Kinsley with an offer to make him the editor of Microsoft's on-line magazine based in Redmond. The company would pay him "two-thirds of what I was getting just to do *Crossfire*," Kinsley says, refusing to confirm his previous salary. (His pay, a Microsoft insider says, is just under two hundred thousand dollars, not counting a bonus.) He would be an employee, not an owner, but if the magazine was successful he would be likely to see the results reflected in bonus checks and stock options. Time Warner also came through with an offer. Unlike Microsoft, it proposed an employment contract that would run for three years, with a heftier salary. But Time Warner would not pledge to back his magazine and promised only to take a hard look at it. Kinsley weighed the two offers, agonizing much as he had done when he turned down *New York* magazine. The safe thing to do was to stay at CNN. But Kinsley was ready now to change his life.

Microsoft announced the appointment on November 6. Within Microsoft, executives congratulated themselves that by hiring Kinsley they had achieved "name branding." He was welcomed as a "media star" by *The Seattle Times*. In D.C., the reaction was that he was simply mad to walk away from a visible and prosperous platform. Friends like Lemann praised his courage. "It's a kind of present-at-the-creation feeling," Lemann said. "I'm intensely curious about what he comes up with." Others were dubious. "There are decent on-line publications like *Word* and *Suck* and *Salon*," said *New York*'s founding editor, Clay Felker, who directs the Felker Magazine Center at the Graduate School of Journalism at Berkeley. "Nobody wants to read very much. If Kinsley is thinking he can put on a version of *The New Republic* or his column, they won't read it. . . . Our eyes get tired quicker on a computer screen. Plus it isn't very practical. People look at a screen as work, not pleasure. It doesn't have the portability of print. As a result, you have to write differently in order to keep people interested." The new medium, Felker added, requires more graphics, more interactivity, more diversions. In a discussion on the WELL, a bulletin board based in San Francisco, the news was greeted with some of the "attitude" that Kinsley had deplored: "It's very VERY nice to know that Mr. Kinsley will be coming in to save us from the veddy tedious web magazines we are forced to sludge through . . . all tepid pieces of crap for cyberfreaks" (kieran); "Somehow, despite his dweebish nature, I don't picture Kinsley as someone who hangs out online and is up on digital culture" (srhodes).

And Kinsley was soon reminded that his project was only a wavelet in the Microsoft ocean. The on-line magazine was just one of more than fifty product lines reporting to the Interactive Media Division. On December 19, five days after Microsoft and NBC announced a joint effort to fashion a twenty-four-hour cable-news channel and on-line news service, Kinsley told me, "I had no idea about it. I didn't know they were planning to become a direct rival to CNN. That was a little embarrassing to me." Not enough to dampen his enthusiasm, however. "Even if it's a total flop, it's an adventure," he said. In a matter-of-fact tone, he added, "There are lots of people who are good writers, many of them are better than me. I think there are fewer people who are better magazine editors than I am."

The first thing Microsoft employees noticed about Michael Kinsley when he appeared for work in Building 25 on January 2, 1996, was not the mail-order khakis or the green four-door Honda Accord, though these stood out as somewhat preppy. They noticed that he was wearing a white "Department of Justice" baseball cap. This was a flashing-neon reminder of the four-year federal probe of Microsoft, as well as a reminder of Kinsley's independence. They might have noticed, as well, that his closet-size one-module office contained no pictures of family or friends. Like many of the driven employees on this sprawling campus, Kinsley has Spartan tastes. A lifelong bachelor, he is thrilled to live in a self-contained world, where there is food and everyone can look out a window and see nothing but grass—and Microsoft buildings.

Physically, Kinsley has about him a languid, professorial air. His long fingers rarely move as he talks, his arms stiffly by his side; his eyes seem stretched open, for he seldom blinks, and from behind his oversized McNamara glasses his dark eyes stay glued to the person to whom he is speaking. Unlike the performer on CNN, Kinsley the editor speaks slowly, deliberately, quietly. Nothing about him—not the short, neatly parted black hair, not the parsimonious gestures, not the bland clothes—detracts from the feature that people at Microsoft or at *The New Republic* usually mention first: his intelligence. It is in this respect that Kinsley fits right in at Redmond. "It's a little like a summer math camp for high-school kids," Steve Ballmer says of the Microsoft culture. "People have to prove they have the right answer—'I challenge that!' "

To walk the mazelike corridors of Building 25, or any of the other buildings on the campus, is to encounter a sameness, where everyone has the same glassed entrance, the same space in which to toil. From his first day at work, Kinsley had the same office with the same adjustable desk and Compaq notebook computer with docking station used by many of his colleagues. What Kinsley didn't have was a plan for his magazine. "All I had was this vague memo," Kinsley remembers. "I had no business plan, no staff, no anything. All I had was an office."

He didn't even have a name for the magazine. He spent the first few days writing a memo that added flesh to the bones of his skeletal periodical. The memo—ten and a half single-spaced pages, which he finished on January 9—began by defining his magazine's audience as the kind of "politically and culturally engaged people" who read *The Economist, The New Yorker,* and *The New Republic.* For those without a computer, there would be a weekly mailed version of the magazine to "reinforce the idea that we are putting out a *real* magazine." He remained opposed to getting swept away by the technology: "Our guiding maxim should be, 'It's the journalism, stupid.' . . . In much on-line stuff now, the tail wags the dog."

The magazine, he continued, would not "push any particular political viewpoint." Nor would it obsess about "objectivity":

> This will be a weekly magazine. Obviously, "weekly" is a metaphor, just like "magazine." Material will come and go all the time. But, at least in the early days, we need to reinforce the metaphor, to make traditional magazine readers feel comfortable in this new medium. . . .
>
> There should be a notional moment each week when we "go to press" and "hit the stands" (one and the same in this medium). I would say Friday midnight. This will allow us to summarize the week, and allow people to read us "fresh" over the weekend. It will also highlight our advantages over traditional weeklies. Time and Newsweek "close" on Saturday; they reach subscribers and newsstands Monday at the earliest, and often Wednesday, Thursday or later.

Even with a Friday closing, Kinsley stressed, the magazine—"the site"—would "not be 'dead' for a week." Here he demonstrated some movement toward Microsoft colleagues who had been pushing him to use their on-line advantages and also to accept the industry view that these "Webzines" be updated with reliable frequency: Kinsley proposed to change features during the week. Again, however, he raised a cautionary flag: "Obsessing too much about being up-to-the-minute is, I suspect, a potential trap and distraction. We are selling analysis and commentary, not news."

As for a title, "this is the biggest single editorial decision we make, and we must make it soon," he wrote. "We want something that doesn't scream 'cyberspace' (or, worse, 'Microsoft'), but suggests it with a wink." The contents of the magazine, Kinsley said, would require "a combination of little appetizers, solid meat-and-potatoes, and dessert." He expanded the definitions of the features presented in his July memo to Time Warner, starting with a cover, or home page:

> We need a signature look, like The New Yorker's illustrations. My idea would be a political cartoon, either by the same artist every week, if we can discover a great new one, or from various artists. . . . Also, would it be possible, with a click, for readers to install the cover cartoon as their screen saver for the next week? That might appeal to a lot of people, and of course would be a constant free ad for us. And

what about a musical logo: a theme song (say, a few bars of jazz piano, or of Bach) that plays whenever our cover pops up?

A section he was particularly excited about—a "mediated forum," which he envisioned as a cross between the *McLaughlin Group* and the letters column of *The New York Review of Books*—would also benefit from technology. With "written, not spoken" responses composed overnight, the dialogue could extend through the week, producing "a far higher level of debate, especially if the participants are chosen for genuine knowledge and intellectual honesty rather than TV charisma." An ideal "first topic," he wrote, "would be antitrust: 'Is Microsoft a Dangerous Monopoly?' " The editor, who had frequently lambasted "spin doctors," added this spin: "A lively and balanced forum on that would get us lots of attention and prove our editorial independence from the start. Maybe Gates would even participate?"

Other departments would include a "short, highly judgmental profile of some important figure in the news"; a "kinetic visual" feature that might offer "an interactive map"; perhaps "interactive fiction," though he confessed to a "strong, ignorant prejudice that it must be crap." The memo pleaded for help from Microsoft to locate a publisher to handle business matters. He noted that he had budget authorization for eight people, including himself and an administrative assistant. If the publisher worked full-time, and if he needed two technical people, that left him with three editorial slots: a deputy working alongside him in Redmond, and an editor in both D.C. and New York. Defending this geographic division, he wrote, "Does it make sense to try to put out a national magazine from Seattle, without *any* presence in the cultural and political capitals? Even in cyberspace?"

As for a great mystery of cyberspace—how to make such ventures viable businesses—Kinsley envisaged revenues coming from advertising, subscribers, and the syndication of pieces. Another mystery: Who was reading the stuff? On-line services often boast of the number of "hits" they receive, equating this with readers. On many sites, however, when a visitor clicks on the title page it is counted as a hit. So is each click on an article or a piece of art. A single reader may be credited with twenty separate hits. Kinsley was also discovering truths that others had encountered before. Last November, David Zweig, a former Time Warner magazine executive, helped launch *Salon* (www.salon1999.com), a magazine close in spirit to the one Kinsley envisioned. Zweig met resistance from advertisers, which he considered normal for any new medium. The tougher obstacle may be the emotional resistance of subscribers. "It's too early to charge," Zweig told me. "I'd love to charge," but customers are "allergic." Kinsley, in his January memo, anticipated this resistance, and separated it into three categories:

> a) Technical. Can you safely give a credit-card number over the Internet? I gather this problem is rapidly being solved. b) Materialist. It's hard for people to feel that they should pay real money for something that has no physical existence. (Software itself originally had—and has largely overcome—the same problem.) c) Spiritual. A widespread feeling that charging for content amounts to "fencing off" cyberspace, and violates the ideology—almost the religion—of the Web.

Kinsley explained why he wanted to charge readers: "One purpose of this whole exercise should be to help people over the conceptual hump of thinking that when they buy a magazine, they're buying the paper and ink rather than the words and pictures. You do that by offering a product so good that people are willing to pay for it." He hoped the first issue would be ready by late April. And, because the first issue would be treated as a Broadway opening, "we don't have the luxury of an out-of-town tryout. We've got to be good from the start."

After his first month at Microsoft, Kinsley was still thought to harbor certain Luddite tendencies. "When he first came here, he was very—I won't say 'anti,' but he didn't want to do 'technology for technology's sake,'" recalls John Williams, who is thirty-one and had been a Microsoft manager for five years before becoming Kinsley's publisher in January 1996. Bill Barnes, a twenty-nine-year-old self-described "geek," who had been at Microsoft as a programmer for seven years and transferred in January to serve as the program manager for Kinsley's project, prodded Kinsley to confront the consequences of his new medium. In an e-mail Barnes expressed enthusiasm for his editorial ideas but warned that to deliver a printed version of the magazine posed problems, including the poor quality of "print on a typical computer paper."

Barnes also cautioned that the technology they were reliant on was still antiquated. Their billing system, for example, would initially depend on "old technology like users entering passwords for authentication" and customers phoning an 800 number to subscribe.

"Prediction: billing on the Internet is the hardest problem. And one we're not chartered to solve, though we need a solution from MSN (or Microsoft, or somebody)."

All these issues, and more, were pressing on Kinsley as January expired. He felt swamped. He still had not met, or even exchanged e-mail, with Bill Gates. He felt like a Martian. He would go to Russ Siegelman's weekly staff meetings and listen to the people chatter about "P&Ls," "boot," "DPI," "IMD," "Schedule +," "domains," "Gantt." He once whispered to me, "I have no idea what they're talking about." He couldn't even figure out how to make his office less hot, and ended up sending an e-mail to "Furnfac Account" on January 26 which read, "Is this the place to write about office temperature? If so, could mine be cooled down a bit?"

Kinsley had not yet had a chance to see Seattle by daylight. He had rented an apartment in Redmond after telling a real-estate agent that he didn't want to look at more than one. The place, with its bare white walls, had an unlived-in look, except for a fussy arrangement of mostly nonfiction books and CDs, ranging from Bach to Fats Waller. Aside from a Falcone piano, most of the functional furniture was new, since Kinsley had sold much of his before leaving Washington. Over morning coffee, he would read *The New York Times* and *The Wall Street Journal*, glancing out the bay window at Lake Washington and, on a clear day, at the Olympic Mountains. Then he would get his Accord from one of the rows of parking spaces allotted to the primarily youthful residents of this homogenized suburban village, drive to Building 25, work, eat lunch and usually dinner in his office, and get home just in time to go to bed. Because the frugal Kinsley would be content living in a medium-sized closet, he was not unhappy, but he told me, "I miss my friends. I miss the buzz a little bit. They're not talking around here about who wrote *Primary Colors.*"

Bill Barnes e-mailed Kinsley on January 24, saying that they needed:

> A codename for the product. This is almost certainly different from the actual product name, which is helpful for discussions with the outside world before we've gotten trademark on the real name, etc. This also helps confuse the media before we're ready to announce. Do you have any preferences? Your favorite bird? The name of your sled as a child? A literary or journalistic idol?

Kinsley e-mailed back:

How do you like "Boot"? Short and snappy. Besides the obvious computer reference, and the image of a kick in the pants, Boot is the hero of the most famous novel about journalism, "Scoop," by Evelyn Waugh.

Despite his frustrations, Kinsley was happy with the career choice he had made. There was an easy rapport in the office. On January 30, Kinsley e-mailed Bill Flora, his art director, whose title is Visual Interface Designer:

Bill, there's a site called Naked Pavement (found at www.razor-fish.com/bluedot) that seems to have a version of yr idea about page numbers. Also a lot of naked people. Either of these two elements (or both) may be of interest.

By the end of January, Kinsley felt he was making some progress. He had come up with the name *Slate*. It was "crisp, clean, short, easy to remember," and suggested "toughness," Kinsley e-mailed Russ Siegelman. With the hiring of John Williams as publisher, he was able to spend more time contemplating the editorial content. Williams had worked as group manager of Content Strategy and Business Development for the Microsoft Network, as lead product manager for Microsoft Works for Windows, and as a product manager in the Consumer Division. He knew how to borrow bodies and push buttons; he assumed responsibility for steering the lawyers and delving into business matters, and, unlike most publishers, he reported to the editor.

By February 1, the magazine was coming alive in Kinsley's mind, and, as his thoughts traveled from idea to execution, the new medium took him to unexpected places. To help navigate, he wrote a third memo, code-named Boot. The memo bulged with practical questions, which he directed to the two Bills—Barnes and Flora. Kinsley had selected a theme song he wanted played when someone got the magazine's home page or cover cartoon on the screen—Fats Waller's "You Meet the Nicest People"—and asked Barnes and Flora if they had to acquire the rights to this song. Should they change it once a month or more often? Should they have a "How to Read This Magazine" section, explaining how to read on-line? How to download in order to read it later and to save the on-line telephone charges? How to order a subscription?

The questions kept coming. Kinsley and his team planned to have hyperlinks, or highlighted words, accompany certain articles, so that readers could click on them and take, say, a detour through the origins of national health insurance. When the link allowed a reader to visit another site—say, an advertiser's—would the technology automatically remove the reader from the magazine's site (undesirable) or could they keep the reader within the magazine's frame? To minimize monotonous scrolling, should they sometimes treat a screen as a page, insisting that some features be squeezed onto a single screen? How much of an impediment was a low-speed modem for a subscriber who wanted to download art or a video clip? In book reviews, should they link a chapter of the book, allowing the reader to sample it? In movie reviews, could they run clips from the movie? How would they maintain an archive of previously published pieces? By topic? By issue?

The memo shifted from technical questions to a further refinement of Kinsley's editorial ideas, which were being infected by this new interactive medium. He wanted a weekly cartoon strip, and wondered whether it could be "moving and/or interactive in some way." Similarly, he hoped that someone like the Democratic media consultant Robert Shrum might do a feature, "Shrum's Ad of the Week," analyzing a political TV commercial and providing a window so that the reader could also watch the ad. Since the magazine offered sound, he proposed to publish a poem every week, and invite the poet to read it.

Much remained to be done before the first issue, including the recruitment of the three editors. To buy more time, in early February it was decided to postpone the inaugural issue a month, to May 31. There were minor setbacks. Kinsley dropped the code name Boot when the natives at Microsoft reported that to Generation X the word meant "barf." When he decided to go with the name *Slate*, Microsoft's lawyers reported that "www.slate.com" was owned by a man named John Slate, who lived in the South and never used it. The lawyers would try to buy the Web address without inflating the price by revealing that they worked for Microsoft. Microsoft executives, not surprisingly, wanted to have the Microsoft brand name as part of the magazine's Web address. Kinsley and his team successfully pressed for *Slate* to be perceived as independent by having an address without the Microsoft name.

In a way, Microsoft was winning a larger battle. By March, those who saw Kinsley every day noticed that he was changing in a fundamental way. The man who had insisted that the technology was only 1 percent of the attraction was finding that he was dazzled by "all the cool things you can do." He oohed when they showed him how a subscriber could read a story on Bosnia and click a map and glimpse how the geography had been altered by the war. He was enthusiastic about using sound so you could read and hear—and feel—a writer's words. He even succumbed to the idea of publishing interactive reader bulletin boards: in one e-mail he borrowed a phrase John Williams used—"the Rubber Room"—and described it as "a padded room where participants can bludgeon one another to their heart's content." Kinsley told me, "I came around that it's not just a sop to readers to have a little interaction. It's one of the most satisfying things about working for this company. You can think of something"—like having poetry read—"and you ask Bill Barnes, 'Hey, can we do that?' "

The Microsoft people were equally enthralled. Russell Siegelman says he welcomed Kinsley's resistance to the lure of technology: "We know how to do that. We needed someone who understood journalism." To them, Kinsley was proving to be a fellow scientist who was open to new ideas. By March, instead of ascribing a 1 percent role to technology, Kinsley said, "It's up about thirty or forty percent."

On some things, Kinsley was implacable. To Esther Dyson's claim that a good Internet editor was "an intellectual bartender" who listened more than he spoke, Kinsley conceded only, "I have come several steps in her direction. Originally, I said no to bulletin boards. But I think Esther has the idea that this technology changes the nature of life itself. I think she's slightly carried away. Even if her vision of the Internet as a community for current users is true, it will be less true for future users. I don't want to sound like an arrogant asshole who's become an Internet fascist, but there is a reason that some people get paid as writers and some don't. I don't want to go to a restaurant and be told, 'This is a community restaurant, and the guy at the next table is cooking for you!' "

In early March, Kinsley conscripted Jodie Allen from *The Washington Post*, where for the past six years she had been the highly regarded editor of the Sunday Outlook section. Allen says she signed on as Washington editor because she admires Kinsley and because "it was new—it was an exciting opportunity." Besides, the on-line element held a particular allure for her: "How many Mondays have I wished we could take back something we printed on Sunday?" Kinsley had already hired much of his Redmond staff, but he had not yet been able to recruit a deputy editor, and was still negotiating for a New York editor. "When you ask me what I've done the last two months, I have a hard time saying," Kinsley told me. "I call my mother every Sunday and she asks me what I've done this week. It's not like *Crossfire*, where you're doing it each night."

He continued to be engrossed by the puzzles that the new enterprise posed. In mid-March, I watched him struggle with several of them during a Tuesday-morning staff meeting in the cafeteria of Building 25. At one point, Kinsley said, "My original idea was that the magazine would close the less urgent stuff first"—the reviews, and other material that often falls under the heading "back of the book." He continued, "My view is that we should post material earlier in the week when it was done. So the idea is that if you log on anytime during the week you'll get a whole week's worth of stuff. Although we publish on Friday night, there will be something different every day. If we post a new book review, it would replace the old review, even if we post it on Monday."

The problem with that, Bill Barnes said, his sandals stretched under the Formica table, is that some readers would miss the review that was yanked.

"I'm not sure the benefits outweigh the confusion," John Williams said.

"It's good if we publish stuff as it's ready," Barnes said. "But it's not good if we erase last week's issue because we have an early close in, say, the arts."

"It will require some new behavior on the part of readers," countered Kinsley, now cast as the nontraditional champion of technology.

"Why not post both—what's the disadvantage?" asked Betsy Davis, the production manager. "If we keep changing stuff in the old edition, how will the reader know what's new?"

"I'm amenable to saying 'Keep the whole edition all week,' " Kinsley said. "But if you add it all up it gets a little long. And what do we do with the table of contents?"

"I may be anal, but I like the idea that I have read the whole magazine," said Davis, who has worked at the company for twelve and a half years.

"It rubs me the wrong way," Barnes said, also chiming in on the side of traditionalists. It offended him that readers might miss something simply because they called up the magazine on a Wednesday.

Kinsley reminded them that they were in a new business: "We're looking for ways to leverage our differences with other magazines. One of our advantages is that the reader can get our reviews 'hot off the press.' "

"Why not a daily magazine, then?" Barnes asked, echoing a question he had asked Kinsley in a March 8 e-mail devoted to whether they were publishing a weekly or a constantly updatable magazine. Instead of addressing this larger issue directly, Kinsley told the staff that he wanted to make use of the magazine's advantages as an electronic publication.

Barnes asked if Kinsley planned to update the printed version of the magazine that would close and be mailed on Fridays.

No, Kinsley said. The projected forty-eight-page printed version would be like a regular weekly magazine.

None of these issues were resolved at this meeting, which lasted a couple of hours, and the discussion only left them in murkier waters. They could, for example, technically update a customer's copy of the magazine by noting whether the reader had already logged on for the new issue. But Barnes asked after the meeting, "How do we know what a person has read or not read?" What if the subscriber scanned the magazine but did not read it? Or didn't finish it, and now it was gone? And if they were constantly updating an issue, as Kinsley asked Barnes in one e-mail exchange, how would they then date the magazine?

Later on the day of the meeting, Kinsley used e-mail to summarize for the staff their discussion about posting daily or just weekly, and he concluded:

> We will still "go to press" once a week, on Friday afternoon, in that (1) this is when our paper version will publish; (2) this is when our print-it-yourself version will be made available, and won't change throughout the week; (3) this is when our most timely features will close and be posted. But certain other features, such as the reviews, will be replaced on other days, on a regular schedule. People who are most interested in, say, the book review will know they can read it "hot" on, say, Monday or Tuesday. People who want to read last Friday's edition exactly as published last Friday can get it out of the archive.

Kinsley knew that this summary would trigger a torrent of e-mail, and he welcomed it. But it would not resolve the larger "philosophical issue" Bill Barnes had referred to: Would what was published be a weekly or a continuous magazine?

The business problems were no less daunting. Most print magazines generate the bulk of their revenues from advertising rather than from circulation. Yet with scant experience to rely on, the publisher John Williams admitted that "it's not clear how the revenue streams will break out." Microsoft was upgrading the software, and Williams expected that by March the company would be able to securely accept credit-card numbers and bill customers from day one. He had no way of knowing whether there would be customers for Kinsley's product, since anything so new involved guesswork. "It's like asking someone who never had a car whether he'd like one," he said. One problem was resolved by March: the magazine would have an ad on each of what would arbitrarily be called a page, and would charge more for ads that appeared on the top of a page.

Then there were the larger, more philosophical things. Kinsley and his team, like others, are roaming the dark side of the moon, trying to avoid the menacing cultural, business, and editorial craters. Apart from the question of charging subscribers, there is another cultural crater—what may be the dueling assumptions of his magazine and the Internet. The two are at war, Jim Albrecht, the deputy editor of *Mr. Showbiz* (www.mrshowbiz.com.), asserts. "The difference between an Internet design and a magazine design is not bells and whistles," he says. "It's about how you get the information. In a magazine, you get what's in that issue. In a Web site, people get what they want." This is an extension of Esther Dyson's argument that an editor must be an intellectual bartender. The Internet is personal. It is about me, and Kinsley's magazine is about him.

Indeed, CompuServe has a new on-line product—called Wow!— that invites users to customize by clicking next to "My News," "My Mail," "My Places." Microsoft Network News, which is housed in a large, open newsroom around the corner from Kinsley's office, is perfecting personalized software tools of a kind that Kinsley abhors. When Steve Forbes made a flat-tax proposal the centerpiece of his quest for the Republican presidential nomination, the Microsoft newsroom, with the help of the accounting firm of Arthur Andersen, devised an interactive form. Individuals could type in a few figures— income, approximate deductions, etc.—and the screen would instantly announce how much you would pay in taxes under the Forbes plan versus the current tax law. Some people see this tool as advancing the trend in journalism toward "news you can use," by giving individuals vital information with which to make intelligent choices. To Kinsley, however, "all these how-it-affects-you things bother me because they play into what's wrong with American politics, which is that too many people take the short-term, selfish view." What if the Forbes plan is good for you, he asks, and bad for the country? "One of the bad things about e-mail is that it's too easy to reach your congressman!" But if the customer insists, Kinsley will become just another talented entrepreneur who guessed wrong.

The dark side of the on-line moon contains other business-editorial craters. Journalism's traditional church-state wall, the one that acts as a stop sign to an ad salesman who would like to suggest a story or urge a reporter to indulge a big advertiser, has not yet been erected in cyberspace. And current trends may work against such walls. "The most popular Web sites are run by corporations," Kinsley notes. "The Pepsi Web site is the most popular. It's like the early days of television." In those days—and, increasingly, in TV's present—advertisers both sponsored and produced programs, weaving their products into the show. So we now bump into hidden ads. If a subscriber summons the magazine *Salon* on the Internet, and clicks the "shop" icon at the end of ten books reviewed in the "Sneak Peeks" section of each issue, he can order one of those books from a Borders bookstore. Since *Salon's* publisher refers to Borders as one of the magazine's "marketing partners," Kinsley asks, "Is that too close a nexus of editorial and advertising? I'm not sure. It makes me a little edgy." And, since Borders wants to sell books, will it exert pressure on *Salon* to publish only positive book reviews?

Another instance of a conflict between technology and journalism may be news packagers like Microsoft Network News, soon to merge its on-line efforts with NBC News. At Microsoft, the rows of editors who work at computer screens labor over news stories supplied by what the news director Andy Beers says are "content providers," which "we edit and then add multimedia elements"—video, audio, music, pictures, personalized interactive features (like how the Forbes tax cut would affect *you*). Every three hours, the editors update stories off the news wires, and in a typical day, he says, they will condense and add "multimedia elements" to a hundred and sixty stories. What they don't provide is original reporting. "The editor operates more in the model of a television producer," Beers says. But his editors don't leave the office. So as the editor gets more distant from the news, and comes to view the news as stories that best lend themselves to a multimedia package, inevitably the package may become more important than the content.

Still another crater: corporate gigantism, which can also vie with journalism. The favored buzzwords at communications companies like Microsoft and Disney and Viacom and Time Warner and AT&T and the News Corporation are "synergy" and "leverage." Microsoft, for example, seems to announce an alliance a day, and has already formed partnerships with NBC to produce news, with MCI to provide Internet access, with Rupert Murdoch's News Corporation to promote Windows 95 in his London newspapers, with John Malone's Tele-Communications, Inc., to offer a new cable channel that is dedicated to the personal-computer market, with America Online to exchange access to Microsoft's Internet Explorer browser in return for making AOL available to users of Microsoft software, with Hughes Electronics to deliver DirecTV satellite services, with DreamWorks SKG to create interactive games. The resultant back-scratching need not take the sinister form of censorship imposed from the top but, rather, the more subtle form of self-censorship: *If we do this story we hurt the company, and maybe ourselves.* Kinsley says he's raised all these questions with Microsoft, and adds, "I got all the right noises. It's in Microsoft's interest. They want to be a media company. They understand you've got to have credibility." They understand now, when the magazine is an abstraction. But what happens, as Kinsley asked before he accepted the job, when he attacks their friends? Or their business interests clash with their editorial interests? This is uncharted terrain for Microsoft.

For Kinsley, pesky business problems kept intruding, and sometimes pitted him against his staff. On March 14, he huddled knee to knee in his office with John Williams to review the budget presentation that Williams, accompanied by Kinsley, would make to Russell Siegelman at one o'clock that day. Looking over the numbers that Williams had prepared, Kinsley appeared placid, except for those unblinking eyes. He skipped over the money set aside for the next three years' editorial budget and homed in on what he thought were profligate expenses for marketing and public relations. "This is what I wanted to get away from," he told Williams.

"You want to put this magazine out there and just expect people to come?" Williams asked. He refused to cede the high ground to Kinsley, believing this part of his brain was still controlled by a reflexive, knee-jerk traditionalist.

"We'll get a lot of free press," Kinsley countered, so why spend the money?

"You've got to spend to get customers," Williams said. To help introduce the magazine, they needed mailings, and targeted e-mail, and ads, and a PR firm.

Kinsley said that he'd like to slim down those numbers, although he realized that there was too little time to do it before their meeting with Siegelman.

"It's a stupid exercise," said Williams, holding his ground. "You can't assume people will find us on their own."

"I'm not pretending we do nothing," Kinsley said.

"You still need someone to make press calls," Williams said. "PR is not enough. It dies after two weeks." Williams said they needed a launch party and ads and people to field press calls so that Kinsley didn't have to do those things himself and could focus on editing and writing.

But on the subject of promotion Kinsley was an unshakable Luddite. "The typical way a magazine fails," he said, "is they create enormous expenses in PR and parties."

Williams thought that *Slate*'s overall budget (a corporate secret) was modest, and said he expected to turn a profit in the third year.

"Every magazine says they're going to turn a profit in the third year," Kinsley said. "It seems to me we're spending too much money for things we can leverage free. And, if we spend, it will be harder to reach a profit."

It was all made-up, conjectural numbers, anyway, Williams cautioned.

"The revenue numbers are made up," Kinsley argued. "The expenditures are not."

They were going around in circles, and Kinsley suggested that they end the discussion. Let Williams present the numbers to Siegelman in a few hours and they could cut the budget later.

"It's important for Russ to understand where our differences are," Williams said, as he took his budget sheets back to his computer, leaving Kinsley at his desk, his khakis pulled up to mid-calf so that his white socks were fully visible. Kinsley sighed, and said, "I don't know to what extent I should give in. It's Microsoft's money. . . . I always dreamed if I ever started a magazine I'd take a pot of money and say, 'Let's put out the first issue and sell ten copies. And let us find our natural audience and save a pot of money to put out more issues.' Maybe that's a crazy idea, but that's what I imagined."

The meeting with Siegelman took place in his third-floor conference room, with him and Williams on one side of a long table and Kinsley sitting across from them. As the general manager for all Microsoft's on-line services, Siegelman was the executive who signed off on Kinsley's budget. Siegelman, with curly hair, orange-and-black-laced hiking boots, and youth and open enthusiasm for this project, was a comforting figure to Kinsley. Every time Siegelman says he cares more about *Slate*'s becoming "a model Web site" than he does about how many subscribers they initially attract, or laughs off what he expects will be a Kinsley punch at Microsoft—"He'll probably do that in issue number one!"—he cements his stature with Kinsley.

Williams began the meeting by presenting their circulation and revenue projections. Within three years, he said, they hoped to about match *The New Republic*'s paid circulation (a hundred thousand), and expected that eventually the bulk of their revenues would come from advertisers, many of whom would advertise not because they could precisely identify the audience but because of their "psychographic" interest in being associated with a brand (Kinsley). But *Slate* was also relying on something else: the largesse of the very folks Kinsley often attacks—lobbyists.

"The biggest issue here is: Can we attract PAC money?" Williams said.

"Not just PACs. Trade associations," added Kinsley, now spinning himself. "You don't mean PACs at all. You mean the trade associations." It was a distinction without a difference.

The subject got around to marketing, including the budget for public relations. As Williams spoke and Kinsley was silent, Siegelman knew enough to tweak a debate. "Speak up, Mr. Kinsley," he said.

"We're not entirely in synch," Williams volunteered.

"What's the issue?" Siegelman asked.

Public relations, Williams answered.

Why pay for free press? Kinsley countered.

Is it in Microsoft's interest to attract intense press interest for the first issue, Siegelman asked, or would it be better to wait until the staff got "the kinks out"?

"You can't say to Jay Leno, 'Put Kinsley on for the second-month anniversary,' " Kinsley said.

"There'll be a lot of buzz on it in May," Williams said. "We're keeping the lid on what this is."

"It's easy," Kinsley said. "We don't know what it is!"

They came back to the marketing costs associated with a May debut. Williams proposed that they launch the magazine with a party in Washington, D.C.

"That's just the kind of thing I don't want to do," Kinsley said, and he argued that it was a waste of time and money. "What you get out of it is a piece in the Style section anyway. And it's embarrassing to me, personally."

Williams said it wasn't a waste, since they would invite journalists, present the magazine's plans, and entertain questions.

"I'd rather send the journalists a mailing," Kinsley said.

They agreed to defer this particular matter and instead returned to the marketing budget, starting with public relations. Siegelman said he thought the figure budgeted was discreet, but Kinsley demurred: "I have a special problem with PR for a journalism product. PR is about spin, and our message is that we are cutting through spin."

"When magazines call and want to find out what this is about, whom do they call?" Siegelman inquired, and he added that they would need statements and facts and some background on Kinsley and the magazine. That took time. And professionals.

"Assign an intern to do it," countered Kinsley, who wouldn't budge.

"So you just want to hire nonprofessional PR people to do PR," Williams responded.

At this point, Siegelman stepped in and ruled: If Kinsley feels a decision "impacts his product," he intoned, "my rule is that Mike rules."

Afterward, Kinsley had reason to be pleased. His twenty-month stay at *Harper's* had featured fractious battles with the chairman of the board. Although under him *Harper's* had won a National Magazine Award for general excellence, the board scolded and briefly suspended him for accepting a policy-oriented junket to Israel, and he embarrassingly whined in a letter to the board about his meager salary and "ordinary one-bedroom apartment in an unfashionable neighborhood." At *The New Republic*, money was always a problem. The magazine continued to lose money under his tutelage, though circulation rose slightly. And although he was a close friend of the owner, Martin Peretz, he had not been free to make marketing decisions, or fully free to set the editorial line (especially on such subjects as Israel), or to make hiring and firing decisions on his own, or to control the back-of-the-book sections. Now he already had enormous freedom, a publisher who reported to him, and a good feeling about Microsoft. "The one thing I don't question is: Is Microsoft behind this?" he told me. It didn't perturb him that he had not yet met Bill Gates. And, reflecting on the meeting in Siegelman's conference room, Kinsley said of John Williams, "I hope I didn't hurt his feelings. I think he's doing a great job. He's been on board a month and a half. Look how far he's progressed."

In truth, they had not, together, progressed far enough, for it was already mid-March and they were behind schedule. Kinsley needed to spend time matching writers and stories, and developing a bank of pieces to fill the magazine in June and July. He told me that he planned to entice writers by claiming that pieces in *Slate* would generate "buzz," that it was a new medium, that some pieces could be published instantly, that technology could enhance their journalism. "And," he added, "in some cases I'm going to beg, which is also an editorial tradition."

By April, the lawyers had secured the Web address for the name *Slate*, and Kinsley had hired his other top editors: Jack Shafer, the former editor of the *City Paper*, in Washington, D.C., and most recently editor of the San Francisco *Weekly*, would become his deputy in Redmond; Judith Shulevitz, the former editor of *Lingua Franca* and deputy editor of *New York*, would supervise the Gotham office and much of the arts coverage; Jodie Allen would start work as the Washington editor on May 1. In addition, he had signed up Herbert Stein, who had been the chairman of the Council of Economic Advisers under President Nixon, as moderator of *Slate's* upscale electronic version of the *McLaughlin Group*, and, Kinsley said, Mark Alan Stamaty, who draws "Washingtoon," was "ninety-nine percent" certain to join as the magazine's cartoonist. To buy two more weeks, Kinsley pushed back the inaugural issue to June 14. And he had scaled down the expensive public-relations campaign and would rely on Microsoft's local PR firm to perform what he called "almost secretarial duties." When he rejected the idea of a big Washington, D.C., party, he said, "It's supposed to be in cyberspace. Why do I have to go to Washington?"

Also in April, Russ Siegelman left the job of running the Microsoft Network, but Kinsley was relieved to know that he would retain responsibility for *Slate*. And, on April 8, Kinsley at last met with Bill Gates, in a pleasant hour-and-a-half discussion of the magazine. Gates was full of questions, according to Siegelman, who accompanied Kinsley to the meeting. The Microsoft chairman focused particularly, both Kinsley and Siegelman recall, on questions relating to whether the magazine was weekly or continuous: If they updated material, how would the reader know what was new and what was old? Would they date it? Would they retain the old information? Would they update the print version as well?

That same week, Kinsley made a fairly momentous decision, one that suggested the distance he had traveled since he wrote his proposal last July for Time Warner. Kinsley remembers that, bothered by the constant back-and-forth over how to date the cover and how to post new pieces in the middle of the week, he had an epiphany: "This is artificial!" The Education of Michael Kinsley was nearly complete. He now decided to abandon the idea of a magazine that went to bed every Friday and replace it with a publication that would never sleep. The decision was announced in an April 9 e-mail Kinsley sent to the staff:

> In short, I propose that we embrace our destiny as a new form of journalism and abandon the conceit that any particular article or feature is attached to a particular weekly "issue." . . . Each article in the ToC [table of contents] could simply indicate the day it was posted and the day we're planning to archive it. . . . As we and the readers get used to this new form of journalism, we could abandon the one-week-up convention completely, and simply have a smorgasbord of stuff to which we add new dishes and remove old ones on no fixed schedule, but simply to keep the whole meal tasting as delicious as possible.

But, as Kinsley's ride through cyberspace cleared one obstacle, others arose. Because of the expense, he was rethinking the idea of a weekly print version of *Slate*, and was inclined to publish it monthly. By May, Kinsley had collided with the reality that this "new form of journalism" was going to remain dependent on old software that made it impossible to assure customers that their credit-card numbers would not be pirated if exchanged over the Internet. To ensure customers complete security, he told me over the telephone, "we're not going to be able to bill people from day one. We're going to have to wait a few months. The software won't be ready." Yet later the same day Kinsley e-mailed:

> The latest . . . we probably WILL be able to charge from the beginning or very near it. At any rate, we still plan to charge as soon as possible.

I then e-mailed:

> Are you "spinning" me? If Carville & Co. sent me such an e-mail I'd interpret it this way: trying to figure way to charge from day one, but unlikely; what is likely is what Kinsley said earlier . . . we'll probably charge by August.

Kinsley e-mailed back:

> Let me "respin." We're going to charge as soon as it's technologically possible. . . . It looks now like we'll have to give it away for a month or so (ie, until July).

A small crater, perhaps, but just one of many lurking in cyber-space.

POSTSCRIPT:

After reporting this piece I better appreciate what useful tools e-mail and the Web offer to the reporter or biographer. The paper trail—of thoughts, correspondence, the steps in a decision-making process, the discussions of strategy or tactics—made it easier for me to re-create Kinsley's journey.

After the piece appeared, both he and I were clubbed by the proprietary fraternal order of pioneer Web users. Some critics honestly believe Kinsley is operating in the wrong medium and will never succeed if he persists in telling his customers: Eat your spinach! They may be right; only time will tell. The electronic magazine *Feed* took me to task for writing about meetings I attended without at least mentioning that the presence of a reporter might have contaminated the data by affecting what people said. That is a fair point. I don't know for sure whether or not my presence changed the result. I believe it did not.

Some complained, How dare Kinsley invade our turf! Media critic Jon Katz, who, when he was a CBS News producer, was a proponent of confecting news "moments" that made viewers feel good, sputtered that Kinsley and Auletta were disgusting primitives, establishment voices who had discovered "the Web suddenly worthy of attention." "Saint Jon," Jack Shafer delightedly dubbed Katz. Steven Johnson, publisher of *Feed*, thoughtfully emoted that "Auletta has lowered the art of CEO porn that would make Larry Flynt blush" (his syntax).

A more serious obstacle to Kinsley's efforts may well be the traffic jams on the Internet. Many customers simply complained that they could not get through because they kept getting the equivalent of busy signals, a problem that intensified in late 1996 when AOL offered unlimited Internet access for twenty dollars per month. It may be, as appears to be the case in early 1997, that the Internet has replaced TV as the interactive medium of choice. Right now, despite the brave pronouncements of believers, the content seems to be ahead of the technology or the economics. In January 1997, Kinsley reluctantly announced that Microsoft had abandoned plans to charge for *Slate*. There are, Kinsley wrote, "too many people who are too damned cheap . . . er, we mean . . . too engaged by the novelty of the medium to feel the need to pay extra for specific content."

This brings to mind something the great urban historian Lewis Mumford once wrote, an oddly appropriate thought on which to end this book: "I'm a pessimist about probabilities, and an optimist about possibilities."

ACKNOWLEDGMENTS

When Tina Brown became the editor of *The New Yorker* in September 1992, I happened to be seated to her right at a small celebratory dinner. She asked what I was working on and I mentioned a book idea. She was obviously unimpressed with the book, for she phoned the next day to see if we might meet that day. We met and she asked if I might like to do a media column that naturally sprang from my last book, *Three Blind Mice: How the TV Networks Lost Their Way.* I said no, I really didn't want to get diverted from doing a book. But we talked about how the communications business was both broadening and changing, and how the beat should be broad enough to encompass not just the networks and the press and the studios but cable and telephone and computers and publishing as well. Think about it, she said.

Can't do it, I said.

Then I thought about it.

To escape the frenzy of deadlines, which prevent most journalists from having time to think or to dive below the surface, I suggested that for the first several months I write nothing. Instead, I would read and conduct background (not-for-attribution) discussions with fifty or so people in the communications industry. No problem. She would subsidize my education. I would guess that these casual, candid conversations planted seeds for half a dozen pieces in this book.

Four years later, the book idea we discussed at dinner is dead and I'm grateful to Tina Brown for providing a front-row seat to witness the information revolution. I'm also grateful to others at *The New Yorker:* Pat Crow and Jeffrey Frank edited these pieces with care and craft; the miraculous fact-checking department spared me from obvious mistakes (any remaining mistakes in this book are mine); Pamela Maffei McCarthy, Dorothy Wickenden, and others too numerous to mention from the editorial, art, and production departments were of immense help.

I am grateful, as always, to Jason Epstein and Random House, who have edited and published six of my seven books, and to the many diligent people there who improved this work, particularly Jason's assistant, Joy de Menil, and copy editor Virginia Avery. I acknowledge another ancient relationship—with my literary agent and friend, Esther Newberg, of ICM. The title *The Highwaymen* was suggested by another friend, Tully Plesser. Khrystine Muldowney spent the summer of her junior year in college cheerfully transferring these pieces onto computer discs and marking those sections of the original manuscript that for space reasons did not appear in *The New Yorker.* This book therefore contains previously unpublished material; most of one piece, "The Power of Shame," appears for the first time in this book.

My wife, Amanda Urban, lent her superb editorial judgments to most of these pieces before they were dispatched to *The New Yorker.*

INDEX

About the Author

KEN AULETTA is the author of two national bestsellers, *Three Blind Mice* and *Greed and Glory on Wall Street*, as well as *The Underclass, The Streets Were Paved with Gold, Hard Feelings,* and *The Art of Corporate Success*. As communications columnist for *The New Yorker*, he has been covering the emerging world of electronic communications for the past five years. Earlier, he was a political columnist for the New York *Daily News*. His pieces have also appeared in *Vanity Fair, The New York Review of Books, The New York Times Magazine, New York, Esquire,* and *The Village Voice*. He appears regularly on National Public Radio and *The NewsHour with Jim Lehrer* and has written and served as correspondent for *Frontline* documentaries; he has been a political commentator for PBS, WCBS, and WNBC.

About the Type

The text of this book was set in Janson, a misnamed typeface designed in about 1690 by Nicholas Kis, a Hungarian in Amsterdam. In 1919 the matrices became the property of the Stempel Foundry in Frankfurt. It is an old-style book face of excellent clarity and sharpness. Janson serifs are concave and splayed; the contrast between thick and thin strokes is marked.